Constitutive Modelling in Geomechanics

Alexander M. Puzrin

Constitutive Modelling in Geomechanics

Introduction

Springer

Prof. Dr. Alexander M. Puzrin
Institute for Geotechnical Engineering
ETH Zurich
Zurich, Switzerland

ISBN 978-3-642-27394-0 e-ISBN 978-3-642-27395-7
DOI 10.1007/978-3-642-27395-7
Springer Heidelberg Dordrecht London New York

Library of Congress Control Number: 2011945686

Printed on acid-free paper

Springer is part of Springer Science+Business Media (www.springer.com)

Table of contents

Preface

This book is based on a course of lectures given over the last 15 years at Technion (Haifa), Georgia Tech (Atlanta), UPC (Barcelona) and ETH Zurich. The purpose of this course has been to bridge the gap between the graduate courses in Geomechanics and those in Numerical Geotechnical Modelling. Traditionally, in many geotechnical programs, Geomechanics is not taught within the rigorous context of Continuum Mechanics and Thermodynamics. There is a good reason for that – the behaviour of soils is very complex: it is more advantageous to explain it at a semi-empirical level, instead of scaring the students away with cumbersome mathematical models. However, when it comes to Numerical Modelling courses, these are often taught using commercially available finite elements (e.g. ABAQUS, PLAXIS) or finite differences (e.g. FLAC) software, which utilize constitutive relationships within the Continuous Mechanics framework. Quite often students and practitioners have to learn the challenging subject of constitutive modelling from a program manual, sometimes ending up with using models which violate the Laws of Thermodynamics!

The book is introductory - by no means does it claim any completeness and state of the art in such a dynamically developing field as numerical and constitutive modelling of soils. Our intention is to achieve a basic understanding of conventional continuum mechanics approaches to constitutive modelling, which can serve as a foundation for exploring more advanced theories. A considerable effort has been invested here into the clarity and brevity of the presentation. We focus on helping the readers to understand how different aspects of complex soil behaviour can be modelled using conventional constitutive models, which can be readily found within the available numerical codes. Another important feature of this book is that it explores thermomechanical consistency of all presented constitutive models in a simple and systematic manner.

The book is built of three parts. Part I gives an introduction into continuum mechanics. Part II deals with the modelling of reversible soil behaviour, while Part III introduces the modelling of irreversible soil behaviour. Finally, Appendix A focuses on modelling the undrained soil behaviour and Appendix B demonstrates an example of incorporation of an irreversible constitutive model into a numerical algorithm for solution of boundary value problems.

We believe that this book can be a useful reference both for researchers and geotechnical engineers, as they face more and more often the necessity of the numerical analysis in their practice. Understanding of the limitations of the built-in constitutive models is crucial for critical assessment of the results of numerical calculations, and, hence, for the safe and cost efficient design of geotechnical structures.

We would like to express our gratitude to Dr. Carlo Rabaiotti, who contributed Appendix B of the book, to Dr. Sophie Messerklinger for her help in preparing the manuscript and to Paul Witteveen for the thorough proofreading of the manuscript.

A.M.Puzrin   Zurich, November 2011

Part I

Introduction to Continuum Mechanics

Chapter 1

Boundary Value Problems in Geotechnics

TABLE OF CONTENTS

Chapter 1

Boundary Value Problems in Geotechnics

1.1 Objectives

The purpose of this book is to introduce the readers to the basic ideas of continuous mechanics (Part I) with a special emphasis on constitutive relationships (Parts II and III). The presentation focuses on constitutive modelling of geomaterials with applications in geotechnical engineering. The main objective here is to bridge the gap between the geomechanics and numerical modelling courses.

1.2 Typical geotechnical problems

The most common geotechnical problems are: analysis and design of foundations, retaining structures, slopes and excavations.

1.2.1 Foundations

In design and analysis of shallow, deep and pile-raft foundations (Figure 1.1) we need to determine their bearing capacity and settlements under working loads.

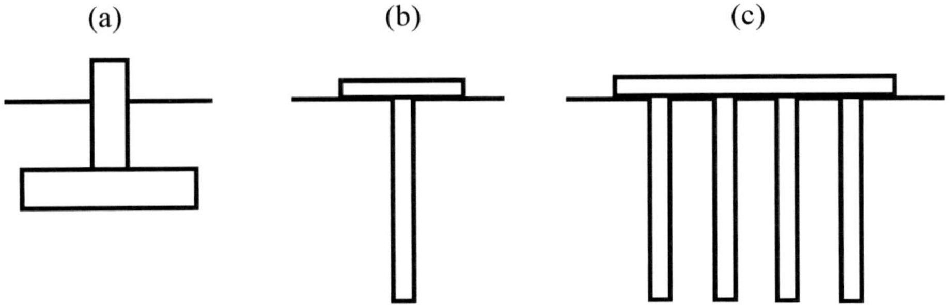

Figure 1.1 Foundations: a) Shallow; b) Deep (piles); c) Pile-raft.

1.2.2 Retaining structures

For gravity, cantilever and anchored retaining walls (Figure 1.2) we need to know the forces acting on the wall, displacements of the wall and the forces in the anchors.

A.M. Puzrin, *Constitutive Modelling in Geomechanics*,

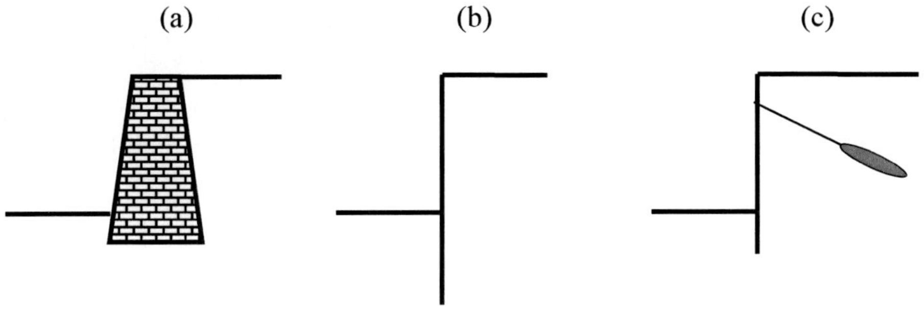

Figure 1.2 Retaining walls: a) Gravity; b) Cantilever; c) Anchored.

1.2.3 Slopes and excavations

In analysis and design of slopes, excavations and tunnels (Figure 1.3) we need to calculate their stability.

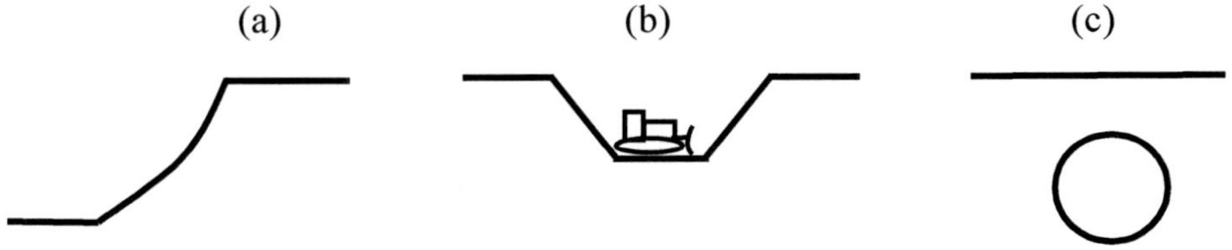

Figure 1.3 a) Slopes; b) Open excavations; c) Tunnels.

What is common between all the geotechnical problems mentioned above? They all require the knowledge of stresses and strains in the soil. To find this stress-strain state, all the above problems can be formulated as boundary value problems (BVP) of the Mechanics of Continuum Media and solved using finite element (FE) or finite difference (FD) methods.

1.3 A boundary value problem of continuum mechanics

In continuum mechanics, any boundary value problem (BVP) is formulated as a set of differential equations (Figure 1.4) supplemented by initial and boundary conditions:

- forces are related to stresses through *equations of motion* based on considerations of equilibrium;
- displacements are related to strains through *compatibility equations*, based on considerations of geometry.

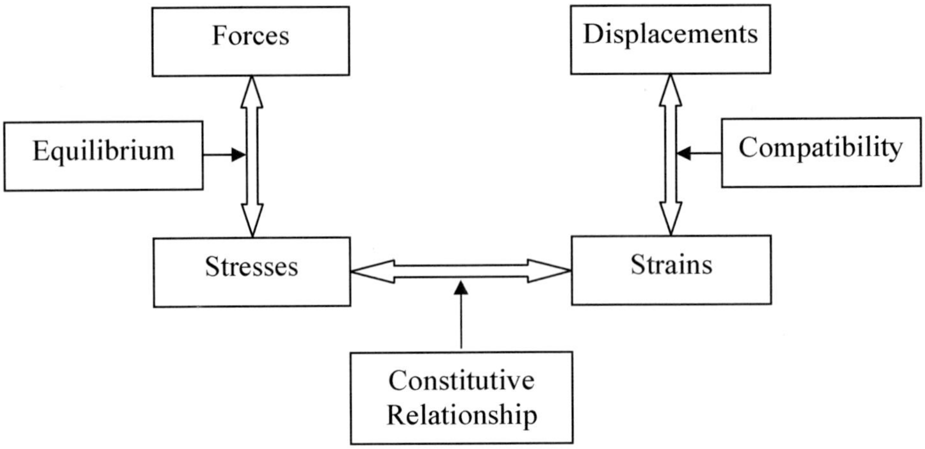

Figure 1.4 Equations of a Boundary Value Problem

These two groups of equations are not sufficient to complete the BVP system of equations. More importantly, these equations are independent of the mechanical properties of materials: they are in principle the same in steel, concrete, wood, soil and even in live tissue.

In order to complete the BVP system of differential equations and introduce the mechanical behaviour of specific materials into the formulation, we have to formulate an additional set of equations (Figure 1.4):

- *constitutive relationships,* relating stresses and strains.

Constitutive relationships are normally derived from the stress-strain data determined in laboratory tests on material samples. A test is called elementary, if the stresses and strains in the test are the same at any point in the sample.

1.4 Typical geotechnical elementary lab tests

Typical geotechnical elementary laboratory tests are: oedometer and triaxial tests. Simple and direct shear tests are also available, but are less popular, and cannot be considered elementary due to the non-homogeneous loading conditions.

1.4.1 Oedometer test

In the oedometer test (Figure 1.5) the sample is subjected to the uniform axisymmetrical loading with controlled (zero) radial strain. The changes in the sample height/volume are monitored as a function of the applied stress and time (at a constant stress).

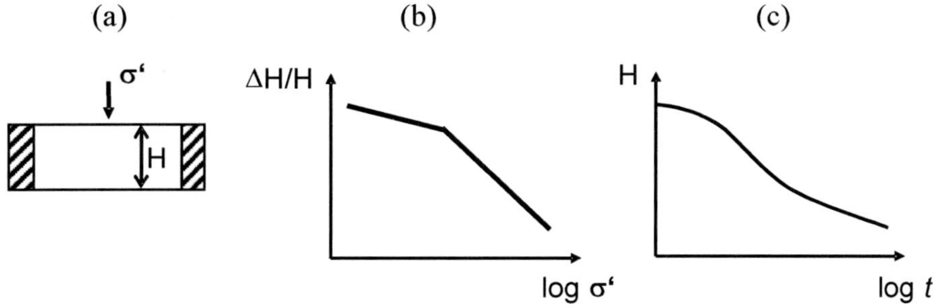

Figure 1.5: Oedometer test: a) Setup; b) Strain-stress, and c) Strain-time curves

1.4.2 Triaxial shear test

In the triaxial shear test (Figure 1.6) the sample is subjected to the uniform axisymmetrical loading with controlled (constant) radial stress. The changes in the sample height and volume (in a drained test) are monitored as a function of the stress. In an undrained test, the volume is kept constant, and the pore water pressure is monitored instead.

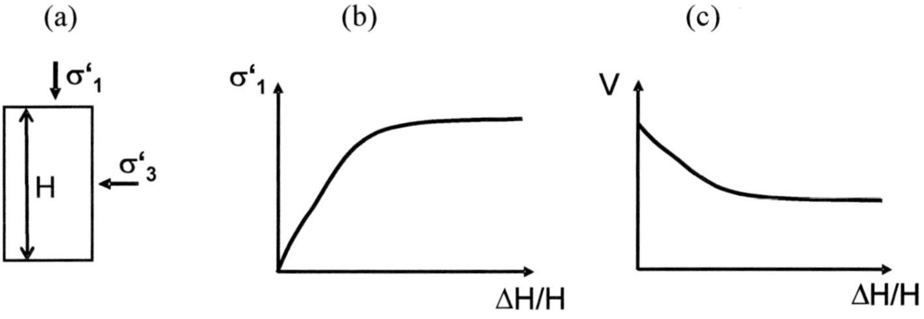

Figure 1.6: Triaxial test: a) Setup; b) Stress-strain and c) Volume-strain curves.

1.5 Complexities of the mechanical behaviour of soils

Soils are very complex materials to describe mathematically. They are:

a) Multiphase (consist of solids, liquids and gases);
b) Granular (built of particles of different sizes and shapes);
c) Non-homogeneous (their mechanical properties vary in space);
d) Anisotropic (their mechanical properties vary with loading direction).

However, even if all the above effects are minimized (e.g., in a drained triaxial compression test on a saturated remoulded isotropically consolidated clay sample), the mechanical behaviour is still rather complex (Figure 1.7). It possesses such features as:

a) Non-linearity (in initial loading, unloading and reloading);
b) Irreversibility (produces residual strains in a closed stress cycle);
c) Hysteresis of unloading-reloading and memory (remembers the highest stress before the unloading and follows the initial loading curve after reaching it in reloading);
d) Stress path dependency (reaches the same stress at different strains);
e) Rate dependency (different stress-strain curves at various strain rates);
f) Time dependency (creep, aging, relaxation).

This book explains how to incorporate all these features into a constitutive model. It has to be kept in mind, however, that in some geotechnical problems certain features will be more important than others. These problems can be solved using simpler, problem oriented constitutive models.

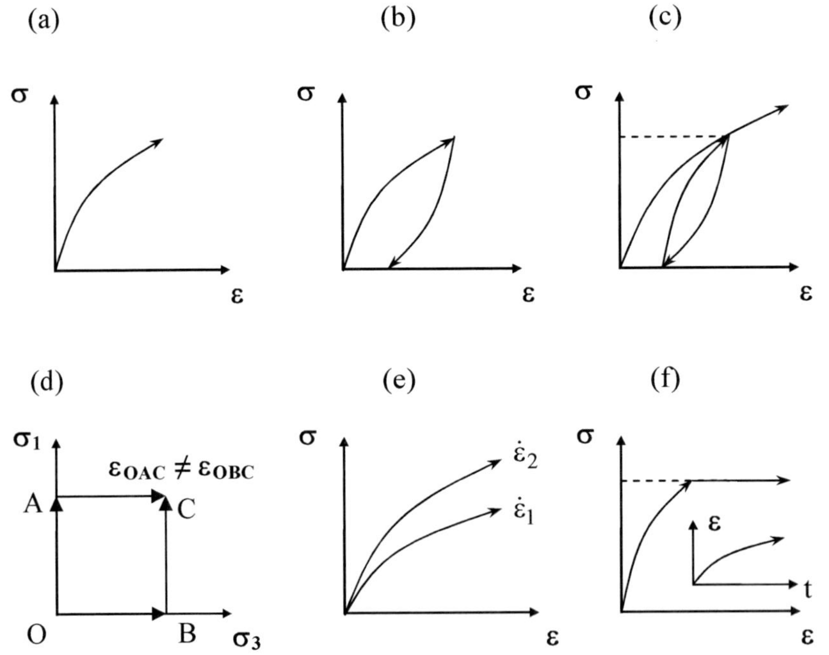

Figure 1.7 Soil behaviour (in a traxial test): a) non-linearity; b) irreversibility; c) material memory; d) stress path dependency; e) rate dependency; f) creep.

1.6 Tensorial form of constitutive relationships

Physical meaning of a constitutive relationship is that it represents a relation between the force applied to the body and the corresponding displacements (Figure 1.8). The simplest one would be a *scalar* equation of a spring or a long rod (Figure 1.8a), where the elongation is determined as a function of the tension force.

If only we could confine our discussion to this case, the book would be a quarter of its present volume. Most of the boundary value problems in geomechanics, however, require the constitutive law to be defined at each point in soil or rock (Figure 1.8b). This implies a relationship between stresses and strains, which are tensorial quantities, hence, leading to *tensor* constitutive equations.

1.7 The structure of Part I: Continuum Mechanics

Tensors may complicate both the presentation and the understanding of the constitutive models considerably, if the readers are not comfortable with the *elementary tensor analysis*. To set up the basic principles of this analysis is the objective of the Chapter 2, which, together with Chapters 3 and 4 introducing *stress and strain tensors*, respectively, closely follows the approach of Desai and Siriwardane (1984). Chapter 5 summarizes *equations of continuum mechanics* and defines a *boundary value problem*. Chapters 6 and 7 demonstrate how this problem can be solved using *finite elements and finite differences methods*, respectively. Chapter 8 describes a *themomechanical framework* within which constitutive models can be validated with respect to their consistency with *the First and the Second Laws of Thermodynamics*. Finally, Chapter 9 introduces important features of soil behaviour which will be modelled in the Parts II and III of this book.

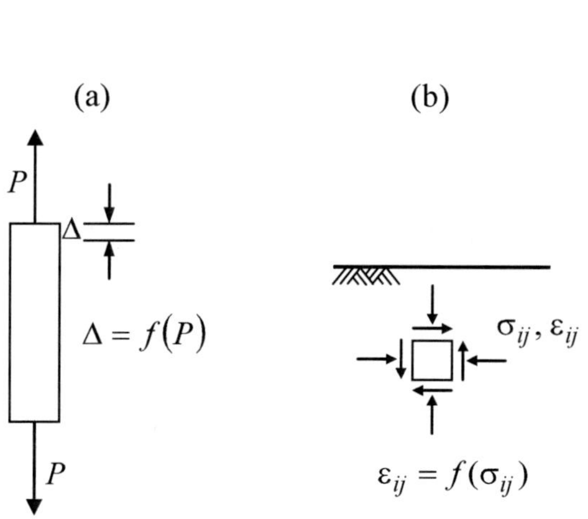

Figure 1.8 Constitutive relationships in: (a) 1D and (b) 3D spaces.

Chapter 2

Elementary Tensor Analysis

TABLE OF CONTENTS

Chapter 2

Elementary tensor analysis

2.1 *Indicial notation*

2.1.1 Importance of indicial notation

Mathematics in tensor analysis for the 3D case (9 components of tensors) is rather complex and extensive. It is necessary to introduce some simplifying notation to make it short and clear (Table 2.1).

Table 2.1 An example: 3D equilibrium equations

Explicit formulation	Indicial notation	Tensor notation
$\frac{\partial \sigma_{xx}}{\partial x}+\frac{\partial \tau_{yx}}{\partial y}+\frac{\partial \tau_{zx}}{\partial z}+F_x=0$ $\frac{\partial \tau_{xy}}{\partial x}+\frac{\partial \sigma_{yy}}{\partial y}+\frac{\partial \tau_{zy}}{\partial z}+F_y=0$ $\frac{\partial \tau_{xz}}{\partial x}+\frac{\partial \tau_{yz}}{\partial y}+\frac{\partial \sigma_{zz}}{\partial z}+F_z=0$	$\sigma_{ij,i}+F_j=0$	$\nabla.\boldsymbol{\sigma}+\mathbf{F}=0$

Clearly both indicial and tensor notations make the presentation exceptionally compact (the tensor notation even more so than the indicial). In the book we adopt the indicial notation, because it achieves a greater clarity, especially when multiplication of a number of multidimensional quantities is involved.

2.1.2 First and second order notation

a) First order objects are denoted by one subscript, e.g. coordinates of the point x, y, z or x_1, x_2, x_3:

$$x_i \quad i=1,2,3 \quad \text{which is equivalent in a matrix notation to } \begin{Bmatrix} x_1 \\ x_2 \\ x_3 \end{Bmatrix}.$$

For convenience, in this chapter the first order objects will be mostly denoted by a small Latin character.

A.M. Puzrin, *Constitutive Modelling in Geomechanics*,

b) Second order objects are denoted by two subscripts, e.g., matrix $\left[X\right]$:

$$X_{ij} \qquad i,j=1,2,3 \qquad \text{which is equivalent to} \qquad \begin{pmatrix} X_{11} & X_{12} & X_{13} \\ X_{21} & X_{22} & X_{23} \\ X_{31} & X_{32} & X_{33} \end{pmatrix} \begin{matrix} i=1 \;\; j=1,2,3 \\ i=2 \;\; j=1,2,3 \\ i=3 \;\; j=1,2,3 \end{matrix} .$$

Second order objects in this chapter will be mostly denoted by a capital Latin character.

2.1.3 Free and dummy indices

a) *Rule 1:* Free index appears once in each term.
Examples: a_i, $a_i = b_i + d_i$, index i is a free index;
$A_{ij} = \alpha_1 B_{ij} + \alpha_2 C_{ij}$, indexes i and j are free indexes.

b) *Rule 2:* Dummy index appears twice and denotes summation.
Examples: $A_{ii} = A_{11} + A_{22} + A_{33}$;
$x_m y_m = x_1 y_1 + x_2 y_2 + x_3 y_3$;
$x_i = y_j Z_{ij} = y_1 Z_{i1} + y_2 Z_{i2} + y_3 Z_{i3}$.

c) *Rule 3:* Dummy indices can be replaced without changing the result.
Example: $A_{ii} = A_{mm} = A_{pp}$, $\quad i = 1,2,3 \quad m = 1,2,3 \quad p = 1,2,3$.

d) *Rule 4:* Index cannot appear three times at the same term.
Example: $\left(A_{11} + A_{22} + A_{33}\right)x_i = A_{ii}x_i$ is wrong, but
$\left(A_{11} + A_{22} + A_{33}\right)x_i = A_{nn}x_i$ is correct.

e) *Rule 5:* Partial differentiation is denoted by a comma followed by the index of differentiation.

Example: $f_{,i} = \dfrac{\partial f}{\partial x_i}$ and $x_{j,i} = \dfrac{\partial x_j}{\partial x_i}$ so that

$$\frac{\partial f}{\partial x_i} = \frac{\partial f}{\partial x_1}\left(\frac{\partial x_1}{\partial x_i}\right) + \frac{\partial f}{\partial x_2}\left(\frac{\partial x_2}{\partial x_i}\right) + \frac{\partial f}{\partial x_3}\left(\frac{\partial x_3}{\partial x_i}\right) \quad \text{becomes} \quad f_{,i} = f_{,j}\, x_{j,i}$$

where i is a free and j is a dummy index.

2.1.4 Kronecker delta

The Kronecker delta is an extremely useful instrument in the tensor analysis. Consider an orthogonal coordinate system (Figure 2.1) defined by unit vectors e_1, e_2, e_3, such that:

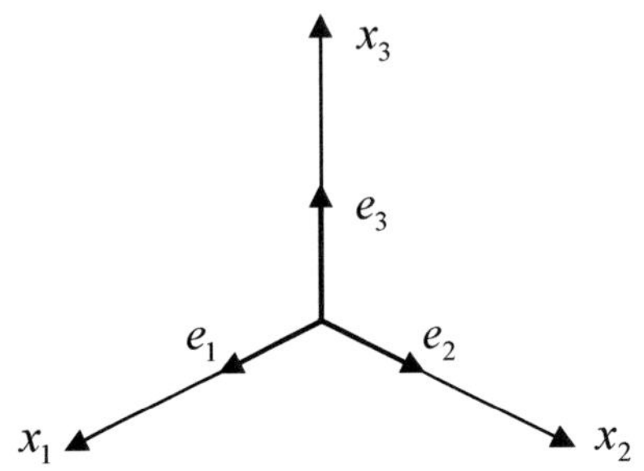

Figure 2.1 Orthogonal coordinate system with unit vectors.

$$e_1e_1 = e_2e_2 = e_3e_3 = 1 \text{ and } e_1e_2 = e_1e_3 = e_2e_3 = 0 \tag{2.1}$$

or in the index notation:

$$e_me_n = \begin{cases} 1 & m = n \\ 0 & m \neq n \end{cases} \tag{2.2}$$

The Kronecker delta is then defined as

$$\delta_{mn} = e_me_n \tag{2.3}$$

It follows that

$$\delta_{ij} \equiv [I] = \begin{bmatrix} 1 & 0 & 0 \\ 0 & 1 & 0 \\ 0 & 0 & 1 \end{bmatrix} \tag{2.4}$$

i.e. is the second order unit matrix.

The properties of the Kronecker delta in a three-dimensional system are:

$$\delta_{ii} = 1 + 1 + 1 = 3 \tag{2.5}$$

$$\delta_{ij}x_i = x_j \text{ and } \delta_{ij}x_j = x_i \tag{2.6}$$

(Proof: $\delta_{ij}x_i = \delta_{1j}x_1 + \delta_{2j}x_2 + \delta_{3j}x_3$, which for $j = 1$ gives $\delta_{i1}x_i = x_1$; for $j = 2$: $\delta_{i2}x_i = x_2$ and for $j = 3$: $\delta_{i3}x_i = x_3$, so that $\delta_{ij}x_i = x_j$).

Similar reasoning is used in proving the following properties:

$$\delta_{ij}\delta_{ji} = \delta_{i1}\delta_{1i} + \delta_{i2}\delta_{2i} + \delta_{i3}\delta_{3i} = \delta_{11} + \delta_{22} + \delta_{33} = 3 \tag{2.7}$$

$$\delta_{ij}A_{jm} = \delta_{i1}A_{1m} + \delta_{i2}A_{2m} + \delta_{i3}A_{3m} = A_{im} \tag{2.8}$$

$$\delta_{ij}A_{ij} = \delta_{i1}A_{i1} + \delta_{i2}A_{i2} + \delta_{i3}A_{i3} = A_{11} + A_{22} + A_{33} = A_{ii} \tag{2.9}$$

In general, multiplication by δ_{ij} replaces a dummy index by a free one.

2.2 Scalars and vectors

2.2.1 Definitions

a) A *scalar* is a zero order object, which in index notation either does not have any index at all, or has only dummy indexes, e.g.: α, A_{nn}, $A_{ij}A_{ij}$, $A_{ij}A_{jk}A_{ki}$.

b) A *vector* is a first order object, which in index notation has only one free index, e.g.: a_i, $A_{ij}b_i$.

2.2.2 Vector transformation

Vector $\overline{a}$ itself is invariant to rotation and translation of coordinate system. Its components a_i, however, will change with the system transformation.

One of the advantages of the index notation is that most of the results obtained for two dimensional cases, after being written in an index form also hold for three dimensional cases. This generalization is achieved by simply letting indexes run from 1 to 3 instead of 1 to 2. This facilitates greatly geometrical interpretation in tensor analysis.

Let us then consider some vector $\overline{a}$ with components a_1 and a_2 in a two dimensional coordinate system x_1Ox_2, defined by the orthogonal unit vectors e_1 and e_2 (Figure 2.2). (Please, pay attention that e_1 and e_2 are not components of some vector *e*, but are vectors themselves). Next, coordinate system x_1Ox_2, undergoes a rotation around the origin *O*, which results in a new coordinate system $x_1^*Ox_2^*$, defined by the orthogonal unit vectors e_1^* and e_2^*. Our task is to find components a_1^* and a_2^* of vector $\overline{a}$ in the new coordinate system.

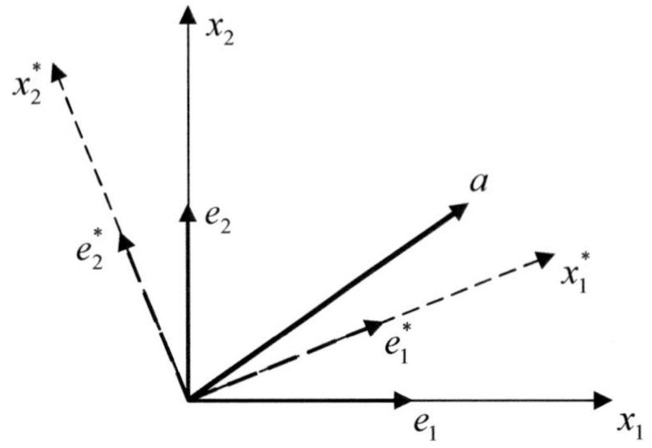

Figure 2.2 Vector transformation.

Let us first express the vector via its components in index notation:

$$a = a_1 e_1 + a_2 e_2 = a_n e_n \qquad a = \overset{*}{a}_1 \overset{*}{e}_1 + \overset{*}{a}_2 \overset{*}{e}_2 = \overset{*}{a}_n \overset{*}{e}_n \tag{2.10}$$

What we are looking here for is the way how to define the new components through the old ones: $\overset{*}{a}_n = f(a_n)$. Because the vector itself is invariant to the transformation, we write:

$$\overset{*}{a}_n \overset{*}{e}_n = a_n e_n \quad \text{or} \quad \overset{*}{e}_n \overset{*}{a}_n = e_n a_n \tag{2.11}$$

Multiplying both sides of second equation (2.11) by a vector $\overset{*}{e}_m$ we obtain $\overset{*}{e}_m \overset{*}{e}_n \overset{*}{a}_n = \overset{*}{e}_m e_n a_n$. Then from the definition of the Kronecker delta (2.3) it follows that $\delta_{mn} \overset{*}{a}_n = T_{mn} a_n$ and replacing the dummy index with the free one (property 2.6) yields the *transformation law*:

$$\overset{*}{a}_m = T_{mn} a_n \tag{2.12}$$

where

$$T_{mn} = \overset{*}{e}_m \cdot e_n = \cos\left(\overset{*}{x}_m, x_n\right) \tag{2.13}$$

is the *transformation matrix*.

Example: $T_{12} = \overset{*}{e}_1 \cdot e_2 = \cos\left(\overset{*}{x}_1, x_2\right)$ is the cosine of the angle between the axes $\overset{*}{x}_1$ and x_2 (Figure 2.2).

At this stage we observe that all the derivations will be also valid for a 3D case if we make $m,n = 1, 2, 3$.

2.2.3 Properties of the transformation

The transformation law $\overset{*}{a}_m = T_{mn} a_n$ (equation 2.12) was obtained by multiplying equation $\overset{*}{e}_n \overset{*}{a}_n = e_n a_n$ by a vector $\overset{*}{e}_m$. If instead it were multiplied by a vector e_m, the *inverse transformation law* would be obtained:

$$a_m = T_{nm} \overset{*}{a}_n \quad \text{or} \quad a_n = T_{in} \overset{*}{a}_i \tag{2.14}$$

we substitute equation (2.14) into (2.12):

$$\overset{*}{a}_m = T_{mn} T_{in} \overset{*}{a}_i \tag{2.15}$$

We notice that $\overset{*}{a}_m = \delta_{mi} \overset{*}{a}_i$, and by comparing it with (2.15) we deduce:

$$\delta_{mi} = T_{mn} T_{in} \tag{2.16}$$

which is an important property of the transformation matrix.

An inverse substitution of (2.12) into (2.14) produces

$$a_m = T_{nm} T_{ni} a_i \tag{2.17}$$

leading to

$$\delta_{mi} = T_{nm} T_{ni} \tag{2.18}$$

which is another important property of the transformation matrix.

Equation (2.16) establishes that any row of the transformation matrix when multiplied by itself gives the unity, while when multiplied by another row it gives zero. Equation (2.18) establishes the same property for a column:

$$[T] = \begin{pmatrix} T_{11} & T_{12} & T_{13} \\ T_{21} & T_{22} & T_{23} \\ T_{31} & T_{32} & T_{33} \end{pmatrix} \qquad \begin{aligned} (Row)^2 &= (Col)^2 = 1 \\ (Row \times Row) &= (Col \times Col) = 0 \end{aligned}$$

Example: Two dimensional transformation matrix ($i = 1, 2$) for rotation of the coordinate system in Figure 2.2 by angle α:

$$T_{ij} = \begin{pmatrix} \cos\left(\overset{*}{x}_1, x_1\right) & \cos\left(\overset{*}{x}_1, x_2\right) \\ \cos\left(\overset{*}{x}_2, x_1\right) & \cos\left(\overset{*}{x}_2, x_2\right) \end{pmatrix} = \begin{pmatrix} \cos\alpha & \sin\alpha \\ -\sin\alpha & \cos\alpha \end{pmatrix} \tag{2.19}$$

Multiplication of each row/column by itself gives $\cos^2\alpha + \sin^2\alpha = 1$, while by another row/column: $-\cos\alpha\sin\alpha + \cos\alpha\sin\alpha = 0$. The components of the vector a_m are transformed into $\overset{*}{a}_m = T_{mn} a_n = \begin{Bmatrix} a_1\cos\alpha + a_2\sin\alpha \\ -a_1\sin\alpha + a_2\cos\alpha \end{Bmatrix}$.

2.3 Second order tensors

2.3.1 Definition

a) Second order *tensor* is a second order object, which in index notation has two free indices e.g.: A_{ij}, $A_{ij}A_{jk}$, $A_{ij}A_{jk}A_{kn}$.

b) V*ector* can then be defined as a first order tensor; s*calar* is a zero order tensor. However, in this book we reserve the term *tensor* only for the second order tensors.

c) *Not every second order object is a tensor. Only those objects are tensors, whose components transform according to the following transformation law.*

2.3.2 Transformation law

For the coordinate system transformation, components $\overset{*}{A}_{ij}$ of a tensor in the new coordinate system can be expressed through components A_{ij} of this tensor in the original coordinate system:

$$\begin{aligned}\overset{*}{A}_{ij} = T_{im}T_{jn}A_{mn} = {} & T_{i1}T_{j1}A_{11} + T_{i1}T_{j2}A_{12} + T_{i1}T_{j3}A_{13} + \\ & T_{i2}T_{j1}A_{21} + T_{i2}T_{j2}A_{22} + T_{i2}T_{j3}A_{23} + \\ & T_{i3}T_{j1}A_{31} + T_{i3}T_{j2}A_{32} + T_{i3}T_{j3}A_{33}\end{aligned} \tag{2.20}$$

Example: For the two dimensional case in Figure 2.2, equation (2.20) becomes:

$$\overset{*}{A}_{ij} = T_{i1}T_{j1}A_{11} + T_{i1}T_{j2}A_{12} + T_{i2}T_{j1}A_{21} + T_{i2}T_{j2}A_{22} \tag{2.21}$$

where $\overset{*}{A}_{ij} = \begin{bmatrix} \overset{*}{A}_{11} & \overset{*}{A}_{12} \\ \overset{*}{A}_{21} & \overset{*}{A}_{22} \end{bmatrix}$ and $T_{ij} = \begin{bmatrix} \cos\alpha & \sin\alpha \\ -\sin\alpha & \cos\alpha \end{bmatrix}$ (from 2.19). This leads to the following equations for the new tensor components:

$$\begin{aligned}\overset{*}{A}_{11} &= A_{11}\cos^2\alpha + A_{22}\sin^2\alpha + \cos\alpha\sin\alpha\left(A_{12} + A_{21}\right) \\ \overset{*}{A}_{22} &= A_{11}\sin^2\alpha + A_{22}\cos^2\alpha - \cos\alpha\sin\alpha\left(A_{12} + A_{21}\right) \\ \overset{*}{A}_{12} &= \left(A_{22} - A_{11}\right)\cos\alpha\sin\alpha + A_{12}\cos^2\alpha - A_{21}\sin^2\alpha \\ \overset{*}{A}_{21} &= \left(A_{22} - A_{11}\right)\cos\alpha\sin\alpha + A_{21}\cos^2\alpha - A_{12}\sin^2\alpha\end{aligned} \tag{2.22}$$

2.3.3 Properties of the transformation

The following properties of the tensor transformation follow from the properties (2.16) and (2.18) of the transformation matrix T_{ij} :

$$A^*_{ii} = T_{im}T_{in}A_{mn} = \delta_{mn}A_{mn} = A_{nn} \tag{2.23}$$

$$A^*_{ij}A^*_{ij} = T_{im}T_{jn}A_{mn}T_{is}T_{jk}A_{sk} = \delta_{ms}\delta_{nk}A_{mn}A_{sk} = A_{mn}A_{mn} \tag{2.24}$$

$$A^*_{ij}A^*_{jk}A^*_{ki} = T_{im}T_{jn}A_{mn}T_{jr}T_{kp}A_{rp}T_{ks}T_{it}A_{st} = \delta_{mt}\delta_{nr}\delta_{ps}A_{mn}A_{rp}A_{st} = A_{mr}A_{rs}A_{sm} \tag{2.25}$$

Clearly, expressions (2.23) - (2.25) do not change with the coordinate transformation, i.e. they are *invariants* of the tensor. This is not at all surprising: in general, tensor expressions, in which all indices are dummy indices, are invariants. Simply because they are scalars! Tensor can have infinitely many invariants, but the above three can be identified as independent – others can always be expressed through them.

2.3.4 Geometric interpretation

The following geometric interpretation demonstrates important property of a tensor. Consider a vector $\overline{n} = n_1e_1 + n_2e_2 + n_3e_3$ defined through its components n_i in the coordinate system, defined by orthogonal unit vectors e_1, e_2, e_3. We require that vector $\overline{n}$ is also a unit vector, so that $n_1^2 + n_2^2 + n_3^2 = 1$. Each vector $\overline{n}$ defines a plane, perpendicular to it (Figure 2.3). When multiplied with tensor A_{ij}, vector $\overline{n}$ gives another vector:

$$a_j^{\overline{n}} = A_{ij}n_i = A_{1j}n_1 + A_{2j}n_2 + A_{3j}n_3 \tag{2.26}$$

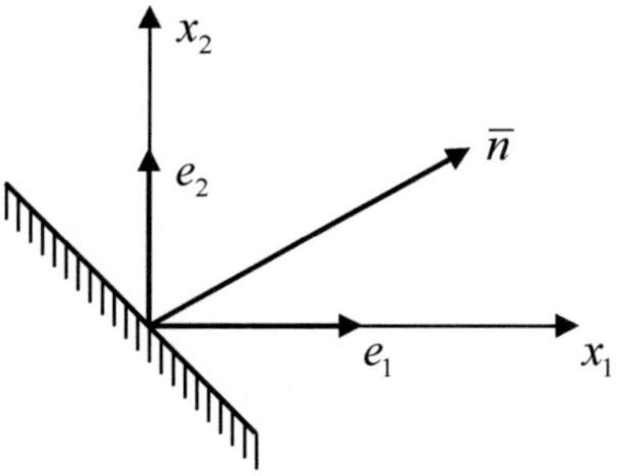

Figure 2.3 Geometric interpretation of tensor.

Let us now assume that vector $\overline{n}$ coincides with one of the basis vectors, e.g. $\overline{n} = e_1$. In this case

$$n_i = \begin{Bmatrix} 1 \\ 0 \\ 0 \end{Bmatrix} \text{ and from (2.26): } a_j^{e_1} = A_{1j} = \begin{Bmatrix} A_{11} \\ A_{12} \\ A_{13} \end{Bmatrix}$$

Repeating the same procedure for $\overline{n} = e_2$ and $\overline{n} = e_3$ we obtain:

$$a_j^{e_i} = A_{ij} \tag{2.27}$$

Clearly, i^{th} row of A_{ij} is a vector conjugate to the direction e_i. Therefore, the tensor A_{ij} is completely defined if three vectors $a_j^{e_i}$ conjugate to e_1, e_2, e_3 are known. Then also a vector $a_j^{\overline{n}}$ conjugate to any other direction $\overline{n}$ can be found from (2.26).

This property of the tensor - to associate a vector with any plane in the continuum - is the major reason why we use tensors in Continuum Mechanics. This property also allows us to define tensor components in mechanical applications.

2.3.5 Characteristic equation

Let us explore a bit this important property of the tensor to associate a vector $a_j^{\overline{n}} = A_{ij} n_i$ with any plane in the continuum, defined by its unit normal $\overline{n}$. Would it be possible to find a plane, where the resulting vector $a_j^{\overline{n}} = A_{ij} n_i$ is perpendicular to this plane, i.e. $a_j^{\overline{n}} = \lambda n_j$ (λ is a scalar)?

This plane exists if we can find a vector n_j such that:

$$A_{ij} n_i = \lambda n_j \tag{2.28}$$

In order to find λ and n_j we transform (2.28):

$$A_{ij} n_i - \lambda n_j = A_{ij} n_i - \lambda \delta_{ij} n_i = \left(A_{ij} - \lambda \delta_{ij}\right) n_i = 0 \tag{2.29}$$

which has a solution for $\overline{n} \neq 0$ only if the determinant of the matrix is zero:

$$\left|A_{ij} - \lambda \delta_{ij}\right| = 0 \quad \text{or} \quad \begin{vmatrix} A_{11} - \lambda & A_{12} & A_{13} \\ A_{21} & A_{22} - \lambda & A_{23} \\ A_{31} & A_{32} & A_{33} - \lambda \end{vmatrix} = 0 \tag{2.30}$$

resulting in the following characteristic equation:

$$\lambda^3 - I_{1A} \lambda^2 + I_{2A} \lambda - I_{3A} = 0 \tag{2.31}$$

with coefficients:

$$I_{1A} = A_{ii} = A_{11} + A_{22} + A_{33} \tag{2.32}$$

$$I_{2A} = \begin{vmatrix} A_{11} & A_{12} \\ A_{21} & A_{22} \end{vmatrix} + \begin{vmatrix} A_{22} & A_{23} \\ A_{32} & A_{33} \end{vmatrix} + \begin{vmatrix} A_{11} & A_{13} \\ A_{31} & A_{33} \end{vmatrix} = \frac{1}{2}\left[A_{ii}A_{jj} - A_{ij}A_{ji}\right] \tag{2.33}$$

$$I_{3A} = \begin{vmatrix} A_{11} & A_{12} & A_{13} \\ A_{21} & A_{22} & A_{23} \\ A_{31} & A_{32} & A_{33} \end{vmatrix} = \frac{1}{6}\left(A_{ii}A_{jj}A_{kk} + 2A_{ij}A_{jk}A_{ki} - 3A_{ij}A_{ji}A_{kk}\right) \tag{2.34}$$

These coefficients are also invariants, because they are functions of the three tensor invariants (2.23) - (2.25).

Once λ is determined from the characteristic equation (2.31), it is substituted back into the equation (2.29) together with the condition $n_1^2 + n_2^2 + n_3^2 = 1$, and the unit normal n_i to the plane is found. These planes are of special importance in Continuum Mechanics.

Chapter 3

Stresses

TABLE OF CONTENTS

Chapter 3

Stresses

3.1 The principle of stresses

Stress is not a measurable quantity - the measurable quantity is force. There are two types of forces acting on a body: volume or mass forces and surface tractions. Stress is the reaction of the body (at each material point) to these forces. This section explores the concept of stress with a special emphasis on its tensorial character.

3.1.1 The principle of stresses (Euler-Cauchy)

We cut a body B by an imaginary plane with normal $\overline{n}$ into two parts B_1 and B_2 (Figure 3.1). These two parts interact on the plane and this interaction is between the surface forces. An elementary area ΔA around the point P of the part B_2 on the plane $\overline{n}$ is subjected to the action of the force ΔF and moment ΔM from the part B_1.

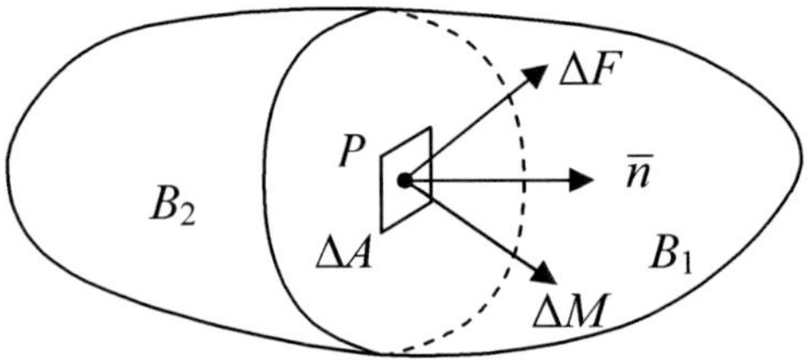

Figure 3.1 The Principle of stresses

According to the principle of stresses of Euler-Cauchy, the following two equations hold:

$$\lim_{\Delta A \to 0} \frac{\Delta F}{\Delta A} = \frac{dF}{dA} = \sigma^{\overline{n}} \quad \text{and} \quad \lim_{\Delta A \to 0} \frac{\Delta M}{\Delta A} = 0 \tag{3.1}$$

where $\sigma^{\overline{n}}$ - is the stress vector at P.

3.1.2 The Third Law of Newton

According to the Third Law of Newton - for every action there is an equal and opposite reaction - an elementary area ΔA around the point P of the part B_1 on the plane $\overline{n}$ is subjected to the action of exactly the same force ΔF

and moment ΔM as the part B_2, only in the opposite direction. Then from equation (3.1) it follows that

$$\sigma^{-\overline{n}} = -\sigma^{\overline{n}} \tag{3.2}$$

3.2 Stress tensor

If only vector $\sigma^{\overline{n}}$ alone could define the state of stress in P, our life would be much easier. Unfortunately, when we repeat the procedure of the previous section for different directions of the plane normal $\overline{n}$ in P, we are going to end up with different $\sigma^{\overline{n}}$. Therefore, in order to define the state of stress in P we need to know $\sigma^{\overline{n}}(P)$ for all possible directions $\overline{n}$, which are infinite in number. This would be an impossible task, if the notion of tensor did not exist. In fact, it was exactly this problem that lead to the discovery of tensors, which in translation from Latin mean "stress."

3.2.1 Equilibrium

The good news is that because of the equilibrium, the vectors $\sigma^{\overline{n}}(P)$ are not independent. For brevity, we consider a two dimensional case (Figure 3.2). All three stress vectors in Figure 3.2: $\sigma^{e_1}(P)$, $\sigma^{e_2}(P)$, $\sigma^{\overline{n}}(P)$ and the normal $\overline{n}$ can be expressed through their components in the coordinate system $(e_1 \perp e_2)$:

$$\sigma^{e_1} = e_1\sigma_1^{e_1} + e_2\sigma_2^{e_1} = e_j\sigma_j^{e_1} \quad \text{and} \quad \sigma^{e_2} = e_1\sigma_1^{e_2} + e_2\sigma_2^{e_2} = e_j\sigma_j^{e_2} \tag{3.3}$$

$$\sigma^n = e_1\sigma_1^n + e_2\sigma_2^n = e_j\sigma_j^n \quad \text{and} \quad n = e_1 n_1 + e_2 n_2 = e_j n_j \tag{3.4}$$

Let us consider equilibrium of forces acting on the wedge in Figure 3.2. The areas of the sides of the wedge are:

$$dA_1 = n_1 dA \quad dA_2 = n_2 dA \quad dA_i = n_i dA \tag{3.5}$$

Then the equilibrium along the axis x_1 yields:

$$\sigma_1^{\overline{n}} dA = \sigma_1^{e_1} dA_1 + \sigma_1^{e_2} dA_2, \text{ so that } \sigma_1^{\overline{n}} = \sigma_1^{e_1} n_1 + \sigma_1^{e_2} n_2 = \sigma_1^{e_i} n_i \tag{3.6}$$

The equilibrium along the axis x_2 yields:

$$\sigma_2^{\overline{n}} dA = \sigma_2^{e_1} dA_1 + \sigma_2^{e_2} dA_2, \text{ so that } \sigma_2^{\overline{n}} = \sigma_2^{e_1} n_1 + \sigma_2^{e_2} n_2 = \sigma_2^{e_i} n_i \tag{3.7}$$

or in general:

$$\sigma_j^{\bar{n}} = \sigma_j^{e_i} n_i \tag{3.8}$$

Therefore, in 2D, in order to calculate $\sigma^{\bar{n}}(P)$ on any plane $\bar{n}$ it is sufficient to know the two vectors $\sigma^{e_1}(P)$ and $\sigma^{e_2}(P)$ on any two mutually orthogonal planes $(e_1 \perp e_2)$. This result can be also extended to a 3D case (see below).

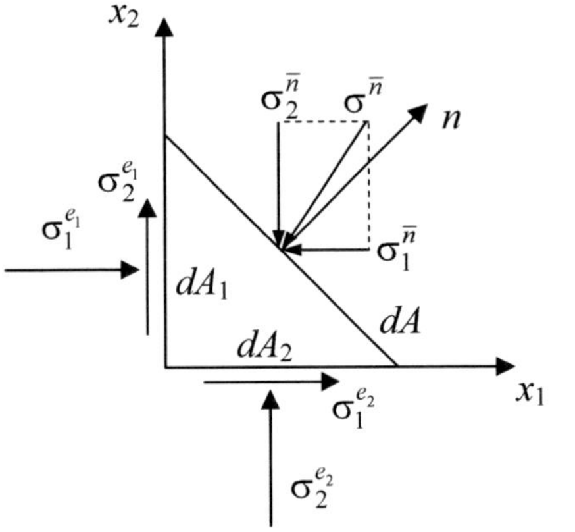

Figure 3.2 Equilibrium

3.2.2 Components of the stress vectors in 3D

Let us, for example, consider vector $\sigma^{e_3}(P)$ in Figure 3.3. It can be expressed through its components in the coordinate system $(e_1 \perp e_2 \perp e_3)$:

$$\sigma^{e_3} = e_1\sigma_1^{e_3} + e_2\sigma_2^{e_3} + e_3\sigma_3^{e_3} = e_k\sigma_k^{e_3} \tag{3.9}$$

and in general:

$$\sigma^{e_i} = e_k\sigma_k^{e_i} = e_k\sigma_{ik} \tag{3.10}$$

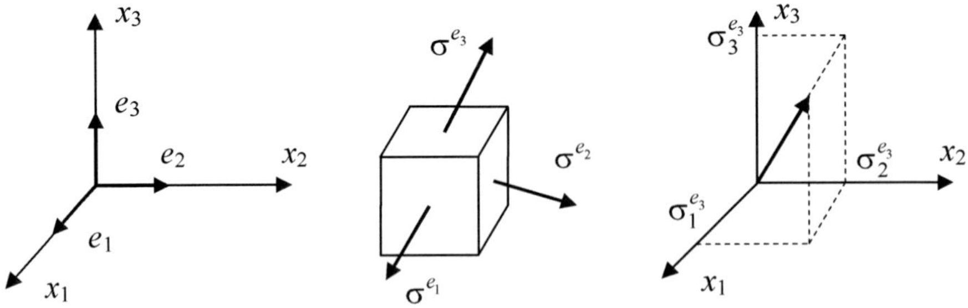

Figure 3.3 Stress vectors

where $\sigma_{ik} = \sigma_k^{e_i}$ is the simplified notation for the components of the stress vector $\sigma^{e_i}(P)$ in the coordinate system $(e_1 \perp e_2 \perp e_3)$. Multiplying equation (3.10) by the basis vector e_j we obtain:

$$e_j \sigma^{e_i} = e_j e_k \sigma_{ik} = \delta_{jk} \sigma_{ik} = \sigma_{ij} \tag{3.11}$$

Equation (3.11) defines the component $\sigma_{ik} = \sigma_k^{e_i}$ of the stress vector $\sigma^{e_i}(P)$ along the direction e_k.

Introducing this simplified notation $\sigma_k^{e_i} = \sigma_{ik}$ into equation (3.8), which also holds for the three dimensional case, we obtain:

$$\sigma_j^{\overline{n}} = \sigma_{ij} n_i \tag{3.12}$$

Equation (3.12) allows for $\sigma^{\overline{n}}(P)$ on any plane $\overline{n}$ to be calculated via the three components $\sigma_{ik} = \sigma_k^{e_i}$ of each of the three vectors $\sigma^{e_1}(P)$, $\sigma^{e_2}(P)$ and $\sigma^{e_3}(P)$ (total $3 \times 3 = 9$) on any three mutually orthogonal planes (e_1, e_2, e_3).

3.2.3 Tensor of stresses

Equation (3.12) can be immediately recognized as the important tensor property (2.26). But is the matrix of vector components σ_{ij} really a tensor?

If we choose a different set of three mutually orthogonal planes $(\overset{*}{e}_1, \overset{*}{e}_2, \overset{*}{e}_3)$, the three stress vectors acting on the new planes and their components $\overset{*}{\sigma}_{ij}$ will be different. However, these components should describe the same state of stress in P. Therefore, the stress vector $\sigma^{\overline{n}}(P) = \sigma_j^{\overline{n}} e_j = \left(\sigma_j^{\overline{n}}\right)^* \overset{*}{e}_j$ acting on a plane $\overline{n} = n_i e_i = \overset{*}{n}_i \overset{*}{e}_i$ should remain the same when calculated for both sets of planes using equation (3.12):

$$\sigma^{\overline{n}} = \sigma_{ij} n_i e_j = \overset{*}{\sigma}_{ij} \overset{*}{n}_i \overset{*}{e}_j \tag{3.13}$$

After replacing dummy indexes $\sigma_{mj} n_m e_j = \overset{*}{\sigma}_{ij} \overset{*}{n}_i \overset{*}{e}_j$, we multiply both parts by $T_{im} \overset{*}{e}_k$, and using equations (2.13) and (2.14) obtain $\sigma_{mj} T_{im} T_{kj} n_m = \overset{*}{\sigma}_{ik} n_m$, which should hold for any $\overline{n}$, so that

$$\overset{*}{\sigma}_{ik} = T_{im} T_{kj} \sigma_{mj} \tag{3.14}$$

which is the tensor transformation law and the definition of tensor.

We conclude that the matrix of vector components σ_{ij} is indeed a tensor:

$$\sigma_{ij} = \begin{bmatrix} \sigma_{11} & \sigma_{12} & \sigma_{13} \\ \sigma_{21} & \sigma_{22} & \sigma_{23} \\ \sigma_{31} & \sigma_{32} & \sigma_{33} \end{bmatrix} = \begin{bmatrix} \sigma_{xx} & \sigma_{xy} & \sigma_{xz} \\ \sigma_{yx} & \sigma_{yy} & \sigma_{yz} \\ \sigma_{zx} & \sigma_{yz} & \sigma_{zz} \end{bmatrix} \tag{3.15}$$

with geometric interpretation given in Figure 3.4, where each component σ_{ij} is represented by a vector. The first subscript in the σ_{ij} denotes the plane on which this vector acts, the second subscript gives the direction of the axis to which this component is parallel. The transformation law for the σ_{ij} is given by equation (3.14).

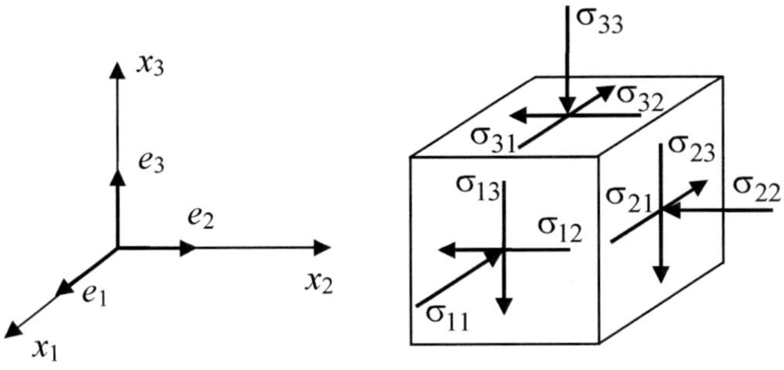

Figure 3.4 The stress tensor

3.2.4 Normal and shear stresses

In order to complete the interpretation of the σ_{ij} tensor components, we note that any vector $\sigma^{\bar{n}}$ can be presented as a sum of two vectors: σ_n - parallel to the normal $\bar{n}$ and called the normal stress; and σ_t - perpendicular to the normal $\bar{n}$ (i.e., parallel to the plane) and called the shear stress, so that:

$$\left|\sigma^{\bar{n}}\right|^2 = \sigma_n^2 + \sigma_t^2 \tag{3.16}$$

In 3D any vector should have 3 components, therefore, these stresses, σ_n and σ_t are not real vector components, unless σ_t is also presented as a sum of two orthogonal components. These two components of σ_t are normally also called shear stresses. It follows (Figure 3.4):

$$\sigma_{ij} = \begin{cases} normal & i = j \\ shear & i \neq j \end{cases}$$

and the strain tensor is often written as:

$$\sigma_{ij} = \begin{bmatrix} \sigma_{11} & \sigma_{12} & \sigma_{13} \\ \sigma_{21} & \sigma_{22} & \sigma_{23} \\ \sigma_{31} & \sigma_{32} & \sigma_{33} \end{bmatrix} = \begin{bmatrix} \sigma_x & \tau_{xy} & \tau_{xz} \\ \tau_{yx} & \sigma_y & \tau_{yz} \\ \tau_{zx} & \tau_{yz} & \sigma_z \end{bmatrix} \quad (3.17)$$

3.3 Equations of motion

Components of the stress tensor σ_{ij} are not independent, but related to each other and to the body forces by equations of motion (or in statics – equilibrium).

3.3.1 Forces

Let us consider the forces acting on an elementary volume with coordinates (x_1, x_2) and dimensions $dx_1 \times dx_2$ (Figure 3.5). The element is subjected to the action of the following forces:

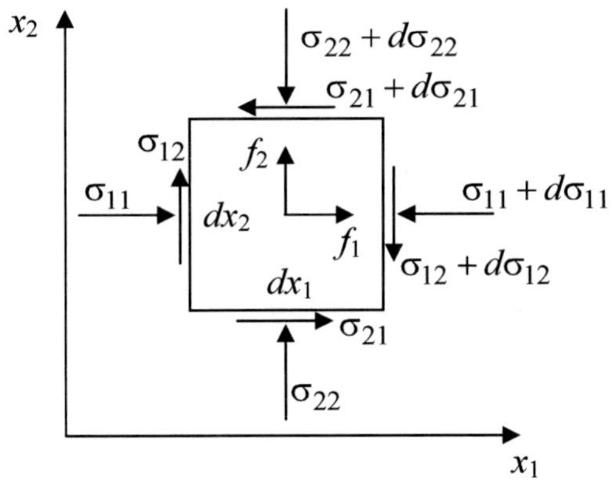

Figure 3.5 Equations of equilibrium.

a) Body forces:

$$\bar{f} = f_1 e_1 + f_2 e_2 = f_j e_j$$

b) Stresses:
The stress state in general is a function of coordinate, but because the element is small, the good approximation to stress increments is provided by the first term in Taylor series:

$$d\sigma_{11} = \sigma_{11}(x_1 + dx, x_2) - \sigma_{11}(x_1, x_2) = \frac{\partial \sigma_{11}}{\partial x_1} dx_1$$

$$d\sigma_{12} = \frac{\partial \sigma_{12}}{\partial x_1} dx_1; \quad d\sigma_{21} = \frac{\partial \sigma_{21}}{\partial x_2} dx_2; \quad d\sigma_{22} = \frac{\partial \sigma_{22}}{\partial x_2} dx_2.$$

3.3.2 Equilibrium of forces

Then the equilibrium in x_1 direction yields:

$$\left(\sigma_{11}+\frac{\partial\sigma_{11}}{\partial x_1}dx_1\right)dx_2-\sigma_{11}dx_2+\left(\sigma_{21}+\frac{\partial\sigma_{21}}{\partial x_2}dx_2\right)dx_1-\sigma_{21}dx_1-f_1dx_1dx_2=0$$

which can be simplified to give

$$\frac{\partial\sigma_{11}}{\partial x_1}+\frac{\partial\sigma_{21}}{\partial x_2}=f_1$$

The equilibrium in x_2 direction yields a similar expression:

$$\frac{\partial\sigma_{12}}{\partial x_1}+\frac{\partial\sigma_{22}}{\partial x_2}=f_2$$

In index notation these two equations can be written as:

$$\sigma_{ij,i}=f_j \tag{3.18}$$

3.3.3 Equilibrium of moments

The equilibrium of moments gives:

$$\left(\sigma_{12}+\frac{\partial\sigma_{12}}{\partial x_1}dx_1+\sigma_{12}\right)dx_2\frac{dx_1}{2}-\left(\sigma_{21}+\frac{\partial\sigma_{21}}{\partial x_2}dx_2+\sigma_{21}\right)dx_1\frac{dx_2}{2}=0$$

This can be simplified to

$$2\sigma_{12}+\frac{\partial\sigma_{12}}{\partial x_1}dx_1=2\sigma_{21}+\frac{\partial\sigma_{21}}{\partial x_2}dx_2$$

and further (because the terms with dx_1 and dx_2 are small relative to the main terms) to $\sigma_{12}=\sigma_{21}$, which in index notation becomes:

$$\sigma_{ij}=\sigma_{ji} \tag{3.19}$$

It follows that the stress tensor is *symmetric*.

3.3.4 Boundary conditions

Static boundary conditions are also determined from equilibrium considerations (Figure 3.6). Surface traction $t^{\overline{n}}$ acts at the boundary at the point with the normal $\overline{n}$. From the Third Law of Newton (equation 3.2): $t^{\overline{n}}=-\sigma^{-\overline{n}}=\sigma^{\overline{n}}$, which after substitution of equation (3.12) gives

$$t_i^{\overline{n}} = \sigma_{ij} n_j \tag{3.20}$$

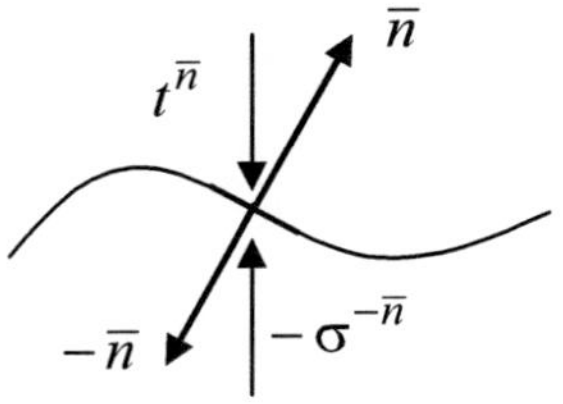

Figure 3.6 Boundary conditions

3.4 Principal stresses

3.4.1 Definition

In previous sections we established that stress state $\sigma_{ij}(P)$ is a tensor. It associates a vector $\sigma^{\overline{n}}(P)$ with any plane $\overline{n}$ passing through the point P (Figure 3.7). Each stress vector $\sigma^{\overline{n}}(P)$ has a normal and a shear component. Do there exist any planes where the stress vector is perpendicular to the plane (i.e., shear stress is zero)? If such a plane exists, it is called a *principal plane*, and its normal is called a *principal direction* (or *axis*). The stress vector (and its length σ) acting on such a plane is called a *principal stress*:

$$\sigma^{\overline{n}} = \sigma \overline{n} \tag{3.21}$$

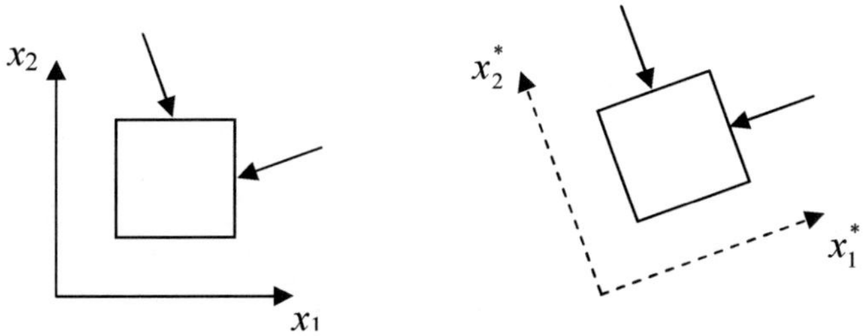

Figure 3.7 Principal stresses

3.4.2 Characteristic equation

Existence of the principal planes and stresses depends on existence of a solution of the corresponding characteristic equation (Section 2.3.5):

$$\sigma_j^{\bar{n}} = \sigma_{ij} n_i = \sigma n_j = \sigma \delta_{ij} n_i, \Rightarrow (\sigma_{ij} - \sigma\delta_{ij}) n_i = 0$$

$$\text{or} \quad \begin{vmatrix} \sigma_{11} - \sigma & \sigma_{12} & \sigma_{13} \\ \sigma_{21} & \sigma_{22} - \sigma & \sigma_{23} \\ \sigma_{31} & \sigma_{32} & \sigma_{33} - \sigma \end{vmatrix} = 0$$

which yields:

$$\sigma^3 - I_{1\sigma}\sigma^2 + I_{2\sigma}\sigma - I_{3\sigma} = 0 \tag{3.22}$$

This third order algebraic equation can be rewritten via its roots $\sigma_1, \sigma_2, \sigma_3$:

$$(\sigma - \sigma_1)(\sigma - \sigma_2)(\sigma - \sigma_3) = 0$$

So that:

$$I_{1\sigma} = \sigma_1 + \sigma_2 + \sigma_3; \quad I_{2\sigma} = \sigma_1\sigma_2 + \sigma_2\sigma_3 + \sigma_3\sigma_1; \quad I_{3\sigma} = \sigma_1\sigma_2\sigma_3. \tag{3.23}$$

3.4.3 Properties of the principle stresses and planes

a) All three roots of the equation (3.22) are *real*:

Proof: One root of the third order equation with real coefficients is always real. Assume that σ_1 is real and transform the axes, so that x_1^* is now the principal direction. Then $\overset{*}{\sigma}_{11} = \sigma_1$, $\overset{*}{\sigma}_{12} = \overset{*}{\sigma}_{21} = \overset{*}{\sigma}_{31} = \overset{*}{\sigma}_{13} = 0$ and the characteristic equation becomes

$$\begin{vmatrix} \sigma_1 - \sigma & 0 & 0 \\ 0 & \overset{*}{\sigma}_{22} - \sigma & \overset{*}{\sigma}_{23} \\ 0 & \overset{*}{\sigma}_{32} & \overset{*}{\sigma}_{33} - \sigma \end{vmatrix} = 0$$

so that

$$\sigma_{2,3} = \frac{1}{2}\left[\left(\overset{*}{\sigma}_{22} + \overset{*}{\sigma}_{33}\right) \pm \sqrt{\left(\overset{*}{\sigma}_{22} - \overset{*}{\sigma}_{33}\right)^2 + 4\overset{*}{\sigma}_{23}{}^2}\right]$$

which are both real, because the square root expression is always non-negative.

b) Principal stresses are *extremal* normal stresses:

Proof. Normal stress is a projection of the stress vector $\sigma^{\bar{n}}$ on the normal $\bar{n}$ ($n_1^2+n_2^2+n_3^2=1$): $\sigma_n=\sigma^{\bar{n}}\bar{n}=\sigma_{ij}n_jn_i$. In order to find extremal values of σ_n, we define the objective function using the constraint $n_1^2+n_2^2+n_3^2-1=0$:

$$F(n_1,n_2,n_3)=\sigma_n-\lambda\left(n_1^2+n_2^2+n_3^2-1\right)$$

where λ is Lagrange multiplier. This function achieves extremal values at

$$\frac{\partial F}{\partial n_i}=2\sigma_{ij}n_j-2\lambda n_i=0$$

which is easily recognizable as the characteristic equation $\left(\sigma_{ij}-\lambda\delta_{ij}\right)n_j=0$. It follows, that the principal stresses $\lambda=\sigma$ provide extremal values for σ_n.

c) Principal directions are *orthogonal*:

Proof. Assume that $\sigma_1\neq\sigma_2\neq\sigma_3$. Then the stress vectors on the principal planes $\bar{n}^{(1)}$ and $\bar{n}^{(2)}$ are given by

$$\sigma^{(1)}=\sigma_{ij}n_j^{(1)}=\sigma_1n_i^{(1)}\text{ and }\sigma^{(2)}=\sigma_{ij}n_j^{(2)}=\sigma_2n_i^{(2)}$$

Multiplying these two equations by $\bar{n}^{(2)}$ and $\bar{n}^{(1)}$, respectively, we obtain:

$$\sigma_{ij}n_j^{(1)}n_i^{(2)}=\sigma_1n_i^{(1)}n_i^{(2)}\text{ and }\sigma_{ij}n_j^{(2)}n_i^{(1)}=\sigma_2n_i^{(2)}n_i^{(1)}$$

From $\sigma_{ij}=\sigma_{ji}$ it follows that $\sigma_{ij}n_j^{(1)}n_i^{(2)}=\sigma_{ij}n_i^{(1)}n_j^{(2)}$, therefore:

$$(\sigma_1-\sigma_2)n_i^{(1)}n_i^{(2)}=0$$

which for $\sigma_1\neq\sigma_2\neq\sigma_3$ is only possible when $n^{(1)}\perp n^{(2)}$.

In order to find the principle directions, we solve 3 independent systems of linear algebraic equations:

$$\left(\sigma_{ij}-\sigma_1\delta_{ij}\right)n_j^{(1)}=0;\ \left(\sigma_{ij}-\sigma_2\delta_{ij}\right)n_j^{(2)}=0;\ \left(\sigma_{ij}-\sigma_3\delta_{ij}\right)n_j^{(3)}=0,$$

each supplemented by the condition $n_1^2+n_2^2+n_3^2=1$.

d) Any stress tensor in any axes can be expressed via the principal stresses:

Proof. In principal axes:

$$\sigma_{ij} = \begin{vmatrix} \sigma_1 & 0 & 0 \\ 0 & \sigma_2 & 0 \\ 0 & 0 & \sigma_3 \end{vmatrix} \qquad \sigma_1 > \sigma_2 > \sigma_3$$

Rotation of the axes will cause the tensor components to change:

$$\sigma_{ij}^* = T_{im} T_{jn} \sigma_{mn} = T_{i1} T_{j1} \sigma_1 + T_{i2} T_{j2} \sigma_2 + T_{i3} T_{j3} \sigma_3$$

From definition of the transformation matrix it follows that any stress tensor can be expressed via the principal stresses and directions:

$$\sigma_{ij} = \sigma_1 n_i^{(1)} n_j^{(1)} + \sigma_2 n_i^{(2)} n_j^{(2)} + \sigma_3 n_i^{(3)} n_j^{(3)}$$

where $n_i^{(k)}$ are the cosines of the principal directions in the chosen coordinate system.

3.4.4 Principal shear stresses

Shear stresses on principal planes are zero. However, on certain planes they also achieve extremal (principal) values.

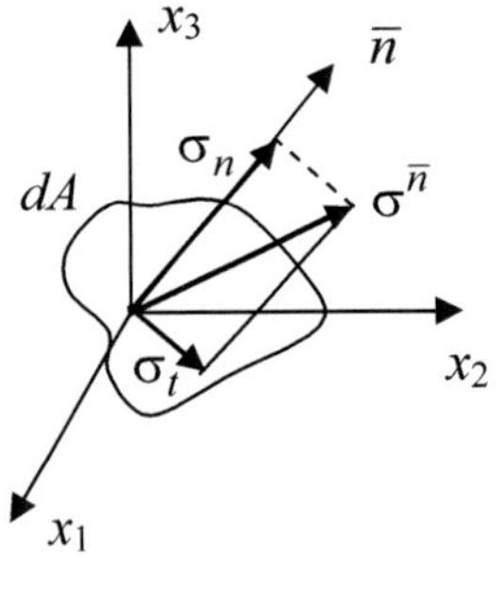

Figure 3.8 Shear stresses

The shear stresses are defined as (Figure 3.8):

$$\sigma_t^2 = \left|\sigma^{\overline{n}}\right|^2 - \sigma_n^2 \tag{3.24}$$

The stress vector is given by $\sigma_i^{\overline{n}} = \sigma_{ij} n_j$ so that in principal axes: $\sigma_1^{\overline{n}} = \sigma_1 n_1$, $\sigma_2^{\overline{n}} = \sigma_2 n_2$ and $\sigma_3^{\overline{n}} = \sigma_3 n_3$:

$$\left|\sigma^{\overline{n}}\right|^2 = (\sigma_1 n_1)^2 + (\sigma_2 n_2)^2 + (\sigma_3 n_3)^2 \tag{3.25}$$

The normal component of the stress vector is given by its projection on the normal $\sigma_n = \sigma^{\overline{n}} \overline{n} = \sigma_{ij} n_i n_j$, or in principal axes:

$$\sigma_n = \sigma_1 n_1^2 + \sigma_2 n_2^2 + \sigma_3 n_3^2 \tag{3.26}$$

Substitution of (3.25) and (3.26) into (3.24) gives:

$$\sigma_t^2 = (\sigma_1 - \sigma_2)^2 n_1^2 n_2^2 + (\sigma_2 - \sigma_3)^2 n_2^2 n_3^2 + (\sigma_3 - \sigma_1)^2 n_1^2 n_3^2 \tag{3.27}$$

which together with the constraint $n_1^2 + n_2^2 + n_3^2 - 1 = 0$ leads to the following objective function:

$$G(n_1, n_2, n_3) = \sigma_t^2 - \lambda\left(n_1^2 + n_2^2 + n_3^2 - 1\right) \tag{3.28}$$

where λ is the Lagrangian multiplier.

The planes of the extremal shear stresses (in principal axes), the principal shear stresses and the corresponding normal stresses are then derived as follows:

$$\frac{\partial G}{\partial n_i} = 0 \qquad \begin{array}{lll} n_1 = 0,\ n_2 = n_3 = \dfrac{\sqrt{2}}{2}, & \sigma_{t1} = \pm\dfrac{(\sigma_2 - \sigma_3)}{2} & \sigma_{n(\tau_1)} = \dfrac{\sigma_2 + \sigma_3}{2} \\ n_2 = 0,\ n_1 = n_3 = \dfrac{\sqrt{2}}{2}, & \sigma_{t2} = \pm\dfrac{(\sigma_1 - \sigma_3)}{2} \ \Rightarrow & \sigma_{n(\tau_2)} = \dfrac{\sigma_1 + \sigma_3}{2} \\ n_3 = 0,\ n_1 = n_2 = \dfrac{\sqrt{2}}{2}, & \sigma_{t3} = \pm\dfrac{(\sigma_1 - \sigma_2)}{2} & \sigma_{n(\tau_3)} = \dfrac{\sigma_1 + \sigma_2}{2} \end{array}$$

It follows that the planes of principal shear stresses contain one of the principal axes and has a $\pi/2$ angle with two other axes. If the principal stresses are chosen so that: $\sigma_1 > \sigma_2 > \sigma_3$, then the maximum shear stress acts on the plane shown in Figure 3.9 and is given by:

$$\tau_{max} = \frac{1}{2}(\sigma_1 - \sigma_3) = |\tau_2|$$

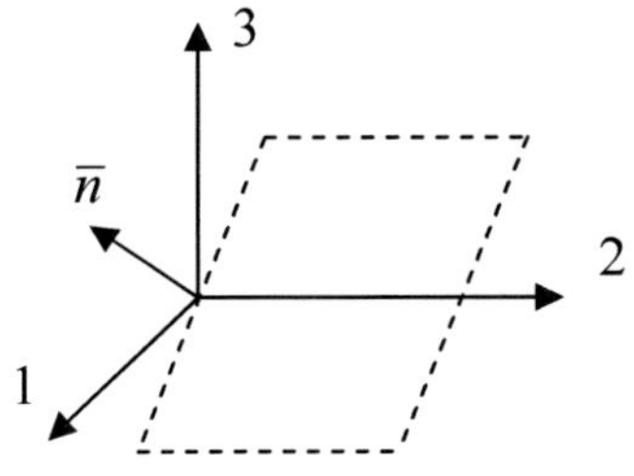

Figure 3.9 The plane of the maximum shear stress.

3.5 Stress tensor invariants

This section summarizes various expressions for the stress tensor invariants and relationships between them.

3.5.1 Stress tensor invariants

$$J_1 = \sigma_{ii}; \quad J_2 = \frac{1}{2}\sigma_{ij}\sigma_{ij}; \quad J_3 = \frac{1}{3}\sigma_{ik}\sigma_{km}\sigma_{mi}. \tag{3.29}$$

3.5.2 Coefficients of the characteristic equation

$$I_{1\sigma} = \sigma_{ii} = J_1 \tag{3.30}$$

$$I_{2\sigma} = \frac{1}{2}\left(\sigma_{ii}\sigma_{jj} - \sigma_{ij}\sigma_{ij}\right) = \frac{1}{2}J_1^2 - J_2 \tag{3.31}$$

$$I_{3\sigma} = \frac{1}{6}\left(\sigma_{ii}\sigma_{jj}\sigma_{kk} + 2\sigma_{ij}\sigma_{jk}\sigma_{ki} - 3\sigma_{ij}\sigma_{ij}\sigma_{kk}\right) = \frac{1}{6}J_1^3 - J_1J_2 + J_3 \tag{3.32}$$

3.5.3 Relations between the invariants

$$J_1 = I_{1\sigma} \tag{3.33}$$

$$J_2 = \frac{1}{2}I_{1\sigma}^2 - I_{2\sigma} \tag{3.34}$$

$$J_3 = -\frac{1}{6}I_{1\sigma}^3 + I_{1\sigma}\left(\frac{1}{2}I_{1\sigma}^2 - I_{2\sigma}\right) + I_{3\sigma} = \frac{1}{3}I_{1\sigma}^3 - I_{1\sigma}I_{2\sigma} + I_{3\sigma} \tag{3.35}$$

3.5.4 Decomposition of the stress tensor

Any stress tensor can be decomposed

$$\sigma_{ij} = s_{ij} + \frac{1}{3}\sigma_{nn}\delta_{ij} \tag{3.36}$$

where $\frac{1}{3}\sigma_{nn} = p$ - the hydrostatic (mean) stress;

$s_{ij} = \sigma_{ij} - \frac{1}{3}\sigma_{nn}\delta_{ij}$ - deviatoric stress tensor.

3.5.5 Invariants of the deviatoric stress tensor

$$J_{1D} = s_{ii} = \sigma_{ii} - \frac{1}{3}\sigma_{nn}\delta_{ii} = 0 \tag{3.37}$$

$$\begin{aligned} J_{2D} &= \frac{1}{2}s_{ij}s_{ij} = \frac{1}{2}\left(\sigma_{ij} - \frac{1}{3}\sigma_{nn}\delta_{ij}\right)\left(\sigma_{ij} - \frac{1}{3}\sigma_{nn}\delta_{ij}\right) \\ &= \frac{1}{2}\left(\sigma_{ij}\sigma_{ij} - 2\sigma_{ij}\frac{\sigma_{nn}}{3}\delta_{ij} + \frac{\sigma_{nn}^2}{9}\delta_{ij}\delta_{ij}\right) \\ &= \frac{1}{6}\left[(\sigma_{11}-\sigma_{22})^2 + (\sigma_{22}-\sigma_{33})^2 + (\sigma_{33}-\sigma_{11})^2\right] + \sigma_{12}^2 + \sigma_{23}^2 + \sigma_{31}^2 \\ &= \frac{1}{6}\left[(\sigma_1-\sigma_2)^2 + (\sigma_2-\sigma_3)^2 + (\sigma_3-\sigma_1)^2\right] = J_2 - \frac{1}{6}J_1^2 \end{aligned} \tag{3.38}$$

$$\begin{aligned} J_{3D} &= \frac{1}{3}s_{ij}s_{jk}s_{ki} = \frac{1}{3}\left(\sigma_{ij} - \frac{J_1}{3}\delta_{ij}\right)\left(\sigma_{jk} - \frac{J_1}{3}\delta_{jk}\right)\left(\sigma_{ki} - \frac{J_1}{3}\delta_{ki}\right) \\ &= \frac{1}{3}\left(\sigma_{ij}\sigma_{jk}\sigma_{ki} - \frac{J_1}{3}\sigma_{ij}\sigma_{ij} - \frac{2}{3}J_1\sigma_{im}\sigma_{im} + \frac{2}{9}J_1^2\sigma_{mm} + \frac{J_1^2}{9}\sigma_{mm} - \frac{J_1^3}{27}\delta_{mm}\right) \\ &= J_3 - \frac{2}{3}J_1J_2 + \frac{2}{27}J_1^3 \end{aligned} \tag{3.39}$$

3.6 Octahedral stresses

There is a special plane, defined in principal axes, which possesses unique properties. The normal stress on it is proportional to the first invariant of the stress tensor J_1, while the shear stress is proportional to the square root of the second invariant of the deviatoric stress tensor $\sqrt{J_{2D}}$.

This plane is called the octahedral plane and is defined by the normal equally inclined to all three principal axes:

$$n_1 = n_2 = n_3 = \frac{1}{\sqrt{3}} \tag{3.40}$$

Then, according to equation (3.26), the normal stress on this plane is:

$$\sigma_{oct} = \sigma_1 n_1^2 + \sigma_2 n_2^2 + \sigma_3 n_3^2 = \frac{1}{3}(\sigma_1 + \sigma_2 + \sigma_3) = \frac{J_1}{3} \tag{3.41}$$

while the shear stress can be calculated from (3.27):

$$\begin{aligned}\tau_{oct}^2 &= (\sigma_1-\sigma_2)^2 n_1^2 n_2^2 + (\sigma_2-\sigma_3)^2 n_2^2 n_3^2 + (\sigma_3-\sigma_1)^2 n_1^2 n_3^2 \\ &= \frac{1}{9}\left[(\sigma_1-\sigma_2)^2 + (\sigma_2-\sigma_3)^2 + (\sigma_3-\sigma_1)^2\right] = \frac{2}{3} J_{2D}\end{aligned} \tag{3.42}$$

These two stresses $\sigma_{oct} = J_1/3$ and $\tau_{oct} = \sqrt{2J_{2D}/3}$ are the generalized measures of the hydrostatic and deviatoric loading of the body at the point.

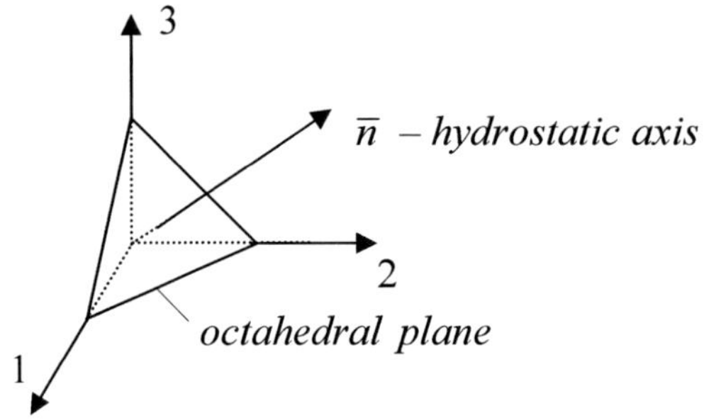

Figure 3.10. Octahedral plane.

Chapter 4

Strains

TABLE OF CONTENTS

Chapter 4

Strains

4.1 Definitions

External forces acting on a non-rigid body cause changes in its shape and volume. These geometric changes are described using the strain tensor.

4.1.1 Assumptions

The following assumptions are made in the Mechanics of Continua:

a) A finite volume cannot disappear;
b) A finite volume cannot become infinite;
c) Continuous body remains continuous during and after deformation;
d) Granular media can be treated as a continuum provided the representative elementary volume is much larger than the grain size.

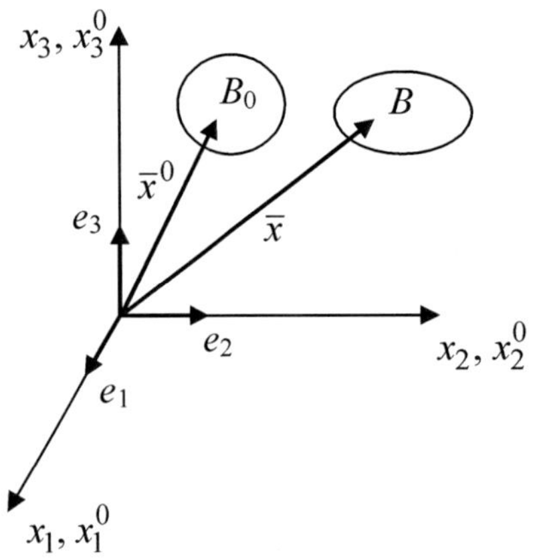

Figure 4.1 Displacements of the body.

Under external loading, the body B_0 moves and changes its shape and volume (Figure 4.1). The new body configuration B can be described using two alternative approaches named after Lagrange and Euler.

4.1.2 Lagrange approach

We attach our local coordinate system to a particle, which is defined by its coordinates in the initial configuration B_0: x_1^0, x_2^0, x_3^0. In the new configuration B after deformation, location of this particle is defined by the vector $\bar{x}$, which is a function of Lagrangian coordinates: $x_i = x_i\left(x_1^0, x_2^0, x_3^0\right)$.

4.1.3 Euler approach

We attach our local coordinate system to a point in space, defined by coordinates: x_1, x_2, x_3, and observe which one of the particles from the initial configuration B_0 has arrived to this point after the deformation.

Location of this particle in the initial configuration B_0 is defined by the vector $\bar{x}^0$, which is a function of Eulerian coordinates: $x_i^0 = x_i^0\left(x_1, x_2, x_3\right)$.

4.2 Strain tensors

4.2.1 Displacements

An elementary vector $d\bar{x}^0 \left(\overline{P_0Q_0}\right)$ in the initial configuration transforms into the vector $d\bar{x} \left(\overline{PQ}\right)$ after deformation (Figure 4.2).

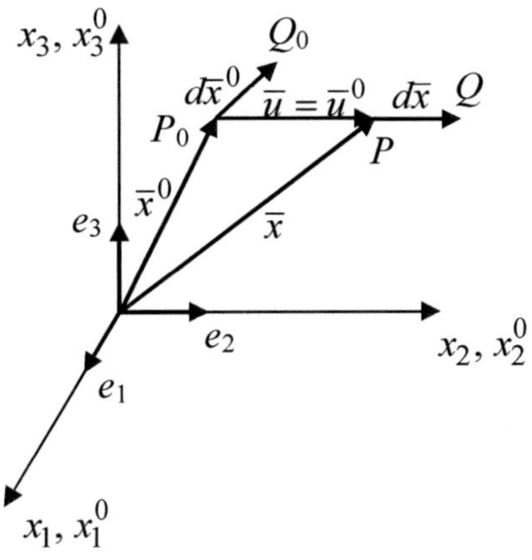

Figure 4.2 Deformations of the body

Displacement of the point P_0 is given by the displacement vector $\left(\overline{P_0P} = \bar{u} = \bar{u}^0\right)$:

$$u_i^0\left(x_1^0, x_2^0, x_3^0\right) = x_i\left(x_1^0, x_2^0, x_3^0\right) - x_i^0 \tag{4.1}$$

$$u_i\left(x_1, x_2, x_3\right) = x_i - x_i^0\left(x_1, x_2, x_3\right) \tag{4.2}$$

in Lagrangian and Eulerian coordinates respectively. Note, that though the displacement vector itself is the same in both coordinates, its function of coordinates is different.

The squared lengths of the elementary vectors $d\overline{x}^0$ and $d\overline{x}$ are, respectively:

$$\left(d\overline{x}^0\right)^2 = dx_i^0 dx_i^0 \text{ and } \left(d\overline{x}\right)^2 = dx_i\, dx_i \tag{4.3}$$

and in order to define strains we are going to consider their difference:

$$\left(d\overline{x}\right)^2 - \left(d\overline{x}^0\right)^2 \tag{4.4}$$

4.2.2 Lagrange-Green strains

In Lagrangian coordinates:

$$\left(d\overline{x}^0\right)^2 = dx_i^0 dx_i^0 = \delta_{mn} dx_m^0 dx_n^0 \tag{4.5}$$

And, since $x_i = x_i\left(x_1^0, x_2^0, x_3^0\right)$, we can write $dx_i = \dfrac{\partial x_i}{\partial x_j^0} dx_j^0$ so that:

$$\left(d\overline{x}\right)^2 = \frac{\partial x_i}{\partial x_m^0}\frac{\partial x_i}{\partial x_n^0} dx_m^0 dx_n^0 \tag{4.6}$$

It follows that:

$$\left(d\overline{x}\right)^2 - \left(d\overline{x}^0\right)^2 = 2e_{mn}^0 dx_m^0 dx_n^0 \tag{4.7}$$

where

$$e_{mn}^0 = \frac{1}{2}\left(\frac{\partial x_i}{\partial x_m^0}\frac{\partial x_i}{\partial x_n^0} - \delta_{mn}\right) \tag{4.8}$$

is the Lagrange-Green strain.

In order to express e_{mn}^0 through displacements (4.1) we write $x_i = x_i^0 + u_i^0$ so that

$$\frac{\partial x_i}{\partial x_j^0} = \frac{\partial x_i^0}{\partial x_j^0} + \frac{\partial u_i^0}{\partial x_j^0} \tag{4.9}$$

For $i = j$: $\dfrac{\partial x_i^0}{\partial x_j^0} = 1$, while for $i \neq j$: $\dfrac{\partial x_i^0}{\partial x_j^0} = 0$, so that (4.9) becomes:

$$\frac{\partial x_i}{\partial x_j^0} = \delta_{ij} + \frac{\partial u_i^0}{\partial x_j^0} \tag{4.10}$$

which after substitution into equation (4.8) gives the Lagrange-Green strain as a function of displacements:

$$e_{mn}^0 = \frac{1}{2}\left(\frac{\partial u_m^0}{\partial x_n^0} + \frac{\partial u_n^0}{\partial x_m^0} + \frac{\partial u_i^0}{\partial x_m^0}\frac{\partial u_i^0}{\partial x_n^0} \right) \tag{4.11}$$

4.2.3 Euler-Almansi strains

Repeating the procedure of the previous section in Eulerian coordinates, we obtain:

$$(d\bar{x})^2 - \left(d\bar{x}^0\right)^2 = 2e_{mn}dx_m dx_n \tag{4.12}$$

where

$$e_{mn} = \frac{1}{2}\left(\frac{\partial u_m}{\partial x_n^0} + \frac{\partial u_n}{\partial x_m^0} - \frac{\partial u_i}{\partial x_m^0}\frac{\partial u_i}{\partial x_n^0} \right) \tag{4.13}$$

is the Euler-Almansi strain.

4.2.4 Is the strain a tensor?

Because $(d\bar{x})^2 - \left(d\bar{x}^0\right)^2$ is a scalar, it should be invariant to the transformation of Lagrange and Euler coordinates. For Euler-Almansi strain this can be expressed as:

$$(d\bar{x})^2 - \left(d\bar{x}^0\right)^2 = 2e_{mn}dx_m dx_n = 2e_{ij}^* dx_i^* dx_j^* \tag{4.14}$$

Because $d\bar{x}$ is a vector, its coordinates transform (equation 2.12) as $dx_m = T_{im}dx_i^*$, which after substitution into (4.14) gives:

$$e_{ij}^* = T_{im}T_{jn}e_{mn} \tag{4.15}$$

which is the transformation law and the definition of the tensor. The same is true for the Lagrange-Green strain.

4.2.5 Cauchy small strain tensor

a) Definition:

Srains are defined as small, when the gradient of all three displacements at any point is small:

$$\frac{\partial u_i^0}{\partial x_j^0} << 1 \text{ and } \frac{\partial u_i}{\partial x_j} << 1 \tag{4.16}$$

b) Small strain tensor:

It follows that at small strains the non-linear terms $\frac{\partial u_i^0}{\partial x_m^0}\frac{\partial u_i^0}{\partial x_n^0}$ and $\frac{\partial u_i}{\partial x_m}\frac{\partial u_i}{\partial x_n}$ in equations (4.11) and (4.13), respectively, become small compared to the linear terms, so that:

$$e_{mn}^0 = \frac{1}{2}\left(\frac{\partial u_m^0}{\partial x_n^0} + \frac{\partial u_n^0}{\partial x_m^0}\right) \text{ and } e_{mn} = \frac{1}{2}\left(\frac{\partial u_m}{\partial x_n} + \frac{\partial u_n}{\partial x_m}\right) \tag{4.17}$$

Thus, at small strains the difference between Lagrangian and Eulerian coordinates vanishes, and the strains can be described using the *symmetric, linear* Cauchy strain tensor:

$$\varepsilon_{mn} = \frac{1}{2}\left(\frac{\partial u_m}{\partial x_n} + \frac{\partial u_n}{\partial x_m}\right) = \frac{1}{2}\left(u_{m,n} + u_{n,m}\right) \tag{4.18}$$

c) Discussion:

In this book we confine our presentation to the case of small strains, therefore, no distinction between Lagrangian and Eulerian coordinates will be made from now on.

What is the small strain in terms of numbers? From (4.16) it follows that even a strain as large as 10% can be considered small, because it makes non-linear terms in equations (4.11) and (4.13), an order of magnitude smaller than the linear terms.

Note, however, that small strains do not necessarily mean small displacements: when integrated over a large area, even very small strains can lead to significant displacements.

4.3 Geometric interpretation

As any second order tensor in 3D, the strain tensor ε_{mn} has 9 components (but only 6 of them are independent, due to the symmetry). What is the physical meaning of these components?

4.3.1 Linear strains

Let us consider a two dimensional case, where a linear element dx_1 undergoes along its axis displacement u_1 at one end, and displacement $u_1 + du_1$ at the other end (Figure 4.3).

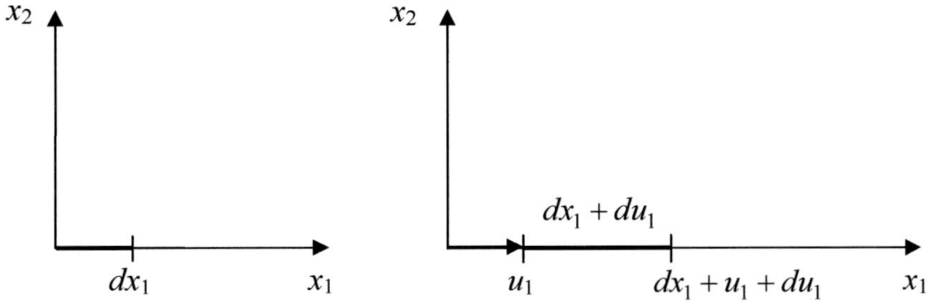

Figure 4.3 Linear strains

The element dx_1 is small, therefore:

$$du_1 = \frac{\partial u_1}{\partial x_1} dx_1 \tag{4.19}$$

and from the definition of the engineering linear strain we obtain:

$$\varepsilon_1 = \frac{l - l_0}{l_0} = \frac{\left(dx_1 + \frac{\partial u_1}{\partial x_1} dx_1 \right) - dx_1}{dx_1} = \frac{\partial u_1}{\partial x_1} = u_{1,1} \tag{4.20}$$

We observe that this expression is equivalent to the following component of the small strain tensor ε_{mn} (equation 4.18):

$$\varepsilon_{11} = \frac{1}{2}\left(u_{1,1} + u_{1,1} \right) \tag{4.21}$$

This allow us to interpret all the components of ε_{mn} with $m = n$ ($\varepsilon_{11}, \varepsilon_{22}, \varepsilon_{33}$) as *linear strains* in the corresponding directions.

4.3.2 Shear strains

Let us now consider a two dimensional case, where an angular element, built of two linear elements dx_1 and dx_2 with a 90° angle between them, undergoes displacements as shown in Figure 4.4.

The elements dx_1 and dx_2 are small, therefore:

$$du_1 = \frac{\partial u_1}{\partial x_2} dx_2 \quad \text{and} \quad du_2 = \frac{\partial u_2}{\partial x_1} dx_1 \tag{4.22}$$

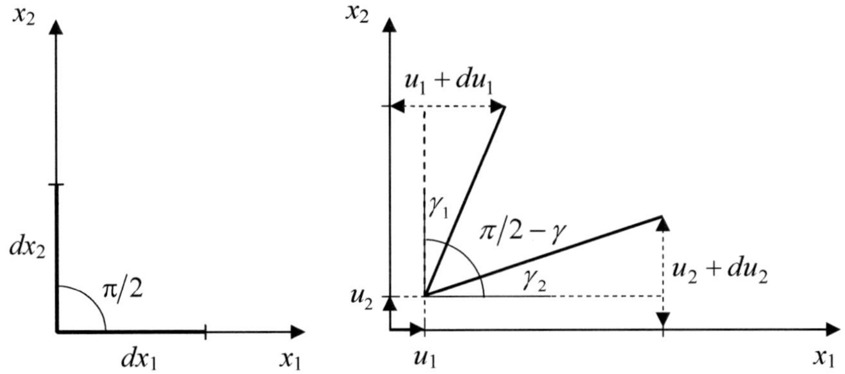

Figure 4.4 Shear strains

and from the definition of the engineering shear strain we obtain:

$$\gamma = \gamma_1 + \gamma_2 = \tan\gamma_1 + \tan\gamma_2 = \frac{\frac{\partial u_1}{\partial x_2} dx_2}{dx_2} + \frac{\frac{\partial u_2}{\partial x_1} dx_1}{dx_1} = \frac{\partial u_1}{\partial x_2} + \frac{\partial u_2}{\partial x_1} \tag{4.23}$$

We observe that a half of this expression is equivalent to the following component of the small strain tensor ε_{mn} (equation 4.18):

$$\varepsilon_{12} = \frac{1}{2}\left(u_{1,2} + u_{2,1}\right) = \frac{1}{2}\gamma_{12} \tag{4.24}$$

This allow us to interpret all the components of ε_{mn} with $m \neq n$ ($\varepsilon_{12}, \varepsilon_{23}, \varepsilon_{31}$) as half of the *shear strains* in the corresponding planes. The strain tensor ε_{mn} is therefore built of linear strains on its diagonal and halves of shear strains off the diagonal:

$$\varepsilon_{mn} = \begin{bmatrix} \varepsilon_{11} & \frac{1}{2}\gamma_{12} & \frac{1}{2}\gamma_{13} \\ \frac{1}{2}\gamma_{21} & \varepsilon_{22} & \frac{1}{2}\gamma_{23} \\ \frac{1}{2}\gamma_{31} & \frac{1}{2}\gamma_{32} & \varepsilon_{33} \end{bmatrix} = \begin{bmatrix} \varepsilon_{xx} & \frac{1}{2}\gamma_{xy} & \frac{1}{2}\gamma_{xz} \\ \frac{1}{2}\gamma_{yx} & \varepsilon_{yy} & \frac{1}{2}\gamma_{yz} \\ \frac{1}{2}\gamma_{zx} & \frac{1}{2}\gamma_{zy} & \varepsilon_{zz} \end{bmatrix} \tag{4.25}$$

4.3.3 Tensor of small rotations

In general, the strain tensor ε_{mn} alone is not sufficient to define the displacement increment du_i. This is because the tensor ε_{mn} is symmetric and has only 6 independent components, while in order to define du_i we need to know all 9 derivatives of the displacement components $u_i = u_i(x_1, x_2, x_3)$:

$$du_i = \frac{\partial u_i}{\partial x_j} dx_j = \beta_{ij} dx_j \tag{4.26}$$

where β_{ij} is the *displacement gradient* built of the symmetric and skew symmetric parts:

$$\beta_{ij} = \varepsilon_{ij} + \omega_{ij} \tag{4.27}$$

The symmetric part is easily recognizable as the small strain tensor $\varepsilon_{ij} = \frac{1}{2}\left(u_{i,j} + u_{j,i}\right)$, while the skew symmetric part:

$$\omega_{ij} = -\omega_{ji} = \frac{1}{2}\left(u_{i,j} - u_{j,i}\right) = \begin{bmatrix} 0 & \omega_{12} & \omega_{13} \\ -\omega_{12} & 0 & \omega_{23} \\ -\omega_{13} & -\omega_{23} & 0 \end{bmatrix} \tag{4.28}$$

is the *tensor of small rotations*, which proves the missing three independent components of the displacement gradient β_{ij}.

Geometric interpretation of the small rotation tensor is given in Figure 4.5. Average (rigid) rotation of the rectangular element $dx_1 \times dx_2$ is given by the mean of the two angles:

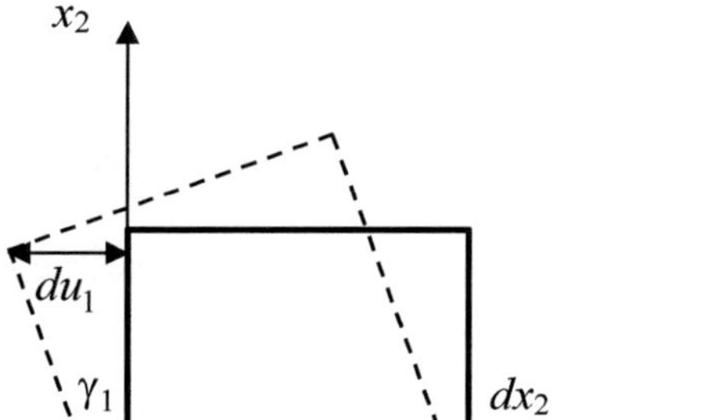

Figure 4.5 Tensor of small rotations.

$$\bar{\gamma}=\frac{1}{2}(\gamma_1+\gamma_2) \tag{4.29}$$

where $\gamma_1=\tan\gamma_1=\frac{-du_1}{dx_2}=-\frac{\partial u_1}{\partial x_2}$ and $\gamma_2=\tan\gamma_2=\frac{du_2}{dx_1}=\frac{\partial u_2}{\partial x_1}$, so that

$$\bar{\gamma}=\frac{1}{2}\left(\frac{\partial u_2}{\partial x_1}-\frac{\partial u_1}{\partial x_2}\right)=\omega_{21}=-\omega_{12} \tag{4.30}$$

As is seen, tensor ω_{ij} describes rotation of a body as a rigid object, and therefore, is independent of the stresses acting in the body. Therefore, the tensor of the small rotations ω_{ij} should not be a part of the constitutive relationship.

4.4 Compatibility equations

4.4.1 Geometric compatibility

The small strain tensor ε_{ij} has 6 independent components, defined, however, by differentiating a displacement vector u_i, which has only 3 independent components. Clearly, there should exist a dependency between the 6 components of ε_{ij}. Geometric interpretation of this dependency is schematically outlined in Figure 4.6.

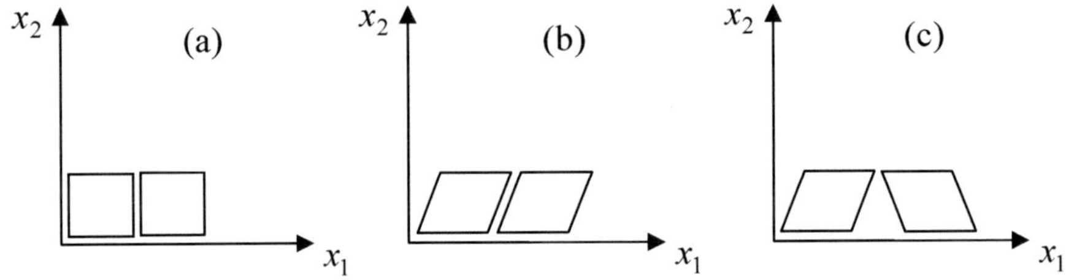

Figure 4.6 Deformation: a) Non-deformed; b) Compatible; c) Incompatible.

Two adjacent elements of the body should deform in a compatible manner so that the body which was continuous before the deformation remains continuous after the deformation. Mathematically, this compatibility condition is expressed via the following two groups of relationships between the second order partial derivatives of the small strain tensor ε_{ij}.

4.4.2 First group of equations

The first group of equations is derived by double differentiation of equations (4.18) for the non-diagonal ($i \neq j$) components of the tensor ε_{ij} and subsequent substitution of the expressions for the diagonal ($i = j$) components of the tensor ε_{ij}, also from equations (4.18):

$$2\frac{\partial^2 \varepsilon_{12}}{\partial x_1 \partial x_2} = \frac{\partial^2 \varepsilon_{11}}{\partial x_2^2} + \frac{\partial^2 \varepsilon_{22}}{\partial x_1^2} \tag{4.31}$$

$$2\frac{\partial^2 \varepsilon_{23}}{\partial x_2 \partial x_3} = \frac{\partial^2 \varepsilon_{22}}{\partial x_3^2} + \frac{\partial^2 \varepsilon_{33}}{\partial x_2^2} \tag{4.32}$$

$$2\frac{\partial^2 \varepsilon_{31}}{\partial x_3 \partial x_1} = \frac{\partial^2 \varepsilon_{33}}{\partial x_1^2} + \frac{\partial^2 \varepsilon_{11}}{\partial x_3^2} \tag{4.33}$$

4.4.3 Second group of equations

The first group of equations is derived by an inverse procedure: double differentiation of equations (4.18) for the diagonal ($i = j$) components of the tensor ε_{ij} and subsequent substitution of the expressions for the non-diagonal ($i \neq j$) components of the tensor ε_{ij}, also from equations (4.18):

$$\frac{\partial^2 \varepsilon_{11}}{\partial x_2 \partial x_3} = \frac{\partial}{\partial x_1}\left(-\frac{\partial \varepsilon_{23}}{\partial x_1} + \frac{\partial \varepsilon_{13}}{\partial x_2} + \frac{\partial \varepsilon_{12}}{\partial x_3}\right) \tag{4.34}$$

$$\frac{\partial^2 \varepsilon_{22}}{\partial x_3 \partial x_1} = \frac{\partial}{\partial x_2}\left(-\frac{\partial \varepsilon_{31}}{\partial x_2} + \frac{\partial \varepsilon_{12}}{\partial x_3} + \frac{\partial \varepsilon_{23}}{\partial x_1}\right) \tag{4.35}$$

$$\frac{\partial^2 \varepsilon_{33}}{\partial x_1 \partial x_2} = \frac{\partial}{\partial x_3}\left(-\frac{\partial \varepsilon_{12}}{\partial x_3} + \frac{\partial \varepsilon_{23}}{\partial x_1} + \frac{\partial \varepsilon_{31}}{\partial x_2}\right) \tag{4.36}$$

4.4.4 Discussion

Now we have 6 equations relating the 6 components of the tensor ε_{ij}. Can it be that by solving them together with the kinematic boundary conditions (i.e. known displacements on the boundaries) we can find the strains and displacements in the body? And this is without even bothering about any stresses? This is too good to be truth. And the truth is that out of the 6 equations (4.31)-(4.36) only 3 are independent. It can be shown that any equation of the second group can be obtained from the 3 equations of the first group (e.g., differentiation of equation (4.34) by $\dfrac{\partial^2}{\partial x_2 \partial x_3}$ and substitution of (4.31)-(4.34) into its left side gives an identity).

4.5 Properties of the strain tensor

Strain tensor has standard properties, which have been explored more in depth in the previous Chapter for the stress tensor, and will be only briefly listed here.

4.5.1 Transformation of the coordinate system T_{mn}

The strain tensor is transformed according to the standard rule:

$$\varepsilon_{ij}^{*} = T_{im} T_{jn} \varepsilon_{mn} \tag{4.37}$$

4.5.2 Characteristic equation and principal strains

In order to solve the characteristic equation:

$$\left(\varepsilon_{ij} - \varepsilon \delta_{ij}\right) n_j = 0 \tag{4.38}$$

we set $\left|\varepsilon_{ij} - \varepsilon \delta_{ij}\right| = 0$, which gives:

$$\varepsilon^3 - I_{1\varepsilon}\varepsilon^2 + I_{2\varepsilon}\varepsilon - I_{3\varepsilon} = 0 \tag{4.39}$$

or in principal strains $\varepsilon_1, \varepsilon_2, \varepsilon_3$:

$$\left(\varepsilon - \varepsilon_1\right)\left(\varepsilon - \varepsilon_2\right)\left(\varepsilon - \varepsilon_3\right) = 0 \tag{4.40}$$

with the following expressions for the coefficients:

$$I_{1\varepsilon} = \varepsilon_1 + \varepsilon_2 + \varepsilon_3; \quad I_{2\varepsilon} = \varepsilon_1\varepsilon_2 + \varepsilon_2\varepsilon_3 + \varepsilon_3\varepsilon_1; \quad I_{3\varepsilon} = \varepsilon_1\varepsilon_2\varepsilon_3 . \tag{4.41}$$

4.5.3 Invariants and their relations

Invariants of the strain tensor are:

$$I_1 = \varepsilon_{ii}; \quad I_2 = \frac{1}{2}\varepsilon_{ij}\varepsilon_{ij}; \quad I_3 = \frac{1}{3}\varepsilon_{ij}\varepsilon_{jk}\varepsilon_{ki}, \tag{4.42}$$

and the coefficients of the characteristic equations can be expressed via these invariants:

$$I_{1\varepsilon} = \varepsilon_{ii} = I_1 \tag{4.43}$$

$$I_{2\varepsilon} = \begin{vmatrix} \varepsilon_{11} & \varepsilon_{12} \\ \varepsilon_{21} & \varepsilon_{22} \end{vmatrix} + \begin{vmatrix} \varepsilon_{22} & \varepsilon_{23} \\ \varepsilon_{32} & \varepsilon_{33} \end{vmatrix} + \begin{vmatrix} \varepsilon_{11} & \varepsilon_{13} \\ \varepsilon_{31} & \varepsilon_{33} \end{vmatrix} = \frac{1}{2} I_1^2 - I_2 \tag{4.44}$$

$$I_{3\varepsilon} = \left| \varepsilon_{ij} \right| = I_3 - I_1 I_2 + \frac{1}{6} I_1^3 \tag{4.45}$$

4.5.4 Decomposition of the strain tensor

The strain tensor can be decomposed

$$\varepsilon_{ij} = e_{ij} + \frac{1}{3} \varepsilon_{kk} \delta_{ij} \tag{4.46}$$

into the volumetric part:

$$\frac{1}{3} \varepsilon_{kk} \delta_{ij} = \begin{bmatrix} \frac{1}{3}\varepsilon_{kk} & 0 & 0 \\ 0 & \frac{1}{3}\varepsilon_{kk} & 0 \\ 0 & 0 & \frac{1}{3}\varepsilon_{kk} \end{bmatrix} \tag{4.47}$$

and the deviatoric part:

$$e_{ij} = \varepsilon_{ij} - \frac{1}{3} \varepsilon_{kk} \delta_{ij} \tag{4.48}$$

which reflect the changes in the volume, and the shape, respectively.

The invariants of the deviatoric strain tensor are:

$$I_{1D} = e_{ii} = 0 \tag{4.49}$$

$$\begin{aligned} I_{2D} &= \frac{1}{6} \left[(\varepsilon_{11} - \varepsilon_{22})^2 + (\varepsilon_{22} - \varepsilon_{33})^2 + (\varepsilon_{33} - \varepsilon_{11})^2 \right] + \varepsilon_{12}^2 + \varepsilon_{23}^2 + \varepsilon_{31}^2 \\ &= I_2 - \frac{I_1^2}{6} \end{aligned} \tag{4.50}$$

$$I_{3D} = I_3 - \frac{2}{3} I_1 I_2 + \frac{2}{27} I_1^3 \tag{4.51}$$

4.5.5 Octahedral strains

Similar to the stress tensor, the linear and shear strains on the octahedral plane $n_1 = n_2 = n_3 = \dfrac{1}{\sqrt{3}}$ (Figure 3.10) are proportional to the first invariant of the strain tensor I_1 and to the square root of the second invariant of the deviatoric strain tensor $\sqrt{I_{2D}}$, respectively:

$$\varepsilon_{oct} = I_1/3 = \varepsilon_v/3 \text{ and } \gamma_{oct} = \sqrt{2I_{2D}/3} \qquad (4.52)$$

which are the generalized measures of the changes in volume and shape.

Chapter 5

Equations of Continuum Mechanics

TABLE OF CONTENTS

Chapter 5

Equations of Continuum Mechanics

Formulation of any boundary value problem in continuum mechanics is based on the following four components:

- Motion (equilibrium);
- Compatibility (geometry);
- Constitutive relationships;
- Initial and boundary conditions.

5.1 Equations of Continuum Mechanics

5.1.1 Motion

The stresses in a continuum cannot simply be specified arbitrarily, but must satisfy certain equilibrium relationships. These are derived in Section 3.3 by examining the equilibrium of a small element in a varying stress field.

Firstly the moment equilibrium of the element establishes the symmetry of the stress tensor:

$$\sigma_{ji} = \sigma_{ij} \tag{5.1}$$

Secondly the direct equilibrium equations give:

$$\sigma_{ij,j} + f_i = \rho \ddot{u}_i \tag{5.2}$$

where f_i is the body force per unit volume (usually equal to ρg_i, where g_i is the gravitational acceleration vector); ρ is the density; u_i is the displacement and $\ddot{u}_i$ is the acceleration. The right hand side of equation (5.2) is the inertia force, which in static (or quasi-static) problems is zero (see equations 3.18).

5.1.2 Compatibility

Similarly, the strains cannot be arbitrarily specified, but have to satisfy certain compatibility relationships. These arise because (in three dimensions) six independent strain components are derived, by definition, from the gradients of only three displacements:

$$\varepsilon_{ij} = \tfrac{1}{2}\left(u_{i,j} + u_{j,i}\right) \tag{5.3}$$

In fact this definition of the strains serves as a complete statement of the compatibility requirements, although it is possible to eliminate the displacements and derive relationships between higher order derivatives of the strains (see equations 4.31-4.36).

5.1.3 Numbers of variables and equations

Within a three-dimensional continuum a solution is required for 15 variables (6 stresses components, 6 strain components and 3 displacements). The equilibrium equation (5.2) provides 3 equations which involve only the stresses, and the strain definitions (5.3) provide 6 equations which involve the strains and displacements. For a solution to the problem we are thus missing 6 equations, and this is where the constitutive relations come in. They provide 6 relationships between the stresses (or more usually their increments) and the strains (or strain increments), thus satisfying the need for 6 further equations and also providing a link between stresses and strains.

5.1.4 Constitutive equations

The constitutive relations are quite different in character from the equilibrium and compatibility relationships.

a) They are algebraic rather than differential equations: $\sigma_{ij} = f(\varepsilon_{ij})$.
b) More importantly, the equilibrium and compatibility relationships are the same for all materials and are universally "true". The constitutive equations are each different for different materials.
c) Furthermore, the constitutive relations are simply approximations to the behaviour of real materials, none of which will behave exactly according to the idealizations employed. Thus constitutive relations are never "true" for a real material, they can only provide solutions which approximate the reality to within a certain degree of precision.

Whoever uses the models should understand this, and must try to assess its implications for the problem in hand.

5.1.5 Initial and boundary conditions

Initial values of the stresses and/or the displacements (and hence through equation (5.3) the strains) need to be specified. The stresses must obey the equilibrium equations (5.1) and (5.2). It is common (but not essential) to treat the displacements and strains as zero at the initial conditions.

The *boundary* conditions fall into four categories:

a) *Kinematic* boundary conditions are specified in terms of displacements on some sections of the boundary.
b) *Static* boundary conditions are specified in terms of the *tractions* on other sections of the boundary. These are often loosely referred to as "stress" boundary conditions, but in fact there is a distinction between stresses (which are second order tensors), and the tractions on a free surface, which are defined as the forces per unit area on the

surface. The traction $t^{\overline{n}}$ is therefore a vector, acting on a surface element with the normal $\overline{n}$. It is the tractions which can be defined at a boundary, and there is then an equilibrium relationship (Section 3.3.4) between these tractions and the stresses in an element immediately within the boundary:

$$t_i^{\overline{n}} = \sigma_{ij} n_j \tag{5.4}$$

c) *Mixed* boundary conditions, in which some displacement components and some traction components are specified *simultaneously*. The most common example is the case where the shear traction on a surface is defined, together with the normal displacement component, to simulate a contact with a smooth rigid body.

d) *Mixed* boundary conditions, in which some *relationship* between displacement and traction is defined. For example, the contact with an elastic support is simulated by expressing some relationship between the increment in displacement and the increment of the traction.

Since we are concerned here principally with constitutive behaviour we do not pursue the issue of boundary conditions further, but note that these need to be specified with some care if the equations to be solved are to form a well-posed mathematical problem.

5.1.6 2D simplifications

Sometimes the geometric and the loading conditions of the boundary value problem under study allow for a simplified 2D analysis:

a) Plane strain problems:

In a plane strain problem, principal strain in one direction is zero. This is the case for long straight structures, where geometric (and loading) conditions are identical in any cross-section perpendicular to the longitudinal axis of the structure. In Geotechnical Engineering, typical examples are: retaining walls, strip foundations, trench excavations, deep horizontal tunnels.

b) Plane stress problems:

In a plane stress problem, principal stress in one direction is zero, e.g., in plates loaded solely in their own plane. A generalized plane stress problem does not require the principal stress to be necessarily zero, it is sufficient that it is uniform across the plate. In Geotechnical Engineering, plane stress problems are less common than the plane strain ones. A typical example would be approximation of a cavity expansion in a deep borehole.

c) Axisymmetric problems:
In an axisymmetric problem, geometric (and loading) conditions have axial symmetry and are identical in any cross-section containing the axis of symmetry. In Geotechnical Engineering, typical examples are: single piles; single circular flat footings and caisson foundations; boreholes and vertical shaft excavations.

5.2 Example of a boundary value problem formulation

As an example we consider a 2D plane strain problem of surface subsidence above a rectangular cut-and-cover tunnel in rock (Figure 5.1a). The trench above the tunnel (with the depth H and width $2B$) has been filled with the backfill. The backfill (with the unit weight γ) settles over time under its own weight. The following simplifying assumptions will help us to obtain analytical solution for the surface subsidence of the backfill (Figure 5.1b):

- the backfill material is isotropic linear elastic (with Young's modulus E and Poisson's ratio ν)
- the shear stress τ_r along the sides of the trench is uniform;
- the tunnel roof provides uniform vertical support p.

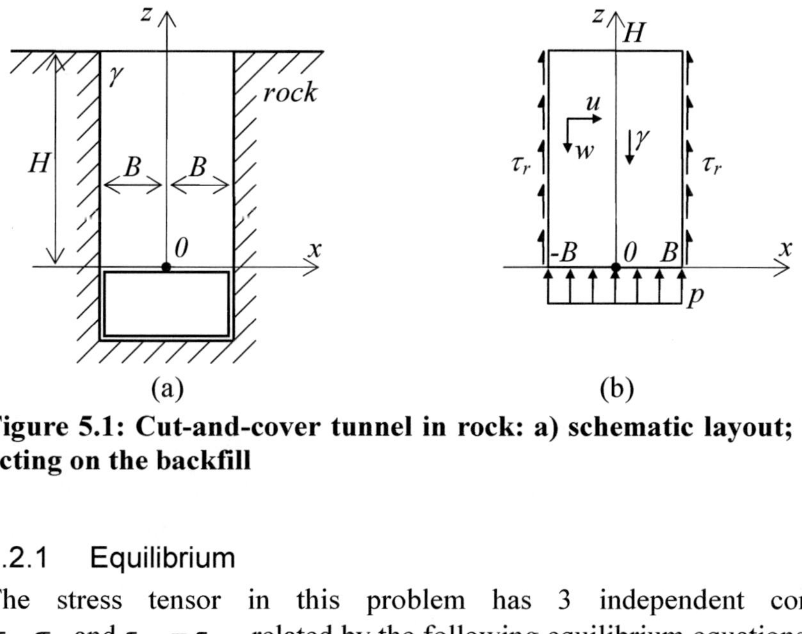

Figure 5.1: Cut-and-cover tunnel in rock: a) schematic layout; b) forces acting on the backfill

5.2.1 Equilibrium

The stress tensor in this problem has 3 independent components σ_x, σ_z and $\tau_{xz} = \tau_{zx}$, related by the following equilibrium equations:

$$\frac{\partial \sigma_x}{\partial x} + \frac{\partial \tau_{xz}}{\partial z} = 0 \qquad \frac{\partial \sigma_z}{\partial z} + \frac{\partial \tau_{xz}}{\partial x} = -\gamma \tag{5.5}$$

where $-\gamma$ is the body force (gravity).

5.2.2 Compatibility

The strain tensor in this problem also has 3 independent components $\varepsilon_x, \varepsilon_z$ and $\frac{1}{2}\gamma_{xz} = \frac{1}{2}\gamma_{zx}$, related to the two displacement components u and v by the following compatibility equations:

$$\varepsilon_x = \frac{\partial u}{\partial x} \qquad \varepsilon_z = \frac{\partial w}{\partial z} \qquad \gamma_{xz} = \frac{\partial u}{\partial z} + \frac{\partial w}{\partial x} \tag{5.6}$$

Alternatively, it is possible to eliminate the displacements and replace 3 equations (5.6) by one compatibility equation:

$$\frac{\partial^2 \gamma_{xz}}{\partial x \partial z} = \frac{\partial^2 \varepsilon_x}{\partial z^2} + \frac{\partial^2 \varepsilon_z}{\partial x^2} \tag{5.7}$$

5.2.3 Numbers of variables and equations

Within a two-dimensional continuum a solution is required for 8 variables (3 stresses components: $\sigma_x, \sigma_z, \tau_{xz}$; 3 strain components: $\varepsilon_x, \varepsilon_z, \gamma_{xz}$; and 2 displacements: u, w). The equilibrium equations (5.5) provide 2 equations which involve only the stresses, and the strain definitions (5.6) provide 3 equations which involve the strains and displacements. For a solution to the problem we are thus missing 3 equations, and this is where the constitutive relations come in.

5.2.4 Constitutive equations

Because we defined the properties of the half-space as linear-elastic, the constitutive relations in this problem are given by the well-known Hooke's Law (see Chapter 10):

$$\begin{Bmatrix} \sigma_x \\ \sigma_z \\ \tau_{xz} \end{Bmatrix} = D \begin{Bmatrix} \varepsilon_x \\ \varepsilon_z \\ \gamma_{xz} \end{Bmatrix} \qquad D = \begin{bmatrix} M & K_o M & 0 \\ K_0 M & M & 0 \\ 0 & 0 & G \end{bmatrix} \tag{5.8}$$

where D is the stiffness matrix. The parameters M, K_0 and G are alternative elastic constants, which can be expressed via the elastic Young's modulus E and Poisson's ratio ν (Chapter 10):

$$M = \frac{E(1-\nu)}{(1+\nu)(1-2\nu)} \tag{5.9}$$

$$K_0 = \frac{\nu}{1-\nu} \tag{5.10}$$

$$G = \frac{E}{2(1+\nu)} \tag{5.11}$$

5.2.5 Initial and boundary conditions

Initial values of the stresses, displacements and strains are zero. These zero values obey both the equilibrium (5.5) and the compatibility (5.6) equations.

Three types of the *boundary* conditions are imposed:

a) *Kinematic* boundary conditions:

Assuming that rock is rigid and that vertical tunnel walls do not deform, neither horizontal nor vertical displacements are allowed at the two points, where a vertical tunnel wall meets the backfill:

$$u(-B,0) = u(B,0) = 0 \qquad w(-B,0) = w(B,0) = 0 \tag{5.12}$$

b) *Static* boundary conditions:

Stress free backfill surface:

$$\sigma_z(x,H) = \tau_{xz}(x,H) = 0 \tag{5.13}$$

Tunnel roof is frictionless and provides a uniform support p:

$$\sigma_z(x,0) = p \qquad \tau_{xz}(x,0) = 0 \tag{5.14}$$

c) *Mixed* boundary conditions:

Assuming that rock is rigid, no horizontal displacement on the rock-backfill interface is allowed:

$$u(-B,z) = u(B,z) = 0 \tag{5.15}$$

while the mobilized shear resistance on this interface is τ_r:

$$\tau_{xz}(B,z) = -\tau_r \qquad \tau_{xz}(-B,z) = \tau_r \tag{5.16}$$

5.2.6 Solution

Let us assume that the backfill experiences no horizontal displacements:

$$u(x,z) = 0 \tag{5.17}$$

This automatically satisfies boundary conditions (5.15) and the symmetry requirement $u(0,z) = 0$. This is, of course, not sufficient: if the assumption (5.17) is wrong, the resulting stress field will violate static boundary conditions.

Assumption (5.17) is substituted into the compatibility equations (5.6) and constitutive relationship (5.8), leading to the following simplifications:

$$\varepsilon_x = 0 \qquad \varepsilon_z = \frac{\partial w}{\partial z} \qquad \gamma_{xz} = \frac{\partial w}{\partial x} \tag{5.18}$$

$$\sigma_x = K_0 M \frac{\partial w}{\partial z} \qquad \sigma_z = M \frac{\partial w}{\partial z} \qquad \tau_{xz} = G \frac{\partial w}{\partial x} \tag{5.19}$$

Substitution of equations (5.19) into the first equilibrium equation (5.5) gives:

$$(K_0 M + G) \frac{\partial^2 w}{\partial x \partial z} = 0 \tag{5.20}$$

which is only possible, if $w(x, z)$ can be decomposed as follows:

$$w(x, z) = f(x) + g(z) \tag{5.21}$$

Then substitution of equations (5.19) and (5.21) into the second equilibrium equation (5.5) gives:

$$M \frac{\partial^2 w}{\partial z^2} + G \frac{\partial^2 w}{\partial x^2} = -\gamma \tag{5.22}$$

and

$$M \frac{\partial^2 g(z)}{\partial z^2} + G \frac{\partial^2 f(x)}{\partial x^2} = -\gamma \tag{5.23}$$

This must be true for any z and x, which is only possible when both terms in the left hand side are constants:

$$\frac{\partial^2 g(z)}{\partial z^2} = A_1 = const \qquad \frac{\partial^2 f(z)}{\partial x^2} = B_1 = const \tag{5.24}$$

$$MA_1 + GB_1 = -\gamma \tag{5.25}$$

Integrating equations (5.24) we obtain:

$$g(z) = \frac{1}{2} A_1 z^2 + A_2 z + A_3 \qquad f(x) = \frac{1}{2} B_1 x^2 + B_2 x + B_3 \tag{5.26}$$

so that

$$w(x, z) = \frac{1}{2} A_1 z^2 + A_2 z + \frac{1}{2} B_1 x^2 + B_2 x + C \tag{5.27}$$

The five coefficients in the above equation are found by substituting expression (5.27) into equations (5.19) and satisfying boundary conditions (5.12)-(5.16) and equation (5.25):

$$w(x, z) = \frac{1}{M}\left(\gamma - \frac{\tau_r}{B}\right)\left(H - \frac{z}{2}\right)z + \frac{1}{2G}\frac{\tau_r}{B}\left(B^2 - x^2\right) \tag{5.28}$$

This results in the following stress field (equations 5.19):

$$\sigma_z = \left(\gamma - \frac{\tau_r}{B}\right)(H - z) \qquad \sigma_x = K_0\sigma_z \qquad \tau_{xz} = -\tau_r\frac{x}{B} \tag{5.29}$$

where the pressure on the tunnel roof is found from the global equilibrium:

$$p = \left(\gamma - \frac{\tau_r}{B}\right)H \tag{5.30}$$

Since both the displacement and stress fields satisfy all the necessary equations and boundary conditions, our initial assumption $u(x, z) = 0$ has been proven to be correct.

The surface subsidence is then given by the following parabola:

$$w(x, H) = \frac{1}{2M}\left(\gamma - \frac{\tau_r}{B}\right)H^2 + \frac{1}{2G}\frac{\tau_r}{B}\left(B^2 - x^2\right) \tag{5.31}$$

with the maximum displacement (Figure 5.2):

$$w_{\max} = w(0, H) = \frac{H^2}{2}\left[\frac{1}{M}\left(\gamma - \frac{\tau_r}{B}\right) + \frac{1}{G}\frac{\tau_r}{B}\left(\frac{B}{H}\right)^2\right] \tag{5.32}$$

For a deep tunnel ($H >> B$) this can be approximated by

$$w_{\max} \approx \frac{H^2}{2M}\left(\gamma - \frac{\tau_r}{B}\right) \tag{5.33}$$

This problem will be used in Chapters 6 and 7 to illustrate application of finite elements and finite differences methods, respectively.

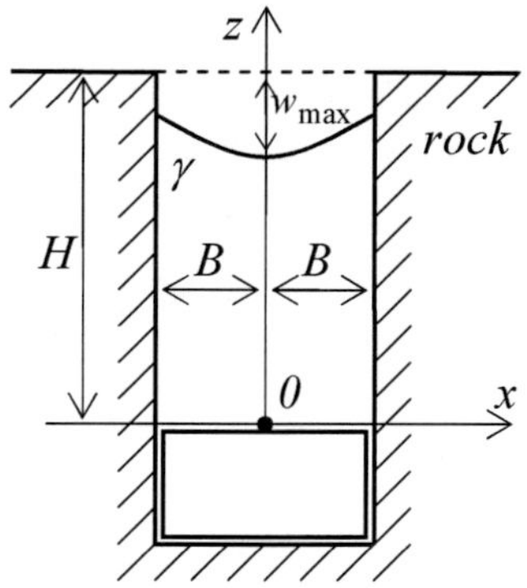

Figure 5.2: Cut-and-cover tunnel in rock: settlement of the backfill.

Chapter 6

Finite Elements

TABLE OF CONTENTS

Chapter 6

Finite Elements

In Chapter 5 we obtained the analytical solution for the plane strain problem of surface subsidence above a cut-and-cover tunnel. This, however, was only possible after introduction of significant simplifications into the problem formulation. In general, analytical solutions are available only for boundary value problems with simple geometry, boundary conditions and constitutive relationships. The majority of the realistic problems have to be solved numerically. Chapter 6 gives a very basic introduction into the most popular numerical approach – Finite Elements Method (FEM). Chapter 7 will explore an alternative approach – Finite Differences Method (FDM), which has advantages for solution of non-linear and unstable boundary value problems.

6.1 Introduction

The 2D FEM method can be briefly described as the following sequence:

1. Discretization of the continuum into a number of "finite elements" (e), interconnected at a discrete number of nodal points n on their boundaries. The displacements $\{u_n\}$ at these nodal points are the main unknowns of the problem. We use matrix notation below.
2. "Shape functions" $\left[N^e(x_i)\right]$ are chosen to define the state of displacement $u_i^e(x_i)$ within each element by interpolating between its nodal displacements $\{u_n\}$: $u_i^e(x_i)=\left[N^e(x_i)\right]\{u_n\}$.
3. Compatibility equations use these displacement functions to define strains $\varepsilon_{ij}^e(x_i)$ in element (e) in terms of $\{u_n\}$: $\varepsilon_{ij}^e(x_i)=\left[B^e(x_i)\right]\{u_n\}$.
4. Constitutive relationships use these strains to define stresses $\sigma_{ij}^e(x_i)$ in each element in terms of $\{u_n\}$: $\sigma_{ij}^e(x_i)=[D]\left[B^e\right]\{u_n\}$.
5. Equilibrium equations for each element use these stresses to define "stiffness matrix" $\left[K^e\right]=\int_{V^e}\left[B^e\right]^T[D]\left[B^e\right]dV$ for element (e), by integrating over its volume V^e, which allows for fictitious "nodal forces" $\left\{q_n^e\right\}$ in this element to be expressed in terms of $\{u_n\}$: $\left\{q_n^e\right\}=\left[K^e\right]\{u_n\}+\left\{f_{bn}^e\right\}+\left\{f_{sn}^e\right\}$, where $\left\{f_{bn}^e\right\}$ and $\left\{f_{sn}^e\right\}$ are body and surface nodal force vectors in element (e), respectively.

6. Equilibrium equations for each node sum these fictitious nodal forces $\{F_n\} = \sum_e \{q_n^e\}$ to assemble element stiffness matrices into the global stiffness matrix $[K]$ for the entire continuum: $\{F_n\} = [K]\{u_n\} + \{f_{bn}\} + \{f_{sn}\}$, $\{f_{bn}\} = \sum_e \{f_{bn}^e\}$, $\{f_{sn}\} = \sum_e \{f_{sn}^e\}$.
7. Static boundary conditions are taken care of in the surface nodal force vector $\{f_{sn}\}$, while the kinematic boundary conditions reduce the number of unknown nodal displacements $\{u_n\}$.
8. Solution of the resulting reduced system of linear algebraic equations produces the vector of nodal displacements $\{u_n\}$, which in turn allows for strains and stresses to be calculated (Steps 3 and 4).

As is seen, FEM uses all the components of the boundary value problem formulation, only in a discretised form. Application of this 8-step procedure to the example problem of surface subsidence above a cut-and-cover tunnel is illustrated below: first in a 1D approximation and then in the complete plain strain formulation.

6.2 Example of a 1D FE solution

The 2D problem formulation and analytical solution are given in Section 5.2. For a deep tunnel, when H >> B, the problem can be approximated in 1D z-space. Equilibrium, compatibility and constitutive equations are reduced to

$$\frac{\partial \sigma_z}{\partial z} = -\gamma + \frac{\tau_r}{B}; \qquad \varepsilon_z = \frac{\partial w}{\partial z}; \qquad \sigma_z = M\varepsilon_z, \tag{6.1}$$

which results in the following differential equation:

$$\frac{\partial^2 w}{\partial z^2} = -\gamma + \frac{\tau_r}{B} \tag{6.2}$$

Equation (6.2) is0 integrated together with the boundary conditions

$$w(0) = 0; \quad \sigma_z(H) = M\frac{\partial w}{\partial z}\bigg|_{z=H} = 0, \tag{6.3}$$

and gives the analytical solution of the 1D problem:

$$w(z) = \frac{1}{M}\left(\gamma - \frac{\tau_r}{B}\right)\left(H - \frac{z}{2}\right)z; \qquad \sigma_z(z) = \left(\gamma - \frac{\tau_r}{B}\right)(H - z). \tag{6.4}$$

Let us now see how the FEM would tackle this problem.

6.2.1 Discretization

The continuum is discretised into two linear elements (1) and (2) of equal length $H/2$ with 2 nodes each (Figure 6.1). Nodal displacements w_1, w_2 and w_3 are the main unknowns of the problem.

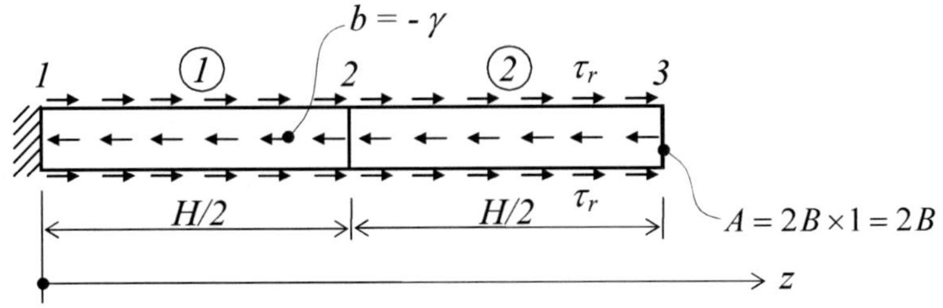

Figure 6.1: Discrete model (rotated by 90°).

6.2.2 Shape functions

Shape functions $\left[N^e\right] = \left[N_i^e(z), N_j^e(z)\right]$ define the state of displacement $w^e(z)$ within element (e) with nodes i and j by interpolating between its nodal displacements w_i and w_j (Figure 6.2):

$$w^e(z) = N_i^e(z) w_i + N_j^e(z) w_j = \left[N^e\right]\{w_n\} \tag{6.5}$$

where $N_i^e(z_i) = 1$ and $N_i^e(z_j) = 0$.

In our example, linear interpolation between the nodal displacements produces the following displacement functions for the two elements:

$$(1):\ w^1(z) = \left(1 - \frac{2z}{H}\right) w_1 + \frac{2z}{H} w_2; \qquad (2):\ w^2(z) = \left(2 - \frac{2z}{H}\right) w_2 + \left(\frac{2z}{H} - 1\right) w_3,$$

so that

$$\left[N^1\right] = \left[1 - \frac{2z}{H}, \frac{2z}{H}\right] \text{ and } \left[N^2\right] = \left[2 - \frac{2z}{H}, \frac{2z}{H} - 1\right] \tag{6.6}$$

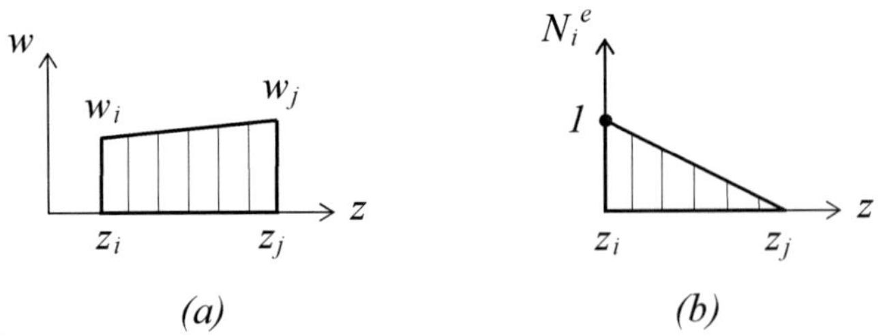

Figure 6.2: Linear shape functions: a) displacement; b) shape function.

6.2.3 Compatibility equations

Strains are defined from displacements via the compatibility equation:

$$\varepsilon_z^e(z) = \frac{\partial w^e(z)}{\partial z} = B_i^e(z)w_i + B_j^e(z)w_j = \left[B^e\right]\{w_n\} \tag{6.7}$$

where $\left[B^e\right] = \left[B_i^e,\, B_j^e\right] = \left[\dfrac{\partial N_i^e}{\partial z}, \dfrac{\partial N_j^e}{\partial z}\right]$.

In our example:

$$(1):\ \varepsilon_z^1(z) = \frac{\partial w^1(z)}{\partial z} = -\frac{2}{H}w_1 + \frac{2}{H}w_2;\quad (2):\ \varepsilon_z^2(z) = \frac{\partial w^2(z)}{\partial z} = -\frac{2}{H}w_2 + \frac{2}{H}w_3,$$

so that

$$\left[B^1\right] = \left[-\frac{2}{H}, \frac{2}{H}\right] \text{ and } \left[B^2\right] = \left[-\frac{2}{H}, \frac{2}{H}\right] \tag{6.8}$$

6.2.4 Constitutive relationship

Stresses are defined from the constitutive relationship:

$$\sigma_z^e(z) = [D]\varepsilon_z^e(z) = [D]\left[B^e\right]\{w_n\}, \qquad [D] = M \tag{6.9}$$

In our example:

$$(1):\ \sigma_z^1(z) = -\frac{2M}{H}w_1 + \frac{2M}{H}w_2;\qquad (2):\ \sigma_z^2(z) = -\frac{2M}{H}w_2 + \frac{2M}{H}w_3,$$

so that

$$[D]\left[B^1\right] = \left[-\frac{2M}{H}, \frac{2M}{H}\right] \text{ and } [D]\left[B^2\right] = \left[-\frac{2M}{H}, \frac{2M}{H}\right] \tag{6.10}$$

6.2.5 Equilibrium of an element

When cut from its surroundings, element (e) is subjected to the action of fictitious forces $\left\{q_n^e\right\} = \begin{Bmatrix} q_i^e \\ q_j^e \end{Bmatrix}$ acting at its nodes i and j (Figure 6.3).

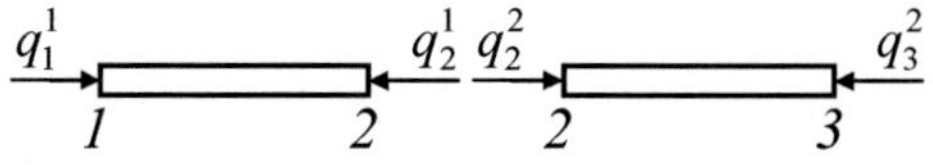

Figure 6.3: Nodal forces

These nodal forces should be in equilibrium with the stresses, body and surfaces forces acting within the element. In the FEM this equilibrium is expressed in a weak form – of the equation of the virtual work.

We assume that the nodes i and j are subjected to virtual displacements $\{\delta w_n\}=\begin{Bmatrix}\delta w_i\\ \delta w_j\end{Bmatrix}$, which produce virtual displacement and strain fields in (e):

$$\delta w^e(z)=\left[N^e\right]\{\delta w_n\};\qquad \delta\varepsilon_z^e(z)=\left[B^e\right]\{\delta w_n\}. \tag{6.11}$$

The virtual work of the nodal forces on the corresponding displacements is

$$\{\delta w_n\}^T\left\{q_n^e\right\}=q_i^e\delta w_i+q_j^e\delta w_j \tag{6.12}$$

If the element is in equilibrium, this external virtual work should be equal to the internal virtual work within the element, done by stresses on the virtual strains and body and surface forces on the corresponding displacements.

Using equations (6.9) and (6.11), work (per unit volume) of stresses and body forces b^e is given by

$$\delta\varepsilon_z^e(z)\sigma_z^e(z)+\delta w^e(z)b^e(z)=\{\delta w_n\}^T\left[\left[B^e\right]^T[D]\left[B^e\right]\{w_n\}+\left[N^e\right]^T b^e(z)\right] \tag{6.13}$$

while the work (per unit area) of surface forces t^e is given by

$$\delta w^e(z)t^e(z)=\{\delta w_n\}^T\left[N^e\right]^T t^e(z) \tag{6.14}$$

Here we use the fact that $([A][B])^T=[B]^T[A]^T$.

The total internal work in the element (e) is calculated by integration of expressions (6.13) over the element volume V^e and (6.14) over the element surface S^e. The result is equated to the external work (6.12) to produce the weak form of the equilibrium equation for the element (e):

$$\begin{aligned}\{\delta w_n\}^T\left\{q_n^e\right\}=&\int_{V^e}\{\delta w_n\}^T\left[\left[B^e\right]^T[D]\left[B^e\right]\{w_n\}+\left[N^e\right]^T b^e(z)\right]dV\\ &+\int_{S^e}\{\delta w_n\}^T\left[N^e\right]^T t^e(z)\,dS\end{aligned} \tag{6.15}$$

Because the above equation should hold for any virtual displacements $\{\delta w_n\}$, and the nodal displacements $\{w_n\}$ are constants, it can be rewritten as:

$$\left\{q_n^e\right\} = \left[K^e\right]\left\{w_n\right\} + \left\{f_{bn}^e\right\} + \left\{f_{sn}^e\right\} \tag{6.16}$$

where

$$\left[K^e\right] = \int_{V^e} \left[B^e\right]^T [D]\left[B^e\right] dV \tag{6.17}$$

is the stiffness matrix for element (e);

$$\left\{f_{bn}^e\right\} = \int_{V^e} \left[N^e\right]^T b^e(z)\, dV \tag{6.18}$$

is the nodal body force vector for element (e);

$$\left\{f_{sn}^e\right\} = \int_{S^e} \left[N^e\right]^T t^e(z)\, dS \tag{6.19}$$

is the nodal surface force vector for element (e).

Let us now return to our example, where the body forces $b(z) = -\gamma$ act over the element volume $V^e = 2B \times H/2 \times 1 = HB$, while surface forces $s(z) = \tau_r$ act on the element surface $S^e = 2 \times H/2 \times 1 = H$. For element (1):

$$\left[K^1\right] = 2B \int_0^{\frac{H}{2}} \begin{bmatrix} -\dfrac{2}{H} \\ \dfrac{2}{H} \end{bmatrix} M \left[-\frac{2}{H}, \frac{2}{H}\right] dz = BM \frac{4}{H} \begin{bmatrix} 1 & -1 \\ -1 & 1 \end{bmatrix}$$

$$\begin{Bmatrix} f_{b1}^1 \\ f_{b2}^1 \end{Bmatrix} = 2B \int_0^{\frac{H}{2}} \begin{bmatrix} 1 - \dfrac{2z}{H} \\ \dfrac{2z}{H} \end{bmatrix} (-\gamma) dz = -2B\gamma \begin{Bmatrix} z - \dfrac{z^2}{H} \\ \dfrac{z^2}{H} \end{Bmatrix}_0^{\frac{H}{2}} = -\frac{1}{2} HB\gamma \begin{Bmatrix} 1 \\ 1 \end{Bmatrix}$$

$$\begin{Bmatrix} f_{s1}^1 \\ f_{s2}^1 \end{Bmatrix} = 2 \int_0^{\frac{H}{2}} \begin{bmatrix} 1 - \dfrac{2z}{H} \\ \dfrac{2z}{H} \end{bmatrix} \tau_r dz = 2\tau_r \begin{Bmatrix} z - \dfrac{z^2}{H} \\ \dfrac{z^2}{H} \end{Bmatrix}_0^{\frac{H}{2}} = \frac{1}{2} H\tau_r \begin{Bmatrix} 1 \\ 1 \end{Bmatrix}$$

$$\left\{q_n^1\right\} = \begin{Bmatrix} q_1^1 \\ q_2^1 \end{Bmatrix} = \frac{4BM}{H} \begin{bmatrix} 1 & -1 \\ -1 & 1 \end{bmatrix} \begin{Bmatrix} w_1 \\ w_2 \end{Bmatrix} - HB\left(\gamma - \frac{\tau_r}{B}\right) \begin{Bmatrix} 1/2 \\ 1/2 \end{Bmatrix} \tag{6.20}$$

For element (2):

$$\left[K^2\right]=2B\int_{\frac{H}{2}}^{H}\begin{bmatrix}-\frac{2}{H}\\ \frac{2}{H}\end{bmatrix}M\left[-\frac{2}{H},\frac{2}{H}\right]dz=BM\frac{4}{H}\begin{bmatrix}1 & -1\\ -1 & 1\end{bmatrix}$$

$$\begin{Bmatrix}f_{b2}^2\\ f_{b3}^2\end{Bmatrix}=2B\int_{\frac{H}{2}}^{H}\begin{bmatrix}2-\frac{2z}{H}\\ \frac{2z}{H}-1\end{bmatrix}(-\gamma)dz=-2B\gamma\begin{Bmatrix}2z-\frac{z^2}{H}\\ \frac{z^2}{H}-z\end{Bmatrix}_{\frac{H}{2}}^{H}=-\frac{1}{2}HB\gamma\begin{Bmatrix}1\\ 1\end{Bmatrix}$$

$$\begin{Bmatrix}f_{s2}^2\\ f_{s3}^2\end{Bmatrix}=2\int_{\frac{H}{2}}^{H}\begin{bmatrix}2-\frac{2z}{H}\\ \frac{2z}{H}-1\end{bmatrix}\tau_r dz=2\tau_r\begin{Bmatrix}2z-\frac{z^2}{H}\\ \frac{z^2}{H}-z\end{Bmatrix}_{\frac{H}{2}}^{H}=\frac{1}{2}H\tau_r\begin{Bmatrix}1\\ 1\end{Bmatrix}$$

$$\left\{q_n^2\right\}=\begin{Bmatrix}q_2^2\\ q_3^2\end{Bmatrix}=\frac{4BM}{H}\begin{bmatrix}1 & -1\\ -1 & 1\end{bmatrix}\begin{Bmatrix}w_2\\ w_3\end{Bmatrix}-HB\left(\gamma-\frac{\tau_r}{B}\right)\begin{Bmatrix}1/2\\ 1/2\end{Bmatrix} \tag{6.21}$$

6.2.6 Equilibrium of the nodes

All the fictitious nodal forces at each node n also have to be in equilibrium:

$$\{F_n\}=\sum_e\left\{q_n^e\right\} \tag{6.22}$$

where F_n is the resulting nodal force at the node n, which in the absence of external nodal forces (such as a reaction to the prescribed displacement) should be equal to zero. By summing up the fictitious nodal forces from each element surrounding the node n, equation (6.22) allows for element stiffness matrices to be assembled into the global stiffness matrix $[K]$ for the entire continuum:

$$\{F_n\}=[K]\{w_n\}+\{f_{bn}\}+\{f_{sn}\} \tag{6.23}$$

where

$$\{f_{bn}\}=\sum_e\left\{f_{bn}^e\right\};\qquad \{f_{sn}\}=\sum_e\left\{f_{sn}^e\right\}. \tag{6.24}$$

In our case:

$$
\begin{array}{llllll}
F_1 = & q_1^1 = & \frac{4BM}{H}w_1 & -\frac{4BM}{H}w_2 & & -\frac{H}{2}B\left(\gamma-\frac{\tau_r}{B}\right) \\
 & q_2^1 = & -\frac{4BM}{H}w_1 & +\frac{4BM}{H}w_2 & & -\frac{H}{2}B\left(\gamma-\frac{\tau_r}{B}\right) \\
F_2 = & + & & & & \\
 & q_2^2 = & & +\frac{4BM}{H}w_2 & -\frac{4BM}{H}w_3 & -\frac{H}{2}B\left(\gamma-\frac{\tau_r}{B}\right) \\
F_3 = & q_3^2 = & & -\frac{4BM}{H}w_2 & +\frac{4BM}{H}w_3 & -\frac{H}{2}B\left(\gamma-\frac{\tau_r}{B}\right)
\end{array}
$$

which produces

$$
\begin{Bmatrix} F_1 \\ F_2 \\ F_3 \end{Bmatrix} = \begin{Bmatrix} F_1 \\ 0 \\ 0 \end{Bmatrix} = \frac{4BM}{H}\begin{bmatrix} 1 & -1 & 0 \\ -1 & 2 & -1 \\ 0 & -1 & 1 \end{bmatrix}\begin{Bmatrix} w_1 \\ w_2 \\ w_3 \end{Bmatrix} - \frac{1}{2}HB\left(\gamma-\frac{\tau_r}{B}\right)\begin{Bmatrix} 1 \\ 2 \\ 1 \end{Bmatrix} \tag{6.25}
$$

Equation (6.25) gives the global system of algebraic equations for the nodal displacements in our 1D boundary value problem.

6.2.7 Boundary conditions

Static boundary conditions are taken care of in the surface nodal force vector $\{f_{sn}\}$ (equations 6.19 and 6.24) and in the vector of external nodal forces $\{F_n\}$, while the kinematic boundary conditions reduce the number of unknown nodal displacements $\{w_n\}$. This allows for the number of equations in the system (6.23) to be reduced by eliminating equations for those F_n, where w_n is prescribed, and by substituting these given values of w_n into the remaining equations. After the reduced system of equations is solved, the derived nodal displacements $\{w_n\}$ can be substituted into the omitted equations to find the nodal reactions F_n.

In our example, in the node $n=1$ the displacement $w_1=0$ is prescribed, while the nodal reaction F_1 is unknown. The first equation in the system (6.25) can then be eliminated, and after substituting $w_1=0$ into other two equations we obtained the reduced system:

$$
\begin{Bmatrix} 0 \\ 0 \end{Bmatrix} = \frac{4BM}{H}\begin{bmatrix} 2 & -1 \\ -1 & 1 \end{bmatrix}\begin{Bmatrix} w_2 \\ w_3 \end{Bmatrix} - \frac{1}{2}HB\left(\gamma-\frac{\tau_r}{B}\right)\begin{Bmatrix} 2 \\ 1 \end{Bmatrix} \tag{6.26}
$$

6.2.8 Solution

The solution of the reduced system (6.26) is provided by

$$w_1 = 0\,; \qquad w_2 = \frac{3H^2}{8M}\left(\gamma - \frac{\tau_r}{B}\right); \quad w_3 = \frac{H^2}{2M}\left(\gamma - \frac{\tau_r}{B}\right) \tag{6.27}$$

which on substitution into the first equation (6.25) gives the reaction

$$F_1 = -2HB\left(\gamma - \frac{\tau_r}{B}\right), \text{ so that } p = -\frac{F_1}{2B} = H\left(\gamma - \frac{\tau_r}{B}\right) \tag{6.28}$$

Note, that the nodal displacements given by equations (6.27) coincide with the exact solution (6.4) at $z = H/2$ and $z = H$, respectively!

Thus, in spite of the fact that we discretised the continuum by only two elements, we managed to obtain exact values of displacements at the nodes. This result is, however, an exception. Normally, good approximation of the exact solution by the FEM is achieved using a relatively large number of finite elements. In our example, the exact solution was obtained thanks to the advantageous choice of the linear shape functions N_n^e, which happened to be the proper weighting functions for the weak Galerkin form of our differential equation (6.2) (see Zienkiewicz and Taylor, 2000).

6.3 Example of a 2D FE solution

In this section, the problem of Section 5.2 is solved in plane strain.

6.3.1 Discretization

Due to symmetry with respect to *z*-axis, it is sufficient to consider just a half of the problem, provided proper boundary conditions are imposed on the new boundary (Figure 6.4). The continuum is discretised into two triangular elements (1) and (2) of equal areas $HB/2$ with 3 nodes each. Nodal displacements $\begin{Bmatrix} w_1 \\ u_1 \end{Bmatrix}$, $\begin{Bmatrix} w_2 \\ u_2 \end{Bmatrix}$, $\begin{Bmatrix} w_3 \\ u_3 \end{Bmatrix}$ and $\begin{Bmatrix} w_4 \\ u_4 \end{Bmatrix}$ are the main unknowns of the problem.

6.3.2 Shape functions

Shape functions $\left[N^e\right]$ define the state of displacement within element (*e*) with nodes *i*, *j* and *k* by interpolating between its nodal displacements (Figure 6.5):

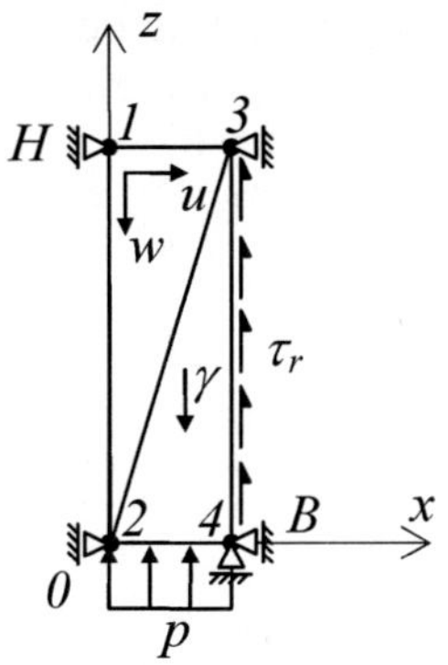

Figure 6.4: Discrete 2D FE model

$$\begin{Bmatrix} w^e(x,z) \\ u^e(x,z) \end{Bmatrix} = N_i^e \begin{Bmatrix} w_i \\ u_i \end{Bmatrix} + N_j^e \begin{Bmatrix} w_j \\ u_j \end{Bmatrix} + N_k^e \begin{Bmatrix} w_k \\ u_k \end{Bmatrix} = \left[N^e\right] \begin{Bmatrix} w_i \\ u_i \\ w_j \\ u_j \\ w_k \\ u_k \end{Bmatrix} \tag{6.29}$$

where $\left[N^e\right] = \begin{bmatrix} N_i^e & 0 & N_j^e & 0 & N_k^e & 0 \\ 0 & N_i^e & 0 & N_j^e & 0 & N_k^e \end{bmatrix}$ (6.30)

where $N_i^e(x_i, z_i) = 1$, $N_i^e(x_j, z_j) = 0$ and $N_i^e(x_k, z_k) = 0$.

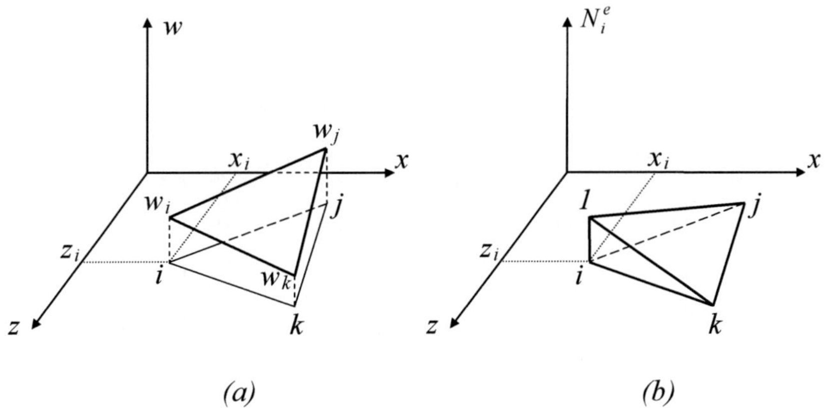

Figure 6.5: Linear shape functions: a) displacement; b) shape function.

In our example, linear interpolation between the nodal displacements produces the following displacement functions for the two elements:

$$(1): \begin{Bmatrix} w^1(x,z) \\ u^1(x,z) \end{Bmatrix} = \left(\frac{z}{H} - \frac{x}{B}\right)\begin{Bmatrix} w_1 \\ u_1 \end{Bmatrix} + \left(1 - \frac{z}{H}\right)\begin{Bmatrix} w_2 \\ u_2 \end{Bmatrix} + \left(\frac{x}{B}\right)\begin{Bmatrix} w_3 \\ u_3 \end{Bmatrix}$$

$$(2): \begin{Bmatrix} w^2(x,z) \\ u^2(x,z) \end{Bmatrix} = \left(1 - \frac{x}{B}\right)\begin{Bmatrix} w_2 \\ u_2 \end{Bmatrix} + \left(\frac{z}{H}\right)\begin{Bmatrix} w_3 \\ u_3 \end{Bmatrix} + \left(\frac{x}{B} - \frac{z}{H}\right)\begin{Bmatrix} w_4 \\ u_4 \end{Bmatrix}$$

so that

$$\begin{aligned} \left[N^1\right] &= \begin{bmatrix} \frac{z}{H} - \frac{x}{B} & 0 & 1 - \frac{z}{H} & 0 & \frac{x}{B} & 0 \\ 0 & \frac{z}{H} - \frac{x}{B} & 0 & 1 - \frac{z}{H} & 0 & \frac{x}{B} \end{bmatrix} \\ \left[N^2\right] &= \begin{bmatrix} 1 - \frac{x}{B} & 0 & \frac{z}{H} & 0 & \frac{x}{B} - \frac{z}{H} & 0 \\ 0 & 1 - \frac{x}{B} & 0 & \frac{z}{H} & 0 & \frac{x}{B} - \frac{z}{H} \end{bmatrix} \end{aligned} \tag{6.31}$$

6.3.3 Compatibility equations

Strains are defined from displacements via the compatibility equation:

$$\left\{\varepsilon^e(x,z)\right\} = \begin{Bmatrix} \varepsilon_z \\ \varepsilon_x \\ \gamma_{xz} \end{Bmatrix} = \begin{bmatrix} \frac{\partial}{\partial z} & 0 \\ 0 & \frac{\partial}{\partial x} \\ \frac{\partial}{\partial x} & \frac{\partial}{\partial z} \end{bmatrix} \begin{Bmatrix} w^e(x,z) \\ u^e(x,z) \end{Bmatrix} = \left[B^e\right] \begin{Bmatrix} w_i \\ u_i \\ w_j \\ u_j \\ w_k \\ u_k \end{Bmatrix} \tag{6.32}$$

where
$$\left[B^e\right] = \begin{bmatrix} \frac{\partial N_i^e}{\partial z} & 0 & \frac{\partial N_j^e}{\partial z} & 0 & \frac{\partial N_k^e}{\partial z} & 0 \\ 0 & \frac{\partial N_i^e}{\partial x} & 0 & \frac{\partial N_j^e}{\partial x} & 0 & \frac{\partial N_k^e}{\partial x} \\ \frac{\partial N_i^e}{\partial x} & \frac{\partial N_i^e}{\partial z} & \frac{\partial N_j^e}{\partial x} & \frac{\partial N_j^e}{\partial z} & \frac{\partial N_k^e}{\partial x} & \frac{\partial N_k^e}{\partial z} \end{bmatrix} \tag{6.33}$$

In our example:

$$\left[B^1\right]=\frac{1}{HB}\begin{bmatrix} B & 0 & -B & 0 & 0 & 0 \\ 0 & -H & 0 & 0 & 0 & H \\ -H & B & 0 & -B & H & 0 \end{bmatrix}$$
$$\left[B^2\right]=\frac{1}{HB}\begin{bmatrix} 0 & 0 & B & 0 & -B & 0 \\ 0 & -H & 0 & 0 & 0 & H \\ -H & 0 & 0 & B & H & -B \end{bmatrix} \tag{6.34}$$

so that

$$\left\{\varepsilon^1\right\}=\left\{\begin{array}{c} \dfrac{w_1-w_2}{H} \\ \dfrac{u_3-u_1}{B} \\ \dfrac{w_3-w_1}{B}+\dfrac{u_1-u_2}{H} \end{array}\right\}; \qquad \left\{\varepsilon^2\right\}=\left\{\begin{array}{c} \dfrac{w_3-w_4}{H} \\ \dfrac{u_4-u_2}{B} \\ \dfrac{w_4-w_2}{B}+\dfrac{u_3-u_4}{H} \end{array}\right\}.$$

6.3.4 Constitutive relationship

Stresses are defined from the constitutive relationship:

$$\left\{\sigma^e(x,z)\right\}=\left\{\begin{array}{c} \sigma_z \\ \sigma_x \\ \tau_{xz} \end{array}\right\}=[D]\left\{\begin{array}{c} \varepsilon_z \\ \varepsilon_x \\ \gamma_{xz} \end{array}\right\}=[D]\left[B^e\right]\left\{\begin{array}{c} w_i \\ u_i \\ w_j \\ u_j \\ w_k \\ u_k \end{array}\right\} \tag{6.35}$$

In our example:

$$[D]=M\begin{bmatrix} 1 & K_0 & 0 \\ K_0 & 1 & 0 \\ 0 & 0 & \xi \end{bmatrix}, \text{ where } \xi=\frac{G}{M} \tag{6.36}$$

$$[D]\left[B^1\right]=\frac{M}{HB}\begin{bmatrix} B & -K_0H & -B & 0 & 0 & K_0H \\ K_0B & -H & -K_0B & 0 & 0 & H \\ -\xi H & \xi B & 0 & -\xi B & \xi H & 0 \end{bmatrix} \tag{6.37}$$

$$[D]\left[B^2\right]=\frac{M}{HB}\begin{bmatrix} 0 & -K_0H & B & 0 & -B & K_0H \\ 0 & -H & K_0B & 0 & -K_0B & H \\ -\xi H & 0 & 0 & \xi B & \xi H & -\xi B \end{bmatrix} \tag{6.38}$$

6.3.5 Equilibrium of the element

When cut from its surroundings, element (e) is subjected to the action of fictitious forces $\{q_i^e\}=\begin{Bmatrix} W_i^e \\ U_i^e \end{Bmatrix}$, $\{q_j^e\}=\begin{Bmatrix} W_j^e \\ U_j^e \end{Bmatrix}$ and $\{q_k^e\}=\begin{Bmatrix} W_k^e \\ U_k^e \end{Bmatrix}$ acting at its nodes i, j and k, respectively (Figure 6.6).

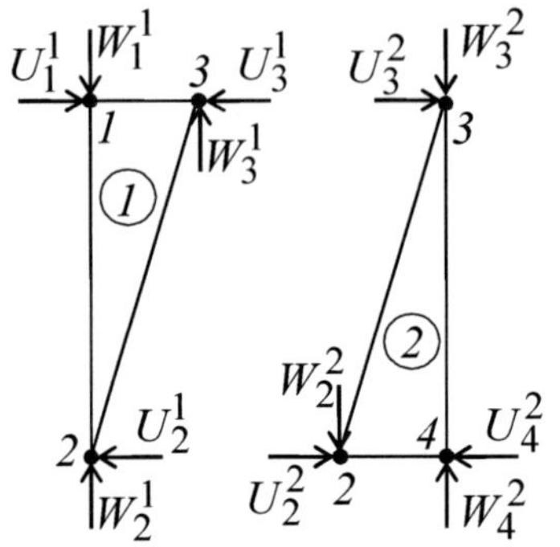

Figure 6.6: Nodal forces

These nodal forces should be in equilibrium with the stresses, body forces $\{b^e(x,z)\}=\begin{Bmatrix} b_z^e \\ b_x^e \end{Bmatrix}$ and surfaces forces $\{t^e(x,z)\}=\begin{Bmatrix} t_z^e \\ t_x^e \end{Bmatrix}$ acting within the element volume V^e and at the element surface S^e, respectively. As in 1D case, this equilibrium is expressed in a weak form via the equation of the virtual work:

$$\begin{Bmatrix} W_i^e \\ U_i^e \\ W_j^e \\ U_j^e \\ W_k^e \\ U_k^e \end{Bmatrix} = \left[K^e\right] \begin{Bmatrix} w_i^e \\ u_i^e \\ w_j^e \\ u_j^e \\ w_k^e \\ u_k^e \end{Bmatrix} + \left\{f_b^e\right\} + \left\{f_s^e\right\} \tag{6.39}$$

where

$$\left[K^e\right] = \int_{V^e} \left[B^e\right]^T [D] \left[B^e\right] dV \tag{6.40}$$

is the stiffness matrix for element (e);

$$\left\{f_b^e\right\} = \int_{V^e} \left[N^e\right]^T \left\{b^e(x,z)\right\} dV \tag{6.41}$$

is the nodal body force vector for element (e);

$$\left\{f_s^e\right\} = \int_{S^e} \left[N^e\right]^T \left\{t^e(x,z)\right\} dS \tag{6.42}$$

is the nodal surface force vector for element (e).

In our example, because all the components of $\left[B^e\right]$ and $[D]$ are constants, the element stiffness matrices can be calculated simply as:

$$\left[K^e\right] = \int_{V^e} \left[B^e\right]^T [D] \left[B^e\right] dV = \frac{HB}{2} \left[B^e\right]^T [D] \left[B^e\right]$$

This gives two symmetrical stiffness matrices:

$$\left[K^1\right] = \frac{M}{2HB} \begin{bmatrix} B^2+\xi H^2 & -(K_0+\xi)HB & -B^2 & HB & -H^2 & K_0HB \\ & H^2+\xi B^2 & -K_0HB & -\xi B^2 & \xi HB & -H^2 \\ & & B^2 & 0 & 0 & -K_0HB \\ & & & \xi B^2 & -\xi HB & 0 \\ & & & & \xi H^2 & 0 \\ & & & & & H^2 \end{bmatrix}$$

$$\left[K^2\right] = \frac{M}{2HB} \begin{bmatrix} \xi H^2 & 0 & 0 & -\xi HB & -\xi H^2 & \xi HB \\ & H^2 & -K_0HB & 0 & K_0HB & -H^2 \\ & & B^2 & 0 & -B^2 & -K_0HB \\ & & & \xi B^2 & \xi HB & -\xi B^2 \\ & & & & B^2+\xi H^2 & -(K_0+\xi)HB \\ & & & & & H^2+\xi B^2 \end{bmatrix}$$

Body forces in our example are:

$$\left\{b^1(x,z)\right\} = \left\{b^2(x,z)\right\} = \begin{Bmatrix} -\gamma \\ 0 \end{Bmatrix}$$

so that

$$\left\{f_b^1\right\}=\int_0^H\left[\int_0^{\frac{B}{H}z}\begin{bmatrix}\frac{z}{H}-\frac{x}{B} & 0\\ 0 & \frac{z}{H}-\frac{x}{B}\\ 1-\frac{z}{H} & 0\\ 0 & 1-\frac{z}{H}\\ \frac{x}{B} & 0\\ 0 & \frac{x}{B}\end{bmatrix}\begin{Bmatrix}-\gamma\\ 0\end{Bmatrix}dx\right]dz=\frac{-1}{2}HB\gamma\begin{Bmatrix}1/3\\ 0\\ 1/3\\ 0\\ 1/3\\ 0\end{Bmatrix}$$

$$\left\{f_b^2\right\}=\int_0^H\left[\int_{\frac{B}{H}z}^{H}\begin{bmatrix}\frac{z}{H}-\frac{x}{B} & 0\\ 0 & \frac{z}{H}-\frac{x}{B}\\ 1-\frac{z}{H} & 0\\ 0 & 1-\frac{z}{H}\\ \frac{x}{B} & 0\\ 0 & \frac{x}{B}\end{bmatrix}\begin{Bmatrix}-\gamma\\ 0\end{Bmatrix}dx\right]dz=\frac{-1}{2}HB\gamma\begin{Bmatrix}1/3\\ 0\\ 1/3\\ 0\\ 1/3\\ 0\end{Bmatrix}$$

We arrived to an obvious result – the weight of the element is distributed equally between its three nodes.

Surface tractions in our example are:

$$\left\{t^1(x,z)\right\}=0\,;\qquad \left\{t^2(x,0)\right\}=\begin{Bmatrix}p\\ 0\end{Bmatrix};\quad \left\{t^2(B,z)\right\}=\begin{Bmatrix}\tau_r\\ 0\end{Bmatrix},$$

so that:

$$\left\{f_s^1\right\}=\begin{Bmatrix}0\\0\\0\\0\\0\\0\end{Bmatrix}\qquad\left\{f_s^2\right\}=\begin{Bmatrix}\dfrac{pB}{2}\\0\\\dfrac{\tau_r H}{2}\\0\\\dfrac{pB}{2}+\dfrac{\tau_r H}{2}\\0\end{Bmatrix}$$

The latter result being obtained from:

$$\left\{f_s^2\right\}=\int_0^B\begin{bmatrix}1-\dfrac{x}{B} & 0\\0 & 1-\dfrac{x}{B}\\\dfrac{z}{H} & 0\\0 & \dfrac{z}{H}\\\dfrac{x}{B}-\dfrac{z}{H} & 0\\0 & \dfrac{x}{B}-\dfrac{z}{H}\end{bmatrix}_{z=0}\begin{Bmatrix}p\\0\end{Bmatrix}dx+\int_0^H\begin{bmatrix}1-\dfrac{x}{B} & 0\\0 & 1-\dfrac{x}{B}\\\dfrac{z}{H} & 0\\0 & \dfrac{z}{H}\\\dfrac{x}{B}-\dfrac{z}{H} & 0\\0 & \dfrac{x}{B}-\dfrac{z}{H}\end{bmatrix}_{x=B}\begin{Bmatrix}\tau_r\\0\end{Bmatrix}dz$$

Clearly, uniform surface loads acting on an element side contribute equal nodal forces to both nodes of this side.

6.3.6 Equilibrium of the nodes

All the fictitious nodal forces at each node *n* also have to be in equilibrium:

$$\{F_n\}=\begin{Bmatrix}F_{zn}\\F_{xn}\end{Bmatrix}=\sum_e\begin{Bmatrix}W_n^e\\U_n^e\end{Bmatrix}\tag{6.43}$$

where $\{F_n\}$ are the resulting nodal forces at the node *n*, which in the absence of external nodal forces (such as a reaction to the prescribed displacement) should be equal to zero. By summing up the fictitious nodal forces from each element surrounding the node *n*, equation (6.43) allows for element stiffness matrices to be assembled into the global stiffness matrix $[K]$ for the entire continuum:

$$\{F_n\}=[K]\{u_n\}+\{f_{bn}\}+\{f_{sn}\}\tag{6.44}$$

where

$$\{f_{bn}\} = \sum_e \{f_{bn}^e\}; \qquad \{f_{sn}\} = \sum_e \{f_{sn}^e\}. \tag{6.45}$$

In our case, this assembly is illustrated in Figure 6.7, where the two elementary matrices $[K^1]$ and $[K^2]$ overlap over the common nodes 2 and 3.

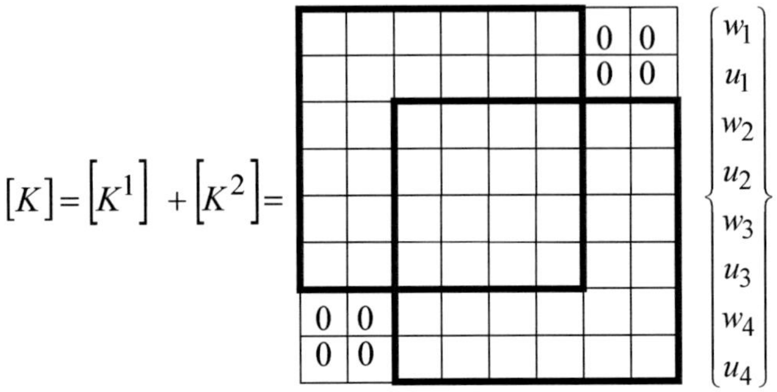

Figure 6.7: Stiffness matrix assembly

The global body and surface nodal forces are:

$$\{f_b\} = -\frac{1}{6} HB\gamma \begin{Bmatrix} 1 \\ 0 \\ 2 \\ 0 \\ 2 \\ 0 \\ 1 \\ 0 \end{Bmatrix}; \qquad \{f_s\} = \frac{1}{2} \begin{Bmatrix} 0 \\ 0 \\ pB \\ 0 \\ \tau_r H \\ 0 \\ pB + \tau_r H \\ 0 \end{Bmatrix}. \tag{6.46}$$

6.3.7 Boundary conditions

Static boundary conditions are taken care of in the surface nodal force vector $\{f_{sn}\}$ and in the vector of external nodal forces $\{F_n\}$, while the kinematic boundary conditions reduce the number of unknown nodal displacements $\{u_n\}$. This allows for the number of equations in the system (6.44) to be reduced by eliminating equations for those F_n, where u_n is prescribed, and by substituting these given values of u_n into the remaining equations. After the reduced system of equations is solved, the derived nodal displacements $\{u_n\}$ are substituted into the omitted equations to find the nodal reactions F_n.

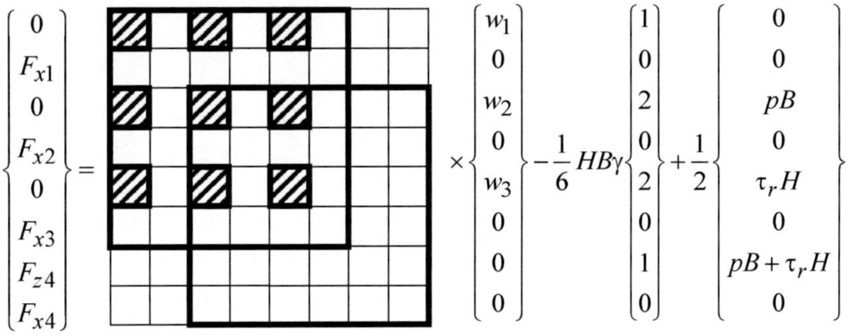

Figure 6.8: Reducing the system using boundary conditions

In our example, the following displacements are prescribed:

$$u_1 = u_2 = u_3 = u_4 = w_4 = 0 \tag{6.47}$$

while the nodal reactions $F_{x1}, F_{x2}, F_{x3}, F_{x4}$ and F_{z4} are unknown. Five corresponding equations can then be eliminated (Figure 6.8), and after substituting expressions (6.47) into the remaining equations we obtain the reduced system:

$$\begin{Bmatrix} 0 \\ 0 \\ 0 \end{Bmatrix} = \frac{MB}{2Ha}\begin{bmatrix} 1 & -a & a-1 \\ -a & 1 & 0 \\ a-1 & 0 & 1 \end{bmatrix}\begin{Bmatrix} w_1 \\ w_2 \\ w_3 \end{Bmatrix} - \frac{\gamma BH}{6}\begin{Bmatrix} 1 \\ 2 \\ 2 \end{Bmatrix} + \frac{1}{2}\begin{Bmatrix} 0 \\ pB \\ \tau_r H \end{Bmatrix} \tag{6.48}$$

$$a = \frac{B^2}{B^2 + \xi H^2} \tag{6.49}$$

where the reduced stiffness matrix is assembled from the highlighted components in Figure 6.8

6.3.8 Solution

When the tunnel support p is calculated from the global equilibrium:

$$p = H\left(\gamma - \frac{\tau_r}{B}\right) \tag{6.50}$$

the solution of the reduced system (6.48) is given by

$$w_1 = \frac{H^2}{2M}\left(\gamma - \frac{\tau_r}{B}\right) + \frac{B^2}{2G}\frac{\tau_r}{B}; \quad w_2 = \frac{B^2}{2G}\frac{\tau_r}{B} + \Delta w; \quad w_3 = \frac{H^2}{2M}\left(\gamma - \frac{\tau_r}{B}\right) + \Delta w, \tag{6.51}$$

where $\Delta w = \dfrac{aH^2\gamma}{6M}$. Substitution of these values into the seventh equation in Figure 6.8:

$$F_{z4} = -\frac{M}{2HB}\xi H^2 w_2 - \frac{M}{2HB}B^2 w_3 - \frac{1}{6}HB\gamma + \frac{1}{2}HB\gamma\,, \qquad (6.52)$$

gives the reaction $F_{z4} = 0$, which confirms the fact that the tunnel support p in equation (6.50) keeps the system in the global equilibrium.

Note, that the nodal displacement w_1 in equations (6.51) coincides with the exact solution (5.28) at $x = 0$ and $z = H$, while the nodal displacements w_2 and w_3 differ from the corresponding exact values by $\Delta w = \dfrac{aH^2\gamma}{6M}$. This difference is due to the fact that in the global nodal body force vector $\{f_b\}$ in the first equation (6.46), the weight of the continuum is not distributed equally between the four nodes. If each node carried one fourth of the weight, all three nodal displacements w_1, w_2 and w_3 would give the exact solution.

Thus, like in the 1D case, in spite of the fact that we discretised the continuum by only two elements, we managed to obtain exact values of displacements at the nodes. This result is, however, again an exception, thanks to the proper choice of the linear shape functions N_n^e (see Zienkiewicz and Taylor, 2000).

Chapter 7

Finite Differences

TABLE OF CONTENTS

Chapter 7

Finite Differences

The version of the FEM described in Chapter 6 is probably the simplest one. The FEM used in the modern commercial FE codes, such as PLAXIS and ABAQUS, benefit from enormous variety of different formulations, element and shape function types. Similarly, there are many different ways to solve a boundary value problem using finite differences. The version of the FDM described in this Chapter follows the one used in FLAC (Fast Lagrangian Analysis of Continua) – the only commercial FD code with broad geotechnical applications. The two special features of this version are: (a) discretization using a non-rectangular mesh, and (b) dynamic approach to solution of static problems. Thanks to these features, this method: (a) is practically identical to the simple version of the FEM described in Chapter 6; (b) is more efficient in terms of computer memory and for solution of non-linear problems.

7.1 Introduction

The 2D FDM method can be briefly described as the following sequence:

1. Discretization of the continuum into "quadrilateral elements" (e) each built of a pair of triangular elements (a) and (b), connected at the nodal points n. At time t, stresses $\sigma_{ij}^{(a)}(t-\Delta t)$ and $\sigma_{ij}^{(b)}(t-\Delta t)$ in the triangular elements and velocities $\{\dot{u}_n(t-\Delta t/2)\}$ at the nodal points are known. Initial conditions for these velocities and stresses have to be specified.
2. FD Compatibility equations use these nodal velocities to define strain rates $\dot{\varepsilon}_{ij}^{(a)}(t)$ and $\dot{\varepsilon}_{ij}^{(b)}(t)$ in triangular elements (a) and (b).
3. Incremental FD constitutive relationships use these strain rates to define stress rates $\dot{\sigma}_{ij}^{(a)}(t)$ and $\dot{\sigma}_{ij}^{(b)}(t)$ in each element. These stress rates are used to update stresses for the current time step: $\sigma_{ij}^{(a)}(t)=\sigma_{ij}^{(a)}(t-\Delta t)+\dot{\sigma}_{ij}^{(a)}(t)\Delta t$ and $\sigma_{ij}^{(b)}(t)=\sigma_{ij}^{(b)}(t-\Delta t)+\dot{\sigma}_{ij}^{(b)}(t)\Delta t$.
4. FD Equilibrium equations for triangular elements use these updated stresses, together with the body and surface forces, to calculate "nodal forces" in each element, which are then summed up to produce the nodal force vector $\left\{q_n^{(e)}(t)\right\}$ for the quadrilateral element (e).
5. Equations of equilibrium for each node sum these fictitious nodal forces from all the quadrilateral elements surrounding the node n to

calculate the vector of “unbalanced forces”: $\{F_n(t)\} = \sum_e \{q_n^{(e)}(t)\}$.

6. Static boundary conditions are taken care of in the nodal force vector $\{q_n^{(e)}(t)\}$, while the kinematic boundary conditions reduce the number of unbalanced forces to be calculated.
7. FD equations of motion for each node use the remaining unbalanced forces, damping forces $\{F_{dn}(t)\}$ and nodal masses to update the nodal velocities: $\{\dot{u}_n(t+\Delta t/2)\} = \{\dot{u}_n(t-\Delta t/2)\} + (\{F_n\} - \{F_{dn}\})\Delta t/m_n$.
8. Solution is achieved by returning to Step 2 with updated nodal velocities $\{\dot{u}_n(t+\Delta t/2)\}$ and repeating the procedure, until unbalanced forces $\{F_n(t)\}$ from Step 5 become negligibly small.

As is seen, apart from using nodal velocities instead of displacements, the described version of the FDM (up to its Step 6) is practically identical to the FEM procedure with linear shape functions presented in Chapter 6. In the FEM, however, the “unbalanced forces” are taken equal to zero, and the resulting system of linear algebraic equations has to be solved with respect to the nodal displacements. FDM avoids this by using the fact that a dynamic system with sufficient damping, when being disturbed, will oscillate around its static state of equilibrium with rapidly decreasing amplitude. This allows for saving a lot of computer memory, but more importantly, it does not require equilibrium to be achieved at each time step, saving a lot of iterations in non-linear problems. This comes, of course, at a price: the time step should be sufficiently small not to allow for the “information” to propagate from an element to its neighbours within one step.

Application of this 8-step FD procedure to the problem of surface subsidence above a cut-and-cover tunnel is illustrated below: first in a 1D approximation and then in the complete plain strain formulation. But before that, let us present the concepts of the FD method on a simplest example.

7.2 Example of a 1D FD solution using local equations

The easiest way to solve the 1D problem described in Section 6.2 is to apply FD procedure directly to the equation of motion:

$$\rho \ddot{w}(z,t) = f(z,t) - f_d(z,t) \tag{7.1}$$

where

$$f(t) = M\frac{\partial^2 w}{\partial z^2} + \gamma - \frac{\tau_r}{B} \tag{7.2}$$

is the unbalanced force per unit volume from equation (6.2);

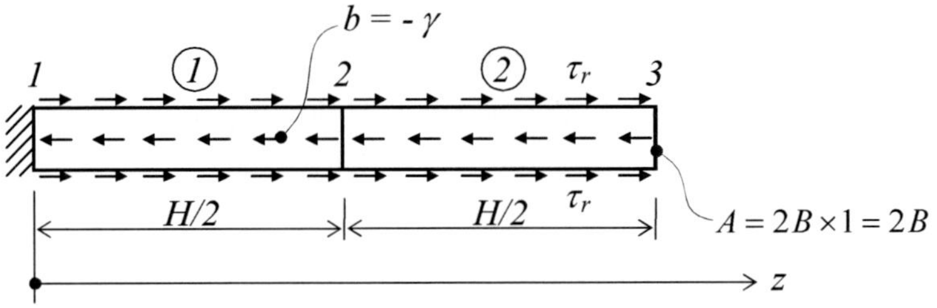

Figure 7.1: Discrete model (rotated by 90°).

$$f_d(z,t) = \alpha \{|f(z,t)|\} \operatorname{sgn}(\dot{w}(z,t)); \quad \operatorname{sgn}(x) = \begin{cases} -1 & for\ x < 0 \\ 0 & for\ x = 0 \\ +1 & for\ x > 0 \end{cases}, \tag{7.3}$$

is the local damping force, chosen to speed up numeric procedure, $\alpha = 0.8$.

Discretising 1D continuum into of intervals of length Δz connected by nodes i (Figure 7.1) and time into steps Δt, we use the following FD approximations:

$$\frac{\partial^2 w(t)}{\partial z^2} = \frac{\frac{w_{i+1} - w_i}{\Delta z} - \frac{w_i - w_{i-1}}{\Delta z}}{\Delta z} = \frac{w_{i+1}(t) - 2w_i(t) + w_{i-1}(t)}{\Delta z^2} \tag{7.4}$$

$$\ddot{w}_i(t) = \frac{\dot{w}_i(t + \Delta t/2) - \dot{w}_i(t - \Delta t/2)}{\Delta t}; \quad \dot{w}_i(t + \Delta t/2) = \frac{w_i(t + \Delta t) - w_i(t)}{\Delta t}. \tag{7.5}$$

The following numerical scheme, supplemented by initial and boundary conditions, solves the problem:

(1): Calculate forces: $f_i(t) = \frac{M}{\Delta z^2}(w_{i+1}(t) - 2w_i(t) + w_{i-1}(t)) + \gamma - \frac{\tau_r}{B}$ (7.6)

(2): Calculate damping: $f_{di}(t) = \alpha\{|f_i(t)|\} \operatorname{sgn}(\dot{w}(t - \Delta t/2))$ (7.7)

(3): Calculate accelerations: $\ddot{w}_i(t) = (f_i(t) - f_{di}(t))/\rho$ (7.8)

(4): Update velocities: $\dot{w}_i(t + \Delta t/2) = \dot{w}_i(t - \Delta t/2) + \ddot{w}_i(t)\Delta t$ (7.9)

(5): Update displacements: $w_i(t + \Delta t) = w_i(t) + \dot{w}_i(t + \Delta t/2)\Delta t$ (7.10)

(6): Return to (7.6) until the unbalanced forces become negligibly small.

This procedure will converge to the static solution provided the chosen time step is sufficiently small: $\Delta t < \Delta z\sqrt{\frac{\rho}{M}}$. This static solution can be obtained directly from equations (7.6), by equating the unbalanced forces to zero and solving the system of differential equations, like in the FEM.

For our example with three nodes (Figure 7.1), equations (7.6) become

$$
\begin{aligned}
f_1 &= \frac{4M}{H^2}\left(w_1 - 2w_2 + w_3\right) + \left(\gamma - \frac{\tau_r}{B}\right) \\
f_2 &= \frac{4M}{H^2}\left(w_1 - 2w_2 + w_3\right) + \left(\gamma - \frac{\tau_r}{B}\right) \\
f_3 &= \frac{4M}{H^2}\left(w_2 - 2w_3 + w_4\right) + \left(\gamma - \frac{\tau_r}{B}\right)
\end{aligned}
\tag{7.11}
$$

The FD versions of the boundary conditions (6.3) are $w_1 = 0$ and $w_2 = w_4$ (Figure 7.2), reducing the system to

$$
\begin{Bmatrix} f_2 \\ f_3/2 \end{Bmatrix} = \frac{4M}{H^2}\begin{bmatrix} -2 & 1 \\ 1 & -1 \end{bmatrix}\begin{Bmatrix} w_2 \\ w_3 \end{Bmatrix} + \frac{1}{2}\left(\gamma - \frac{\tau_r}{B}\right)\begin{Bmatrix} 2 \\ 1 \end{Bmatrix}
\tag{7.12}
$$

When the unbalanced forces are zero, this system becomes identical to the FEM system (6.26), and will produce the same solution (6.27), which happens to be the exact solution.

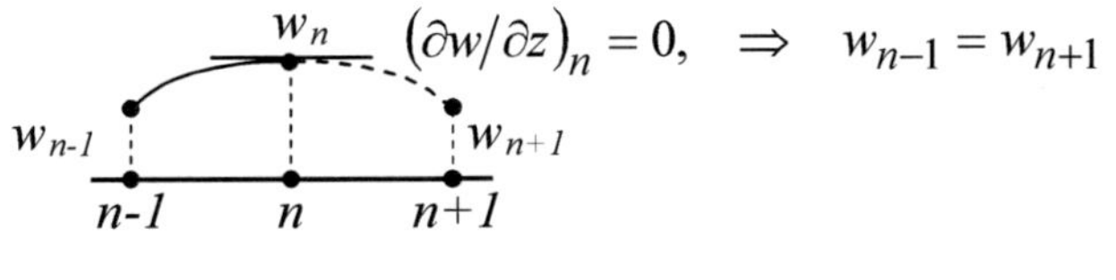

Figure 7.2: Numeric boundary conditions with fictitious nodes

However, the FDM described here avoids solving large systems of algebraic equations by using the fact that a dynamic system with sufficient damping, when being disturbed, will oscillate around its static state of equilibrium with rapidly decreasing amplitude. In this method, equations (7.12) will be used in the above numerical scheme in place of equations (7.6), with corresponding nodal damping, accelerations, velocities and displacements at nodes 2 and 3 being subsequently updated using equations (7.7)-(7.10). The high speed of convergence is ensured by the choice of the local damping (7.3). The first order numerical errors are eliminated thanks to the fact that velocities are shifted by half a step in time with respect to forces.

7.3 A 1D FD solution using element equilibrium

2D FDM employs a more formal 8-step FD procedure described in Section 7.1, which is also useful to follow on our 1D example, in order to facilitate better understanding of the 2D case presented in Section 7.4.

7.3.1 Discretization and initial conditions

The continuum is discretised into two linear elements (1) and (2) of equal length $H/2$ connected by 3 nodes (Figure 7.1). At the previous time step, the values of nodal velocities $\dot{w}_1(t-\Delta t/2)$, $\dot{w}_2(t-\Delta t/2)$ and $\dot{w}_3(t-\Delta t/2)$, as well as the stresses $\sigma_z^{(1)}(t-\Delta t)$ and $\sigma_z^{(2)}(t-\Delta t)$ in linear elements (1) and (2) are assumed to be known. For initial conditions at $t=0$ we assume:

$$\begin{gathered}\dot{w}_1(-\Delta t/2)=\dot{w}_2(-\Delta t/2)=\dot{w}_3(-\Delta t/2)=0\\ \sigma_z^{(1)}(-\Delta t)=\sigma_z^{(2)}(-\Delta t)=0\end{gathered} \tag{7.13}$$

7.3.2 Compatibility equations

Strain rates are defined from velocities via the FD compatibility equation:

$$\dot{\varepsilon}_z^{(e)}(t-\Delta t/2)=\frac{\dot{w}_i(t-\Delta t/2)-\dot{w}_{i-1}(t-\Delta t/2)}{\Delta z} \tag{7.14}$$

In our example:

$$(1)\text{: } \dot{\varepsilon}_z^{(1)}(t-\Delta t/2)=\frac{\dot{w}_2-\dot{w}_1}{H/2}; \quad (2)\text{: } \dot{\varepsilon}_z^{(2)}(t-\Delta t/2)=\frac{\dot{w}_3-\dot{w}_2}{H/2}.$$

7.3.3 Incremental constitutive relationship

Stress rates are defined from strain rates via the incremental constitutive relationship:

$$\dot{\sigma}_z^{(e)}(t-\Delta t/2)=M\dot{\varepsilon}_z^{(e)}(t-\Delta t/2)=M\frac{\dot{w}_i(t-\Delta t/2)-\dot{w}_{i-1}(t-\Delta t/2)}{\Delta z} \tag{7.15}$$

And the stresses are updated for the current time step:

$$\sigma_z^{(e)}(t)=\sigma_z^{(e)}(t-\Delta t)+\dot{\sigma}_z^{(e)}(t-\Delta t/2)\Delta t \tag{7.16}$$

In our example:

$$(1)\text{: } \dot{\sigma}_z^{(1)}(t-\Delta t/2)=M\frac{\dot{w}_2-\dot{w}_1}{H/2}; \quad (2)\text{: } \dot{\sigma}_z^{(2)}(t-\Delta t/2)=M\frac{\dot{w}_3-\dot{w}_2}{H/2},$$

so that the updated stresses are given by

$$\sigma_z^{(1)}(t)=\sigma_z^{(1)}(t-\Delta t)+M\frac{\dot{w}_2-\dot{w}_1}{H/2}\Delta t; \quad \sigma_z^{(2)}(t)=\sigma_z^{(2)}(t-\Delta t)+M\frac{\dot{w}_3-\dot{w}_2}{H/2}\Delta t.$$

7.3.4 Equilibrium of the element

When cut from its surroundings, element (e) is subjected to action of fictitious forces $\left\{q_n^e\right\} = \begin{Bmatrix} q_i^e \\ q_{i-1}^e \end{Bmatrix}$ acting at its nodes i and i-1 (Figure 7.3).

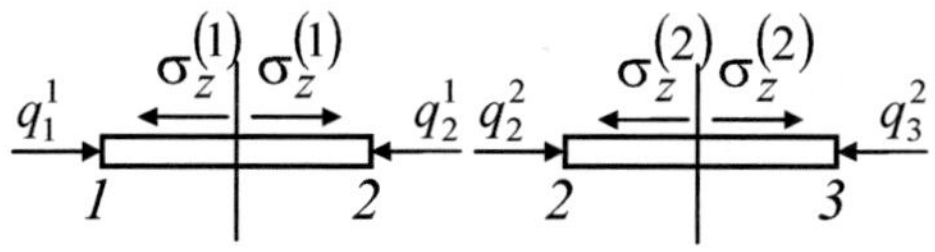

Figure 7.3: Nodal forces

In the FDM, these nodal forces are calculated by cutting element (e) in the middle and summing the forces caused by stresses $\sigma_z^{(e)}$, body forces $b^{(e)}$ and surfaces forces $t^{(e)}$ in the half adjacent to the node:

$$\left\{q_n^{(e)}\right\} = \left\{f_{\sigma n}^{(e)}\right\} + \left\{f_{bn}^{(e)}\right\} + \left\{f_{sn}^{(e)}\right\} \tag{7.17}$$

$$f_{\sigma i}^{(e)} = \sigma_z^{(e)} A^{(e)}; \qquad f_{bi}^{(e)} = b^{(e)} A^{(e)} \Delta z/2; \qquad f_{si}^{(e)} = t^{(e)} S^{(e)}, \tag{7.18}$$

where $A^{(e)}$ is the area of the element cross-section, $S^{(e)}$ is the area of the half of the element surface subjected to action of surface tractions.

In our example, for both elements (1) and (2):

$$A^{(e)} = 2B \times 1 = 2B; \quad S^{(e)} = 2 \times H/4 \times 1 = H/2; \quad b^{(e)} = -\gamma; \quad t^{(e)} = \tau_r,$$

so that

$$\begin{Bmatrix} q_1^{(1)} \\ q_2^{(1)} \end{Bmatrix} = \begin{Bmatrix} 2B\sigma_z^{(1)} + \gamma HB/2 - \tau_r H/2 \\ -2B\sigma_z^{(1)} + \gamma HB/2 - \tau_r H/2 \end{Bmatrix}$$

$$\begin{Bmatrix} q_2^{(2)} \\ q_3^{(2)} \end{Bmatrix} = \begin{Bmatrix} 2B\sigma_z^{(2)} + \gamma HB/2 - \tau_r H/2 \\ -2B\sigma_z^{(2)} + \gamma HB/2 - \tau_r H/2 \end{Bmatrix}$$

7.3.5 Equilibrium of the nodes

All the fictitious nodal forces at each node n also have to be in equilibrium:

$$\{F_n\} = \sum_e \left\{q_n^{(e)}\right\} \tag{7.19}$$

where F_n is the resulting "unbalanced" nodal force at the node n.

For our example:

$$\begin{Bmatrix} F_1 \\ F_2 \\ F_3 \end{Bmatrix}_t = 2B \begin{Bmatrix} \sigma_z^{(1)} \\ \sigma_z^{(2)} - \sigma_z^{(1)} \\ -\sigma_z^{(2)} \end{Bmatrix}_t + \frac{1}{2} HB\left(\gamma - \frac{\tau_r}{B}\right) \begin{Bmatrix} 1 \\ 2 \\ 1 \end{Bmatrix}$$

or, after substituting updated stresses

$$\begin{Bmatrix} F_1 \\ F_2 \\ F_3 \end{Bmatrix}_t = 2B \begin{Bmatrix} \sigma_z^{(1)} \\ \sigma_z^{(2)} - \sigma_z^{(1)} \\ -\sigma_z^{(2)} \end{Bmatrix}_{t-\Delta t} - \frac{4BM\Delta t}{H} \begin{bmatrix} 1 & -1 & 0 \\ -1 & 2 & -1 \\ 0 & -1 & 1 \end{bmatrix} \begin{Bmatrix} \dot{w}_1 \\ \dot{w}_2 \\ \dot{w}_3 \end{Bmatrix} + \frac{1}{2} HB\left(\gamma - \frac{\tau_r}{B}\right) \begin{Bmatrix} 1 \\ 2 \\ 1 \end{Bmatrix} \tag{7.20}$$

7.3.6 Boundary conditions

Static boundary conditions are taken care of in the surface nodal force vector $\{f_{sn}\}$, while the kinematic boundary conditions eliminate equations for those F_n, where $\dot{w}_n$ is prescribed.

In our example, in the node $n = 1$ the velocity $\dot{w}_1 = 0$ is prescribed. The first equation in the system (7.20) can then be eliminated, and after substituting $\dot{w}_1 = 0$ into other two equations we obtained the reduced system:

$$\begin{Bmatrix} F_2 \\ F_3 \end{Bmatrix}_t = 2B \begin{Bmatrix} \sigma_z^{(2)} - \sigma_z^{(1)} \\ -\sigma_z^{(2)} \end{Bmatrix}_{t-\Delta t} - \frac{4BM\Delta t}{H} \begin{bmatrix} 2 & -1 \\ -1 & 1 \end{bmatrix} \begin{Bmatrix} \dot{w}_2 \\ \dot{w}_3 \end{Bmatrix} + \frac{1}{2} HB\left(\gamma - \frac{\tau_r}{B}\right) \begin{Bmatrix} 2 \\ 1 \end{Bmatrix} \tag{7.21}$$

In static formulation, the unbalanced forces and initial stresses are zero and $\{\dot{w}_n\}\Delta t = \{w_n\}$. In this case, the system (7.21) is identical to the one produced by the FEM, and when solved will also give the exact solution at the nodes:

$$w_1 = 0; \qquad w_2 = \frac{3H^2}{8M}\left(\gamma - \frac{\tau_r}{B}\right); \qquad w_3 = \frac{H^2}{2M}\left(\gamma - \frac{\tau_r}{B}\right). \tag{7.22}$$

The FDM, however, instead of solving the system, invokes equations of motion.

7.3.7 FD equations of motion

For node n, equation of motion is given by:

$$m_n \ddot{w}_n(t) = F_n(t) \tag{7.23}$$

where m_n is the mass of the node. FD approximation of this equation allows for the nodal velocities to be updated for the next time step:

$$\dot{w}_n(t+\Delta t/2)=\dot{w}_n(t-\Delta t/2)+F_n(t)\Delta t/m_n \tag{7.24}$$

This equation, however, will lead to the steady state vibrations around the static solution. To avoid this, a special kind of numerical damping is added:

$$\dot{w}_n(t+\Delta t/2)=\dot{w}_n(t-\Delta t/2)+\left(F_n(t)-F_{dn}(t)\right)\Delta t/m_n \tag{7.25}$$

where

$$F_{dn}(t)=\alpha\left|F_n(t)\right|\mathrm{sgn}\left(\dot{w}_n(t-\Delta t/2)\right) \tag{7.26}$$

is the damping force at the node n; $\alpha = 0.8$ for faster convergence.

7.3.8 Solution

Solution is achieved by subsequently updating nodal forces (7.21):

$$\begin{Bmatrix} F_2 \\ F_3 \end{Bmatrix}_t = \begin{Bmatrix} F_2 \\ F_3 \end{Bmatrix}_{t-\Delta t} - \frac{4BM\Delta t}{H}\begin{bmatrix} 2 & -1 \\ -1 & 1 \end{bmatrix}\begin{Bmatrix} \dot{w}_2 \\ \dot{w}_3 \end{Bmatrix}_{t-\Delta t/2} \tag{7.27}$$

and nodal velocities (7.25):

$$\begin{Bmatrix} \dot{w}_2 \\ \dot{w}_3 \end{Bmatrix}_{t+\Delta t/2} = \begin{Bmatrix} \dot{w}_2 \\ \dot{w}_3 \end{Bmatrix}_{t-\Delta t/2} + \left(\begin{Bmatrix} F_2/m_2 \\ F_3/m_3 \end{Bmatrix}_t - \begin{Bmatrix} F_{d2}/m_2 \\ F_{d3}/m_3 \end{Bmatrix}_t \right)\Delta t \tag{7.28}$$

where
$$\begin{Bmatrix} F_{d2} \\ F_{d3} \end{Bmatrix}_t = 0.8\begin{Bmatrix} |F_2| \\ |F_3| \end{Bmatrix}_t \mathrm{sgn}\left(\begin{Bmatrix} \dot{w}_2 \\ \dot{w}_3 \end{Bmatrix}_{t-\Delta t/2} \right); \quad \begin{Bmatrix} m_2 \\ m_3 \end{Bmatrix} = \rho \begin{Bmatrix} 2HB \\ HB \end{Bmatrix} \tag{7.29}$$

using initial conditions:

$$\begin{Bmatrix} F_2 \\ F_3 \end{Bmatrix}_0 = \frac{1}{2}HB\left(\gamma - \frac{\tau_r}{B}\right)\begin{Bmatrix} 2 \\ 1 \end{Bmatrix} \quad \text{and} \quad \begin{Bmatrix} \dot{w}_2 \\ \dot{w}_3 \end{Bmatrix}_{-\Delta t/2} = \begin{Bmatrix} 0 \\ 0 \end{Bmatrix} \tag{7.30}$$

until unbalanced forces become negligibly small. At each step, velocities are used to update displacements:

$$\begin{Bmatrix} w_2 \\ w_3 \end{Bmatrix}_t = \begin{Bmatrix} w_2 \\ w_3 \end{Bmatrix}_{t-\Delta t} + \Delta t \begin{Bmatrix} \dot{w}_2 \\ \dot{w}_3 \end{Bmatrix}_{t-\Delta t/2} \tag{7.31}$$

This procedure will converge to the static solution (7.22), provided the chosen time step is sufficiently small.

$$\Delta t < \Delta z\sqrt{\frac{\rho}{M}} = \frac{H}{2}\sqrt{\frac{\rho}{M}} \tag{7.32}$$

Here ρ is the density defining the fictitious nodal masses in (7.29), which can be adjusted to increase the speed of convergence, or just taken $\rho = \gamma/g$. The high speed of convergence is also ensured by the local damping (7.26), while the first order numerical errors are eliminated thanks to the fact that velocities are shifted by half a step in time with respect to forces.

7.4 Example of a 2D FD solution

FD approximations for the 2D case are based on the finite difference expression for Gauss' divergence theorem, which allows an integral over an area A to be replaced by an integral along the curve S bounding this area:

$$\int_A \frac{\partial f}{\partial x} dA = \int_S f n_x ds \; ; \qquad \int_A \frac{\partial f}{\partial z} dA = \int_S f n_z ds \, , \tag{7.33}$$

where n_x and n_z are the corresponding components of the outwardly directed unit normal to the surface S.

If the integrals (7.33) are taken over the area of a triangle, then the average values $\left\langle \frac{\partial f}{\partial x} \right\rangle$ and $\left\langle \frac{\partial f}{\partial z} \right\rangle$ of the gradient components over the triangle area are given by

$$\left\langle \frac{\partial f}{\partial x} \right\rangle = \frac{1}{A} \sum_S \langle f \rangle n_x \Delta S \; ; \qquad \left\langle \frac{\partial f}{\partial z} \right\rangle = \frac{1}{A} \sum_S \langle f \rangle n_z \Delta S \, , \tag{7.34}$$

where summation is performed for the tree sides of the triangle and $\langle f \rangle$ is the average value of the function over the corresponding side; n_x and n_z are the components of the unit normal to this side; ΔS is the length of the side.

7.4.1 Discretization and initial conditions

The continuum is discretised into a single quadrilateral element with four nodes (Figure 7.4). This quadrilateral element is built out of two alternative sets of two triangular elements: (a) and (b), and (c) and (d).

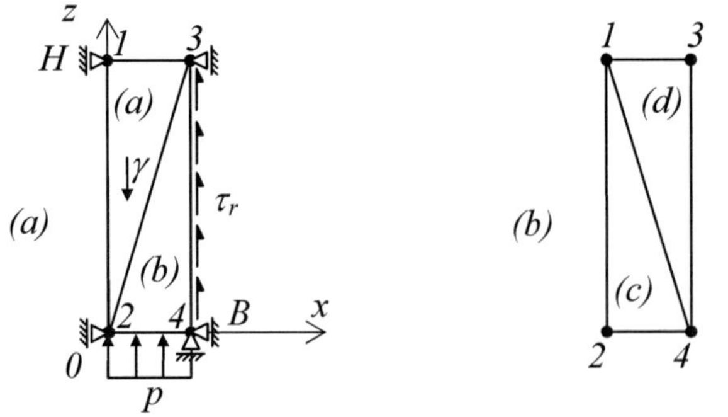

Figure 7.4: Discrete 2D FD model: a) First set of constant strain triangular elements; b) second set.

All the nodal forces are calculated separately for each set, and then are averaged. We are going to demonstrate the procedure for the first set: elements (a) and (b). At the previous time step, the values of nodal velocities $\dot{w}_1(t-\Delta t/2)$, $\dot{w}_2(t-\Delta t/2)$ and $\dot{w}_3(t-\Delta t/2)$, as well as the stresses $\{\sigma^{(a)}(t)\}$ and $\{\sigma^{(b)}(t)\}$ in triangular elements are assumed to be known. For initial conditions at $t=0$ we assume:

$$\begin{gathered}\dot{w}_1(-\Delta t/2)=\dot{w}_2(-\Delta t/2)=\dot{w}_3(-\Delta t/2)=0\\ \{\sigma^{(a)}(-\Delta t)\}=\{\sigma_z^{(b)}(-\Delta t)\}=0\end{gathered} \tag{7.35}$$

7.4.2 Compatibility equations

Strain rates in a triangular element (Figure 7.5) are defined from nodal velocities $\begin{Bmatrix}\dot{w}_n\\ \dot{u}_n\end{Bmatrix}$ via the FD compatibility equations, based on (7.34):

$$\dot{\varepsilon}_z=\frac{\partial \dot{w}}{\partial z}=\frac{1}{A}\left[\frac{\dot{w}_i+\dot{w}_j}{2}n_z^{(ij)}\Delta S^{(ij)}+\frac{\dot{w}_i+\dot{w}_k}{2}n_z^{(ik)}\Delta S^{(ik)}+\frac{\dot{w}_j+\dot{w}_k}{2}n_z^{(jk)}\Delta S^{(jk)}\right]$$

$$\dot{\varepsilon}_x=\frac{\partial \dot{u}}{\partial x}=\frac{1}{A}\left[\frac{\dot{u}_i+\dot{u}_j}{2}n_x^{(ij)}\Delta S^{(ij)}+\frac{\dot{u}_i+\dot{u}_k}{2}n_x^{(ik)}\Delta S^{(ik)}+\frac{\dot{u}_j+\dot{u}_k}{2}n_x^{(jk)}\Delta S^{(jk)}\right]$$

$$\begin{aligned}\dot{\gamma}_{zx}&=\frac{\partial \dot{w}}{\partial x}+\frac{\partial \dot{u}}{\partial z}\\ &=\frac{1}{A}\left[\frac{\dot{w}_i+\dot{w}_j}{2}n_x^{(ij)}\Delta S^{(ij)}+\frac{\dot{w}_i+\dot{w}_k}{2}n_x^{(ik)}\Delta S^{(ik)}+\frac{\dot{w}_j+\dot{w}_k}{2}n_x^{(jk)}\Delta S^{(jk)}\right]\\ &+\frac{1}{A}\left[\frac{\dot{u}_i+\dot{u}_j}{2}n_z^{(ij)}\Delta S^{(ij)}+\frac{\dot{u}_i+\dot{u}_k}{2}n_z^{(ik)}\Delta S^{(ik)}+\frac{\dot{u}_j+\dot{u}_k}{2}n_z^{(jk)}\Delta S^{(jk)}\right]\end{aligned}$$

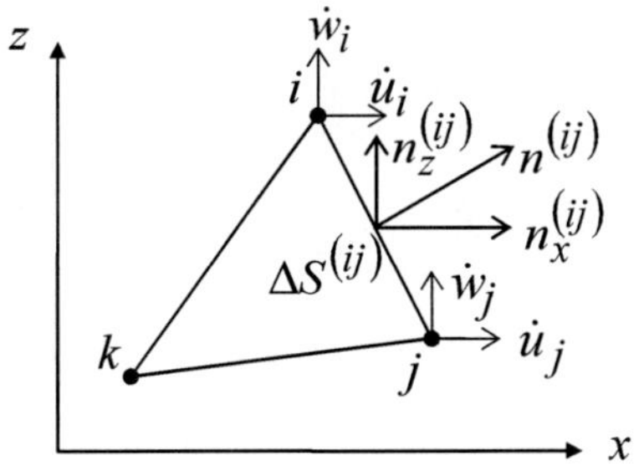

Figure 7.5: Velocities and strain rates

For our example (Figure 7.4) for element (a):

$$n_x^{(12)} = -1; \qquad n_x^{(13)} = 0; \qquad n_x^{(23)} = \frac{H}{\sqrt{H^2 + B^2}};$$

$$n_z^{(12)} = 0; \qquad n_z^{(13)} = 1; \qquad n_z^{(23)} = \frac{-B}{\sqrt{H^2 + B^2}};$$

$$\Delta S^{(12)} = H; \qquad \Delta S^{(13)} = B; \qquad \Delta S^{(23)} = \sqrt{H^2 + B^2}.$$

For element (b):

$$n_x^{(34)} = 1; \qquad n_x^{(24)} = 0; \qquad n_x^{(23)} = \frac{-H}{\sqrt{H^2 + B^2}};$$

$$n_z^{(34)} = 0; \qquad n_z^{(24)} = -1; \qquad n_z^{(23)} = \frac{B}{\sqrt{H^2 + B^2}};$$

$$\Delta S^{(34)} = H; \qquad \Delta S^{(24)} = B; \qquad \Delta S^{(23)} = \sqrt{H^2 + B^2}.$$

The area of both elements: $A = HB/2$. Invoking the boundary conditions:

$$\dot{w}_4 = 0; \quad \dot{u}_1 = \dot{u}_2 = \dot{u}_3 = \dot{u}_4 = 0,$$

we obtain the strain rates for both triangular elements:

$$\dot{\varepsilon}_z^{(a)}(t - \Delta t/2) = \frac{\dot{w}_1 - \dot{w}_2}{H}; \qquad \dot{\varepsilon}_x^{(a)}(t - \Delta t/2) = 0; \qquad \dot{\gamma}_{xz}^{(a)}(t - \Delta t/2) = \frac{\dot{w}_3 - \dot{w}_1}{B};$$

$$\dot{\varepsilon}_z^{(b)}(t - \Delta t/2) = \frac{\dot{w}_3}{H}; \qquad \dot{\varepsilon}_x^{(b)}(t - \Delta t/2) = 0; \qquad \dot{\gamma}_{xz}^{(b)}(t - \Delta t/2) = -\frac{\dot{w}_2}{B}.$$

7.4.3 Incremental constitutive relationship

Stress rates in plain strain are defined from strain rates via the following incremental elastic constitutive relationship:

$$\begin{Bmatrix} \dot{\sigma}_z \\ \dot{\sigma}_x \\ \dot{\tau}_{xz} \end{Bmatrix}_{(t-\Delta t/2)} = M \begin{bmatrix} 1 & K_0 & 0 \\ K_0 & 1 & 0 \\ 0 & 0 & \xi \end{bmatrix} \begin{Bmatrix} \dot{\varepsilon}_z \\ \dot{\varepsilon}_x \\ \dot{\gamma}_{xz} \end{Bmatrix}_{(t-\Delta t/2)} \tag{7.36}$$

And the stresses are updated for the current time step:

$$\{\sigma^{(e)}(t)\} = \{\sigma^{(e)}(t - \Delta t)\} + \{\dot{\sigma}^{(e)}(t - \Delta t/2)\}\Delta t \tag{7.37}$$

In our example, the updated stresses are:

$$\begin{Bmatrix} \sigma_z^{(a)} \\ \sigma_x^{(a)} \\ \tau_{zx}^{(a)} \end{Bmatrix}_{(t)} = \begin{Bmatrix} \sigma_z^{(a)} \\ \sigma_x^{(a)} \\ \tau_{zx}^{(a)} \end{Bmatrix}_{(t-\Delta t)} + M\Delta t \begin{Bmatrix} \dfrac{\dot{w}_1 - \dot{w}_2}{H} \\ K_0 \dfrac{\dot{w}_1 - \dot{w}_2}{H} \\ \xi \dfrac{\dot{w}_3 - \dot{w}_1}{B} \end{Bmatrix}_{(t-\Delta t/2)} \tag{7.38}$$

$$\begin{Bmatrix} \sigma_z^{(b)} \\ \sigma_x^{(b)} \\ \tau_{zx}^{(b)} \end{Bmatrix}_{(t)} = \begin{Bmatrix} \sigma_z^{(b)} \\ \sigma_x^{(b)} \\ \tau_{zx}^{(b)} \end{Bmatrix}_{(t-\Delta t)} + M\Delta t \begin{Bmatrix} \dfrac{\dot{w}_3}{H} \\ K_0 \dfrac{\dot{w}_3}{H} \\ \xi \dfrac{\dot{w}_2}{B} \end{Bmatrix}_{(t-\Delta t/2)} \tag{7.39}$$

7.4.4 Equilibrium of the element

When cut from its surroundings, element (e) is subjected to the action of fictitious forces $\left\{q_i^{(e)}\right\} = \begin{Bmatrix} W_i^{(e)} \\ U_i^{(e)} \end{Bmatrix}$, $\left\{q_j^{(e)}\right\} = \begin{Bmatrix} W_j^{(e)} \\ U_j^{(e)} \end{Bmatrix}$ and $\left\{q_k^{(e)}\right\} = \begin{Bmatrix} W_k^{(e)} \\ U_k^{(e)} \end{Bmatrix}$ acting at its nodes *i*, *j* and k, respectively (Figure 7.6). These nodal forces should be in equilibrium with the stresses $\left\{\sigma^{(e)}\right\}$, body forces $\left\{b^{(e)}(x,z)\right\} = \begin{Bmatrix} b_z^{(e)} \\ b_x^{(e)} \end{Bmatrix}$ and surfaces forces $\left\{t^{(e)}(x,z)\right\} = \begin{Bmatrix} t_z^{(e)} \\ t_x^{(e)} \end{Bmatrix}$ acting within the element volume V^e and at the element surface S^e, respectively:

$$\left\{q_n^{(e)}\right\} = \left\{f_{\sigma n}^{(e)}\right\} + \left\{f_{bn}^{(e)}\right\} + \left\{f_{sn}^{(e)}\right\} \tag{7.40}$$

Nodal forces $\left\{f_{\sigma i}^{(e)}\right\} = \begin{Bmatrix} W_{\sigma i}^{(e)} \\ U_{\sigma i}^{(e)} \end{Bmatrix}$, which have to be in equilibrium with the stresses, are calculated by integrating normal and shear stresses acting on the sides of the triangle (Figure 7.6). Only half of each adjacent side contributes to the nodal force:

$$W_{\sigma i}^{(e)} = \frac{1}{2}\left(\left[\sigma_z^{(e)} n_z^{(ik)} + \tau_{xz}^{(e)} n_x^{(ik)}\right]\Delta S^{(ik)} + \left[\sigma_z^{(e)} n_z^{(ij)} + \tau_{xz}^{(e)} n_x^{(ij)}\right]\Delta S^{(ij)}\right)$$

$$U_{\sigma i}^{(e)} = \frac{1}{2}\left(\left[\sigma_x^{(e)} n_x^{(ik)} + \tau_{xz}^{(e)} n_z^{(ik)}\right]\Delta S^{(ik)} + \left[\sigma_x^{(e)} n_x^{(ij)} + \tau_{xz}^{(e)} n_z^{(ij)}\right]\Delta S^{(ij)}\right)$$

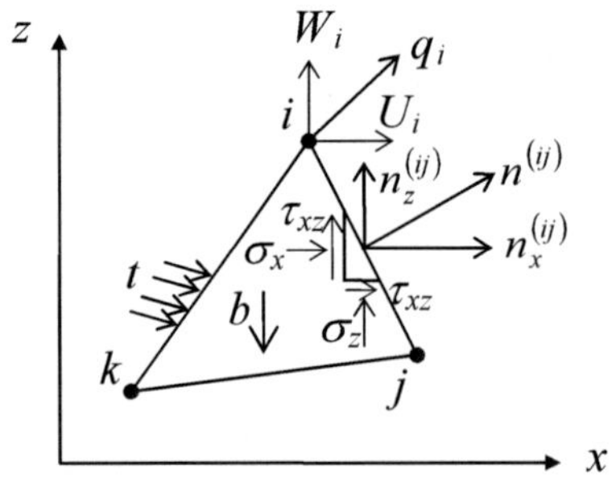

Figure 7.6: Stresses and nodal forces

In our example,

$$\dot{w}_4 = 0\,; \quad \dot{u}_1 = \dot{u}_2 = \dot{u}_3 = \dot{u}_4 = 0\,.$$

Therefore, only the components $W_{\sigma 1}^{(e)}$, $W_{\sigma 2}^{(e)}$, $W_{\sigma 3}^{(e)}$ have to be calculated:

(a): $W_{\sigma 1}^{(a)} = \frac{1}{2}\left(B\sigma_z^{(a)} - H\tau_{zx}^{(a)}\right)$; $\quad W_{\sigma 2}^{(a)} = \frac{1}{2}B\sigma_z^{(a)}$; $\quad W_{\sigma 3}^{(a)} = \frac{1}{2}H\tau_{zx}^{(a)}$,

(b): $W_{\sigma 2}^{(b)} = \frac{1}{2}H\tau_{zx}^{(b)}$; $\quad W_{\sigma 3}^{(b)} = \frac{1}{2}B\sigma_z^{(b)}$,

where the values for the components normal vectors n_i and side lengths ΔS have been taken from Section 7.4.2.

Nodal forces $\left\{f_{bi}^{(e)}\right\} = \begin{Bmatrix} W_{bi}^{(e)} \\ U_{bi}^{(e)} \end{Bmatrix}$, which have to be in equilibrium with body forces, are calculated in the FDM by lumping one third of the weight of a triangular element at each of the nodes:

$$W_{b1}^{(a)} = W_{b2}^{(a)} = W_{b3}^{(a)} = W_{b2}^{(b)} = W_{b3}^{(b)} = \frac{1}{6}HB\gamma$$

Nodal forces $\left\{f_{si}^{(e)}\right\} = \begin{Bmatrix} W_{bi}^{(e)} \\ U_{si}^{(e)} \end{Bmatrix}$, which have to be in equilibrium with surface forces, are calculated in the FDM as static reactions of the nodes to the surface tractions acting on the side of triangle connecting these two nodes:

$$W_{s1}^{(a)} = W_{s2}^{(a)} = W_{s3}^{(a)} = 0\,;\; W_{s2}^{(a)} = -\frac{1}{2}Bp = -\frac{1}{2}HB\left(\gamma - \frac{\tau_r}{B}\right);\; W_{s3}^{(a)} = -\frac{1}{2}H\tau_r\,.$$

Then the total nodal forces for element (a) are:

$$W_1^{(a)} = \frac{1}{2}\left(B\sigma_z^{(a)} - H\tau_{zx}^{(a)}\right) + \frac{1}{6}HB\gamma$$

$$W_2^{(a)} = \frac{1}{2}B\sigma_z^{(a)} + \frac{1}{6}HB\gamma\,; \quad W_3^{(a)} = \frac{1}{2}H\tau_{zx}^{(a)} + \frac{1}{6}HB\gamma\,,$$

for element (b):

$$W_2^{(b)} = \frac{1}{2}H\tau_{zx}^{(b)} + \frac{1}{6}HB\gamma - \frac{1}{2}HB\left(\gamma - \frac{\tau_r}{B}\right); \quad W_3^{(b)} = \frac{1}{2}B\sigma_z^{(b)} + \frac{1}{6}HB\gamma - \frac{1}{2}H\tau_r\,.$$

7.4.5 Equilibrium of the nodes

All the fictitious nodal forces at each node n also have to be in equilibrium:

$$\{F_n\} = \begin{Bmatrix} F_{zn} \\ F_{xn} \end{Bmatrix} = \sum_e \{q_n^e\} \tag{7.41}$$

where F_{xn} and F_{zn} are the resulting "unbalanced" nodal force at the node n.

For our example:

$$\begin{Bmatrix} F_{z1} \\ F_{z2} \\ F_{z3} \end{Bmatrix} = \begin{Bmatrix} W_1^{(a)} \\ W_2^{(a)} + W_2^{(b)} \\ W_3^{(a)} + W_3^{(b)} \end{Bmatrix} = \frac{1}{2}\begin{Bmatrix} B\sigma_z^{(a)} - H\tau_{zx}^{(a)} \\ B\sigma_z^{(a)} + H\tau_{zx}^{(b)} \\ B\sigma_z^{(b)} + H\tau_{zx}^{(a)} \end{Bmatrix} + \frac{1}{6}HB\gamma\begin{Bmatrix} 1 \\ 2 \\ 2 \end{Bmatrix} - \frac{1}{2}H\begin{Bmatrix} 0 \\ B\gamma - \tau_r \\ \tau_r \end{Bmatrix}$$

or, after substituting updated stresses (7.38) and (7.39) and using notation:

$$a = \frac{B^2}{B^2 + \xi H^2} \tag{7.42}$$

we obtain:

$$\begin{aligned} \begin{Bmatrix} F_{z1} \\ F_{z2} \\ F_{z3} \end{Bmatrix}_{(t)} &= \frac{1}{2}\begin{Bmatrix} B\sigma_z^{(a)} - H\tau_{zx}^{(a)} \\ B\sigma_z^{(a)} + H\tau_{zx}^{(b)} \\ B\sigma_z^{(b)} + H\tau_{zx}^{(a)} \end{Bmatrix}_{(t-\Delta t)} + \frac{1}{6}HB\gamma\begin{Bmatrix} 1 \\ 2 \\ 2 \end{Bmatrix} - \frac{1}{2}H\begin{Bmatrix} 0 \\ B\gamma - \tau_r \\ \tau_r \end{Bmatrix} \\ &\quad - \frac{MB\Delta t}{2Ha}\begin{bmatrix} 1 & -a & a-1 \\ -a & 1 & 0 \\ a-1 & 0 & 1 \end{bmatrix}\begin{Bmatrix} \dot{w}_1 \\ \dot{w}_2 \\ \dot{w}_3 \end{Bmatrix}_{(t-\Delta t/2)} \end{aligned} \tag{7.43}$$

7.4.6 Boundary conditions

Static boundary conditions are taken care of in the surface nodal force vector $\{f_{sn}\}$, while the kinematic boundary conditions eliminate equations for those F_n, where $\dot{w}_n$ is prescribed. In our example, we used kinematic conditions $\dot{w}_4 = 0$ and $\dot{u}_1 = \dot{u}_2 = \dot{u}_3 = \dot{u}_4 = 0$ early in the development in order to reduce the amount of calculations, which resulted in three unbalanced forces (7.43) instead of eight.

In static formulation, the unbalanced forces and initial stresses are zero and $\{\dot{w}_n\}\Delta t = \{w_n\}$. In this case, the system (7.43) is identical to (6.48) produced by the FEM, and when solved will also give:

$$w_1 = \frac{H^2}{2M}\left(\gamma - \frac{\tau_r}{B}\right) + \frac{B^2}{2G}\frac{\tau_r}{B}; \qquad w_2 = \frac{B^2}{2G}\frac{\tau_r}{B} + \Delta w; \qquad w_3 = \frac{H^2}{2M}\left(\gamma - \frac{\tau_r}{B}\right) + \Delta w.$$

where $\Delta w = \frac{aH^2\gamma}{6M}$. This result differs from the exact solution by Δw. This difference is due to the fact that in the global nodal body force vector $\{f_{bn}\}$ in (7.43), the weight of the continuum is not distributed equally between the four nodes of the quadrilateral element. The FD procedure, however, calculates the unbalanced forces for each quadrilateral element twice: once for the set of triangles (a) and (b), and the second time for the set (c) and (d) (Figure 7.4), and then averages them. In our example, this gives:

$$\begin{Bmatrix} F_{z1} \\ F_{z2} \\ F_{z3} \end{Bmatrix}_{(t)} = \frac{1}{4}\begin{Bmatrix} B\sigma_z^{(a)} - H\tau_{zx}^{(a)} \\ B\sigma_z^{(a)} + H\tau_{zx}^{(b)} \\ B\sigma_z^{(b)} + H\tau_{zx}^{(a)} \end{Bmatrix}_{(t-\Delta t)} + \frac{1}{4}\begin{Bmatrix} B\sigma_z^{(c)} - H\tau_{zx}^{(d)} \\ -B\sigma_z^{(c)} - H\tau_{zx}^{(c)} \\ B\sigma_z^{(d)} - H\tau_{zx}^{(d)} \end{Bmatrix}_{(t-\Delta t)}$$
$$+\frac{1}{4}HB\gamma\begin{Bmatrix} 1 \\ 1 \\ 1 \end{Bmatrix} - \frac{1}{2}H\begin{Bmatrix} 0 \\ B\gamma - \tau_r \\ \tau_r \end{Bmatrix} - \frac{MB\Delta t}{2Ha}\begin{bmatrix} 1 & -a & a-1 \\ -a & 1 & 0 \\ a-1 & 0 & 1 \end{bmatrix}\begin{Bmatrix} \dot{w}_1 \\ \dot{w}_2 \\ \dot{w}_3 \end{Bmatrix}_{(t-\Delta t/2)} \qquad (7.44)$$

which in the static case will produce the exact solution:

$$w_1 = \frac{H^2}{2M}\left(\gamma - \frac{\tau_r}{B}\right) + \frac{B^2}{2G}\frac{\tau_r}{B}; \qquad w_2 = \frac{B^2}{2G}\frac{\tau_r}{B}; \qquad w_3 = \frac{H^2}{2M}\left(\gamma - \frac{\tau_r}{B}\right). \qquad (7.45)$$

The FDM, however, instead of solving the system, invokes equations of motion.

7.4.7 FD equations of motion

For node n, equations of motion are given by:

$$
\begin{aligned}
m_n \ddot{w}_n(t) &= F_{zn}(t) \\
m_n \ddot{u}_n(t) &= F_{xn}(t)
\end{aligned}
\tag{7.46}
$$

where m_n is the mass of the node. FD approximation of this equation allows for the nodal velocities to be updated for the next time step:

$$
\begin{aligned}
\dot{w}_n(t+\Delta t/2) &= \dot{w}_n(t-\Delta t/2) + F_{zn}(t)\Delta t/m_n \\
\dot{u}_n(t+\Delta t/2) &= \dot{u}_n(t-\Delta t/2) + F_{xn}(t)\Delta t/m_n
\end{aligned}
\tag{7.47}
$$

This equation, however, will lead to the steady state vibrations around the static solution. To avoid this, a special kind of numerical damping is added:

$$
\begin{aligned}
\dot{w}_n(t+\Delta t/2) &= \dot{w}_n(t-\Delta t/2) + \left(F_{zn}(t) - W_{dn}(t)\right)\Delta t/m_n \\
\dot{u}_n(t+\Delta t/2) &= \dot{u}_n(t-\Delta t/2) + \left(F_{xn}(t) - U_{dn}(t)\right)\Delta t/m_n
\end{aligned}
\tag{7.48}
$$

where

$$
\begin{aligned}
W_{dn}(t) &= \alpha\left|F_{zn}(t)\right|\mathrm{sgn}\left(\dot{w}_n(t-\Delta t/2)\right) \\
U_{dn}(t) &= \alpha\left|F_{xn}(t)\right|\mathrm{sgn}\left(\dot{u}_n(t-\Delta t/2)\right)
\end{aligned}
\tag{7.49}
$$

is the damping force at the node n; $\alpha = 0.8$ for faster convergence.

7.4.8 Solution

Solution is achieved by subsequently updating nodal forces (7.44):

$$
\begin{Bmatrix} F_{z1} \\ F_{z2} \\ F_{z3} \end{Bmatrix}_{(t)} = \begin{Bmatrix} F_{z1} \\ F_{z2} \\ F_{z3} \end{Bmatrix}_{(t-\Delta t)} - \frac{MB\Delta t}{2Ha} \begin{bmatrix} 1 & -a & a-1 \\ -a & 1 & 0 \\ a-1 & 0 & 1 \end{bmatrix} \begin{Bmatrix} \dot{w}_1 \\ \dot{w}_2 \\ \dot{w}_3 \end{Bmatrix}_{(t-\Delta t/2)}
\tag{7.50}
$$

and nodal velocities (7.47):

$$
\dot{w}_n(t+\Delta t/2) = \dot{w}_n(t-\Delta t/2) + \left(F_{zn}(t) - W_{dn}(t)\right)\Delta t/m_n \qquad n = 1,2,3
\tag{7.51}
$$

where $m_n = \rho HB/4$ and

$$
W_{dn}(t) = \alpha\left|F_{zn}(t)\right|\mathrm{sgn}\left(\dot{w}_n(t-\Delta t/2)\right) \qquad n = 1,2,3
\tag{7.52}
$$

using initial conditions:

$$\begin{Bmatrix} F_{z1} \\ F_{z2} \\ F_{z3} \end{Bmatrix}_{(0)} = \frac{1}{4} HB\gamma \begin{Bmatrix} 1 \\ 1 \\ 1 \end{Bmatrix} - \frac{1}{2} H \begin{Bmatrix} 0 \\ B\gamma - \tau_r \\ \tau_r \end{Bmatrix} \quad \text{and} \quad \begin{Bmatrix} \dot{w}_1 \\ \dot{w}_2 \\ \dot{w}_3 \end{Bmatrix}_{(-\Delta t/2)} = \begin{Bmatrix} 0 \\ 0 \\ 0 \end{Bmatrix} \tag{7.53}$$

until unbalanced forces become negligibly small. At each step, velocities are used to update stresses (7.38) and (7.39) and displacements:

$$w_n(t) = w_n(t - \Delta t) + \dot{w}_n(t - \Delta t/2)\Delta t \qquad n = 1,2,3 \tag{7.54}$$

This procedure will converge to the static solution (7.45) provided the chosen time step is sufficiently small:

$$\Delta t < \frac{1}{2} \min(H, B) \sqrt{\frac{\rho}{M}} \tag{7.55}$$

Here ρ is the density defining the fictitious nodal masses, which can be adjusted to increase the speed of convergence, or just taken $\rho = \gamma/g$. The high speed of convergence is also ensured by the local damping (7.52), while the first order numerical errors are eliminated thanks to the fact that velocities are shifted by half a step in time with respect to forces.

Chapter 8

Equations of Continuum Thermodynamics

TABLE OF CONTENTS

Chapter 8

Equations of Continuum Thermodynamics

Before we move on to the constitutive modelling, there is an important issue to be considered. As we learned in Chapter 5, formulation of any boundary value problem in continuum mechanics is based on the following four components:

- Motion (equilibrium);
- Compatibility (geometry);
- Constitutive relationships;
- Initial and boundary conditions.

There are, however, two other conditions, which do not enter the boundary value problem formulation explicitly, and yet have to be satisfied by any reliable constitutive relationship. These are the basic universal Laws governing the energy conservation and dissipation:

- The First Law of Thermodynamics;
- The Second Law of Thermodynamics.

The reader may ask: why do we need thermodynamics in this book? Indeed, the processes in geotechnical engineering are either so slow (e.g., during construction) that they can be considered isothermal, or so fast (e.g., during an earthquake) that they can be considered adiabatic. In any case, the temperature and heat play a little role in the vast majority of the geotechnical problems (apart, probably, from those dealing with the ground freezing and nuclear waste disposal).

The answer to this question is that Thermodynamics is not just about the temperature and heat. It is mainly about the fundamental laws of energy conservation and dissipation, which all existing materials obey, in all processes, including isothermal and adiabatic ones. And because the real materials obey these laws, it seems logical to require that their mathematical and mechanical models, i.e. constitutive relationships, should also obey the Laws of Thermodynamics.

In the following we present the thermodynamical concepts in terms of the Classical Thermodynamics and after that interpret them in terms of the Thermodynamics of Continua. This framework will allow us to explore thermomechanical consistency of the constitutive models introduced in the following chapters.

A.M. Puzrin, *Constitutive Modelling in Geomechanics,*

8.1 The Laws of Thermodynamics

8.1.1 The First Law of Thermodynamics

In Classical Thermodynamics we consider a closed system which is isolated from its surroundings by certain walls. A *process* involves interaction between the system and its surroundings, and can involve transfer of two types of energy: a heat flow $\dot{Q}$ into the system from the surroundings and a mechanical power input $\dot{W}$, also from the surroundings. *The First Law* is usually stated in the following form: *for a system in thermodynamic equilibrium, there is a property of the system, called the internal energy U, such that:*

$$\dot{Q} + \dot{W} = \dot{U} \tag{8.1}$$

The above is in fact a somewhat simplified form of the First Law, in that it ignores the fact that some of the input power may, in general, cause an increase in kinetic and potential energies. However, equation (8.1) expresses the essential principle of conservation of energy: the sum of all the sources of power input to a body is equal to the rate of increase of the energy of the body. Furthermore the most important sources of power we must consider are mechanical power and heat supply.

Important consequence of the first Law is that internal energy is path independent: no matter by which combination of heating and loading has been the current thermodynamic state achieved, the internal energy should be the same. It follows that in a closed loading and/or heating cycle the change in internal energy should be zero. If the First Law is violated, the energy in the closed cycle would either disappear or be generated. In the latter case we would be able to construct a so-called perpetual engine of the first kind.

8.1.2 The Second Law of Thermodynamics

The Second Law is considerably more subtle than the First. It applies certain restrictions to the processes that can occur. For instance, one of the basic consequences of the Second Law is that heat cannot spontaneously flow from a colder place to a hotter one.

There are a number of ways by which the Law can be expressed, but the most useful is in the form of the Clausius-Duhem inequality. To express the First law we had to introduce the concept of a property called the internal energy U. The Second Law is also best expressed by making the hypothesis that there is a further property, called the entropy S. Most people find the entropy a more abstract concept to understand than the internal energy. The interpretation of entropy in Statistical Mechanics is the measure of uncertainty, which remains about a system after its observable macroscopic properties, such as temperature, pressure and volume, have been taken into account. For a given set of macroscopic variables, the entropy measures the degree to which the probability of the system is spread out over different possible microstates, which specify all molecular details about the system including the position and velocity of every molecule. The more such states are available to the system with appreciable probability, the greater is the entropy. The Second Law states that the total entropy of any system will not decrease other than by increasing the entropy of some other system.

In Classical Thermodynamics, however, perhaps the most useful approach is to treat entropy simply as a mathematical abstraction, without seeking a physical meaning. A part of the system surroundings which is sufficiently large that its properties can be regarded as unchanged by any interactions with the system is said to be a *reservoir*. The Clausius-Duhem inequality states that, for a system which exchanges heat with n reservoirs at temperatures θ_i, the change of entropy is such that:

$$\dot{S} \geq \sum_{i=1}^{n} \frac{\dot{Q}_i}{\theta_i} \tag{8.2}$$

where $\dot{Q}_i$ is the rate of heat input from reservoir i. It can readily be seen that inequality (8.2) does not permit the heat to spontaneously flow from a colder place to a hotter one. Indeed, consider the process shown in Figure 8.1 in which an unchanged system exchanges heat with two reservoirs at different temperatures. (The system is *unchanged* by a certain process when all state variables and properties of the system are the same after completion of the process as they were at the beginning). The First Law requires that $\dot{Q}_1 = \dot{Q}_2$, and the Second Law requires, since $\dot{S} = 0$ for the unchanged system:

$$0 \geq \frac{\dot{Q}_1}{\theta_1} + \frac{-\dot{Q}_2}{\theta_2} \tag{8.3}$$

Rearranging this, making use of the fact that θ_1, θ_2 and $\dot{Q}_1$ are all positive gives $\theta_1 \geq \theta_2$, so that in a process of pure heat transfer, heat can only flow from a hotter place to a colder one.

Another basic consequence of the Second Law is that work can be dissipated in the form of heat, but that heat cannot be changed into work without some side-effects occurring too. If the second Law is violated, we would be able to construct a so-called perpetual engine of the second kind.

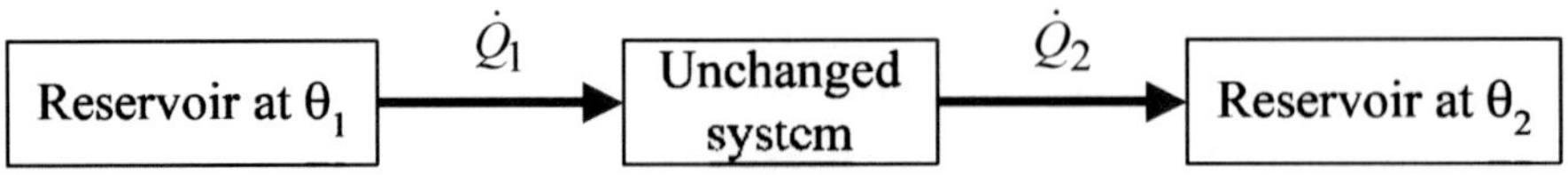

Figure 8.1: An unchanged system exchanging heat with two reservoirs

8.1.3 Reversibility and dissipation

An important concept in thermodynamics is that of *reversibility*. A process is reversible if all the directions of the work and heat flow can be reversed simultaneously, and the resulting process still obeys the Second Law. It is straightforward to show that reversibility is only possible when equality rather than inequality holds for the Clausius-Duhem relationship for the process. So for all reversible processes:

$$\dot{S} = \sum_{i=1}^{n} \frac{\dot{Q}_i}{\theta_i} \tag{8.4}$$

In practice, real geomaterials do not exhibit purely reversible behaviour, in that they are *dissipative*.

8.2 Equations of Continuum Thermodynamics

In the Classical Thermodynamics, the thermodynamic properties of materials are only defined when the material is in a state of *thermodynamic equilibrium*. This requires that all processes occur very slowly. In the Thermodynamics of Continua this is generally not the case. In practice, however, the application of "equilibrium thermodynamics" even to quite rapidly evolving processes is very successful. It is common, therefore, to assume that quantities strictly defined for thermodynamic equilibrium conditions also have applicability to the states of "frozen" inequilibrium, which occur in rate-independent continua.

8.2.1 Work conjugacy

A very important concept in continuum mechanics is that of work conjugacy of stresses and strains. A proper set of definitions satisfy the condition that the stresses and strains are work conjugate in the sense that an increment of work input to the material, per unit volume, is given by the product of the stresses with the strain increments:

$$\dot{W} = \sigma_{ij}\dot{\varepsilon}_{ij} \tag{8.5}$$

It is straightforward to show that the conventional small-strain definitions of the Cauchy stress and the linear strain satisfy this condition.

8.2.2 Thermodynamic state

In Thermodynamics of Continua it is assumed that the local *state* of the material is completely defined by knowledge of (a) the strain ε_{ij} (measured from some reference configuration), (b) the specific (per unit volume) entropy s and (c) certain internal variables α_{ij} (most often plastic strains).

8.2.3 The First Law

The First Law of Thermodynamics states that there is a *property*, termed the specific internal energy, which is a function of state $u = u(\varepsilon_{ij}, s, \alpha_{ij})$ and:

$$\dot{u} = \sigma_{ij}\dot{\varepsilon}_{ij} - q_{k,k} \tag{8.6}$$

where σ_{ij} is the stress that is work-conjugate to the strain rate $\dot{\varepsilon}_{ij}$, and q_k is the specific heat flux vector.

8.2.4 The Second Law

The Second Law of Thermodynamics states that there is a property (the entropy) s such that:

$$\dot{s} \geq -\left(\frac{q_k}{\theta}\right)_{,k} \tag{8.7}$$

where (q_k/θ) is the entropy flux, θ is the non-negative thermodynamic temperature. Equation (8.7) can be restated as:

$$\theta\dot{s} + q_{k,k} - \frac{q_k\theta_{,k}}{\theta} \geq 0 \tag{8.8}$$

The first two terms $\theta\dot{s} + q_{k,k} = d$ are called the *mechanical dissipation*. The third term $-q_k\theta_{,k}/\theta$ is called the *thermal dissipation*. The thermal dissipation is always non-negative by virtue of the fact that the heat flux is always in the direction of the negative thermal gradient. Requiring both $\theta\dot{s} + q_{k,k} = d \geq 0$ and $-(q_k\theta_{,k}/\theta) \geq 0$ is a slightly more stringent condition than the Second Law (8.2), but is widely accepted. Therefore, The Second Law can be stated as:

$$\theta\dot{s} + q_{k,k} = d \geq 0 \tag{8.9}$$

where $d = 0$ is for reversible material behaviour only.

8.3 Thermodynamic restrictions on constitutive equations

How can we judge whether our constitutive relationship satisfies the laws of thermodynamics?

8.3.1 Derivation

Adding equations (8.6) and (8.9) we obtain

$$\dot{u} + d = \sigma_{ij}\dot{\varepsilon}_{ij} + \theta\dot{s} \tag{8.10}$$

Assuming that $u = u(\varepsilon_{ij}, s, \alpha_{ij})$ is differentiable, we write:

$$\dot{u} = \frac{\partial u}{\partial \varepsilon_{ij}}\dot{\varepsilon}_{ij} + \frac{\partial u}{\partial \alpha_{ij}}\dot{\alpha}_{ij} + \frac{\partial u}{\partial s}\dot{s} \tag{8.11}$$

which on substitution into (8.10) yields:

$$\left(\frac{\partial u}{\partial \varepsilon_{ij}} - \sigma_{ij}\right)\dot{\varepsilon}_{ij} + \left(\frac{\partial u}{\partial s} - \theta\right)\dot{s} + \left(\frac{\partial u}{\partial \alpha_{ij}}\dot{\alpha}_{ij} + d\right) = 0 \tag{8.12}$$

Assuming that

$$\frac{\partial u}{\partial \alpha_{ij}}\dot{\alpha}_{ij} + d = 0 \tag{8.13}$$

and that the processes of straining and change of entropy are mutually independent, we obtain from equation (8.12):

$$\sigma_{ij} = \frac{\partial u}{\partial \varepsilon_{ij}} \tag{8.14}$$

$$\theta = \frac{\partial u}{\partial s} \tag{8.15}$$

and from equation (8.13) and inequality (8.9):

$$\frac{\partial u}{\partial \alpha_{ij}}\dot{\alpha}_{ij} = -d \le 0 \tag{8.16}$$

Equation (8.14) is easily recognizable as a constitutive relationship. It follows that if our constitutive relationship can be expressed in the form of equation (8.14) using some potential function $u = u(\varepsilon_{ij}, s, \alpha_{ij}) \ge 0$ satisfying inequality (8.16), then this relationship automatically satisfies both the First and the Second Laws of Thermodynamics.

Note, however, that for a *dissipative material*, even if our constitutive relationship does not satisfy conditions (8.14) - (8.16) it can still satisfy the First and the Second Laws of Thermodynamics. This is because in our derivation we imposed the assumption (8.13), also known as Ziegler's orthogonality condition, which limits applicability of the above criteria. Therefore, for a *dissipative material*, conditions (8.14) - (8.16) are *sufficient but not necessary* for the constitutive relationship to satisfy the laws of Thermodynamics. For *reversible* (non-dissipative) materials, however, conditions (8.14) and (8.15) are both *sufficient and necessary*.

8.3.2 Alternative energy functions

The thermodynamic state of material can be defined in several alternative ways through the conjugate variables: stresses σ_{ij} or strains ε_{ij}, and temperature θ or entropy s. These four possible combinations result in four alternative energy functions: internal energy u, Helmholtz free energy f, enthalpy h and Gibbs free energy g, all related to each other by Legendre transformations (Table 8.1).

Sufficient conditions (8.14) - (8.16) for the constitutive relationship to satisfy the Laws of Thermodynamics change correspondingly (Table 8.1). For isothermal problems (which constitute a majority in the Geotechnical Engineering), it is more convenient to use the Helmholtz and Gibbs free energy functions, for strain and stress controlled loading, respectively.

Table 8.1: Energy definitions for small strain continuum mechanics

Internal energy	Helmholtz free energy	Enthalpy	Gibbs free energy
$u = u(\varepsilon_{ij}, \alpha_{ij}, s)$	$f = f(\varepsilon_{ij}, \alpha_{ij}, \theta)$ $f = u - s\theta$	$h = h(\sigma_{ij}, \alpha_{ij}, s)$ $h = u - \sigma_{ij}\varepsilon_{ij}$	$g(\sigma_{ij}, \alpha_{ij}, \theta)$ $g = h - s\theta$ $= f - \sigma_{ij}\varepsilon_{ij}$
$\sigma_{ij} = \dfrac{\partial u}{\partial \varepsilon_{ij}}$ $\theta = \dfrac{\partial u}{\partial s}$	$\sigma_{ij} = \dfrac{\partial f}{\partial \varepsilon_{ij}}$ $s = -\dfrac{\partial f}{\partial \theta}$	$\varepsilon_{ij} = -\dfrac{\partial h}{\partial \sigma_{ij}}$ $\theta = \dfrac{\partial h}{\partial s}$	$\varepsilon_{ij} = -\dfrac{\partial g}{\partial \sigma_{ij}}$ $s = -\dfrac{\partial g}{\partial \theta}$
$\dfrac{\partial u}{\partial \alpha_{ij}} \dot{\alpha}_{ij} \le 0$	$\dfrac{\partial f}{\partial \alpha_{ij}} \dot{\alpha}_{ij} \le 0$	$\dfrac{\partial h}{\partial \alpha_{ij}} \dot{\alpha}_{ij} \le 0$	$\dfrac{\partial g}{\partial \alpha_{ij}} \dot{\alpha}_{ij} \le 0$

Chapter 9

Modelling Soil Behaviour

TABLE OF CONTENTS

Chapter 9

Modelling Soil Behaviour

9.1 Introduction

Compared to other natural and man-made materials, soils are not as spectacular as, e.g., the mother-of-pearl or liquid crystals. However, in terms of the mathematical description of their mechanical behaviour, soils are probably the most complex materials in the world. For instance, the most important advances in the Theory of Plasticity (originally developed for metals) have been achieved in an attempt to describe the irreversible mechanical behaviour of soils. As has been already mentioned in Chapter 1 of this book, soils are:

- Multiphase (i.e., consist of solids, liquids and gases),
- Granular (i.e., build of particles of different sizes and shapes),
- Non-homogeneous (i.e., their mechanical properties vary in space),
- Anisotropic (their mechanical properties vary with loading direction).

However, even if all the above effects are minimized by special sample preparation and testing procedures, the mechanical behaviour of soils is still rather complex due to the following features:

- non-linearity (in initial loading, unloading and reloading),
- stress path dependency (reaches different strains at the same stress),
- stress level dependency (changes properties with confinement stress),
- irreversibility (produces residual strains in a closed stress cycle),
- material memory (remembers the highest stress before the unloading and follows the initial loading curve after reaching it in reloading),
- dilatancy (changes in volume during shearing),
- hardening (changes in the yield stress with plastic straining),
- rate dependency (different stress-strain curves at various strain rates),
- time dependency (creep, aging, relaxation).

In the Parts II and III of this book it will be shown how these and other features of soil behaviour can be described mathematically. In order to facilitate the explanations, the soil behaviour will be demonstrated and modelled using a simple triaxial test described in Chapter 1.

A.M. Puzrin, *Constitutive Modelling in Geomechanics*,

9.2 Triaxial stress-strain space

In the triaxial shear test (Figure 9.1) the sample is subjected to the uniform axisymmetrical loading with controlled (constant) radial stress $\sigma_2 = \sigma_3$. The axial loading can be either stress (σ_1) or strain (ε_1) controlled, with changes in the sample height or the axial force being monitored, respectively. In a drained test the changes in the sample volume are also monitored. In an undrained test, the volume is kept constant, and the pore water pressure is monitored instead.

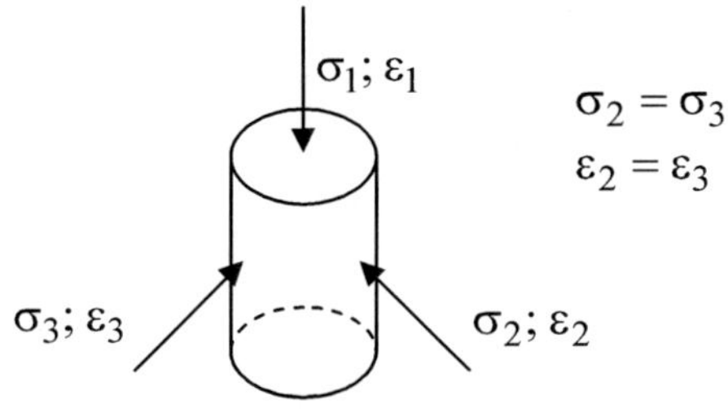

Figure 9.1: Triaxial test setup.

The setup allows for two principal stresses to be controlled independently, and the stress state of the sample can be expressed through their combination:

$$p = \frac{\sigma_1 + 2\sigma_3}{3} = \frac{J_1}{3} \qquad q = \sigma_1 - \sigma_3 = \sqrt{3J_{2D}} \tag{9.1}$$

where p and q are the mean and deviatoric stresses, which are proportional to the first invariant of the stress tensor J_1 and the square root of the deviatoric stress tensor $J_{2D} = \frac{1}{6}\left[(\sigma_1 - \sigma_2)^2 + (\sigma_2 - \sigma_3)^2 + (\sigma_3 - \sigma_1)^2\right]$, respectively. The principal strains can also be combined:

$$\varepsilon_v = \varepsilon_1 + 2\varepsilon_3 = I_1 \qquad \varepsilon_s = \frac{2}{3}(\varepsilon_1 - \varepsilon_3) = \frac{2}{3}\sqrt{3I_{2D}} \tag{9.2}$$

where ε_v and ε_s are the volumetric and shear strains, which are proportional to the first invariant of the strain tensor I_1 and the square root of the deviatoric stress tensor I_{2D}, respectively.

Equations (9.1) - (9.2) define the so called triaxial stress-strain space. The stress and strain components have been chosen to satisfy the work conjugacy condition (8.5):

$$\dot{W} = \sigma_{ij}\dot{\varepsilon}_{ij} = \sigma_1\dot{\varepsilon}_1 + 2\sigma_3\dot{\varepsilon}_3 = p\dot{\varepsilon}_v + q\dot{\varepsilon}_s \tag{9.3}$$

In accordance with Terzaghi's principle of effective stresses, the constitutive behaviour of soils depends on the mean effective stress $p' = p - u$, where u is the pore water pressure.

An important advantage of the presented definition of the triaxial stress-strain space is the proportionality of the stress and strain components to the corresponding tensor invariants. This allows for a straightforward generalization of the triaxial models to the general six-dimensional stress space required for applications in 3D boundary value problems.

9.3 A typical triaxial soil behaviour

Typical results of a drained triaxial compression test on slightly overconsolidated clay are presented in Figure 9.2. The sample preparation includes isotropic consolidation up to the pre-consolidation pressure p_c' with subsequent swelling (unloading) down to the initial mean effective stress p_0' (dashed lines in Figure 9.2a). The volumetric behaviour during this consolidation and swelling stage is shown by dashed lines in Figure 9.2b. This behaviour is *non-linear* and *irreversible* (because loading and unloading follow different stress-strain curves).

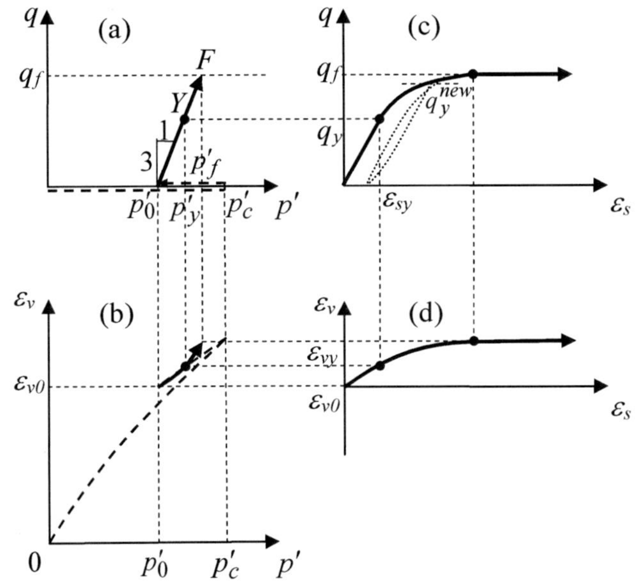

Figure 9.2: Results of a triaxial compression test on a slightly overconsolidated clay: (a) the stress path; (b) volumetric behaviour; (c) deviatoric behaviour; (d) the strain path.

After the sample is brought to certain small overconsolidation ratio $OCR = p'_c / p'_0 \leq 2$, the next step begins: the drained triaxial compression. The axial stress increases in increments $\Delta\sigma_1$, while the radial stresses are kept constant $\Delta\sigma_3 = 0$. From equations (9.1) it follows that $\Delta p' = \Delta\sigma_1/3$ and $\Delta q = \Delta\sigma_1$, so that the drained stress path is straight with the inclination $\Delta q / \Delta p' = 3$ (Figure 9.2a).

In the initial part of this stress path, both the volumetric (Figure 9.2b) and the deviatoric (Figure 9.2c) behaviours of the sample are almost *reversible* (for example, in Figure 9.2b the stress-strain curve initially follows the swelling line). This reversible behaviour is rather complex. For example, if the drained stress path in Figure 9.2a was inclined at a different angle, the soil stiffness in both the volumetric (Figure 9.2b) and the deviatoric (Figure 9.2c) plots would also change, reflecting *stress-path dependency*, related to material *anisotropy*. Also, this stiffness would be different, if the shearing started from a different initial stress p'_0, reflecting *pressure dependency* of stiffness. Finally, even though this behaviour is reversible, it may exhibit a high degree of *small strain nonlinearity*.

The soil behaviour continues to be reversible only until the stress path reaches the yield stress state $Y(p'_y, q_y)$, which depends on the *pre-consolidation* pressure p'_c. Starting from this stress state the material is yielding plastically and its behaviour becomes *irreversible*: for example, the descending dotted line in Figure 9.2c, representing unloading, does not follow the stress-strain curve for initial loading.

When the sample is reloaded (the ascending dotted line in Figure 9.2c), the yield stress q_y^{new} is higher than the one reached in the initial loading, i.e. the material is *strain-hardening*. In fact, this new yield stress q_y^{new} will be equal to the maximum stress reached before the unloading, i.e. the soil has *material memory* of the maximum stress.

Another important feature of this plastic yielding: the shearing is accompanied by increasing volumetric plastic strains (Figure 9.2d), i.e. the soil is *contracting*. If the clay was strongly overconsolidated, the volumetric plastic strain would be negative, i.e. the material would be *dilating*.

The plastic yielding continues until the soil sample reaches the failure stress state $F(p'_f, q_f)$. At *failure*, the plastic flow continues at constant stresses (Figures 9.2a-c). If the volumetric strain also stays constant (Figures 9.2d), the soil has reached the *critical state*.

A more advanced analysis of the stress cycle in Figure 9.2c indicates that even before the yield stress q_y^{new} is reached, the reloading stress-strain curve does not follow the unloading curve exactly, i.e., the cyclic soil behaviour is *irreversible* even before the large scale yielding takes place.

Finally, if the test was performed at a higher stress or strain rate, both the volumetric (Figure 9.2b) and the deviatoric (Figure 9.2c) behaviours would be different due to their *rate dependency*.

In the Parts II and III of the book these and other features of soil behaviour will be modelled using the modern theories of elasticity, plasticity and rheology. The model descriptions will follow the path from more simple models to more complex ones, also reflecting their historical development.

9.4 The structure of Part II: Modelling Reversible Behaviour

In Part II, Chapters 10-13 deal with the *reversible* soil behaviour. They introduce models that can describe not only the pre-yielding soil behaviour, but also the reversible component of the total strain during yielding and failure. Chapter 10 introduces the theories of *elasticity* and *hyperelasticity*, which are the main tools in describing the reversible behaviour of materials. Chapter 11 deals with the *anisotropy* and *stress path dependency*. Chapter 12 explores the modelling of the *stiffness dependency on the mean effective stress*. Chapter 13 describes the most popular models for the *small-strain non-linearity*. In all these chapters a special emphasis has been made on ensuring that the models do not violate *the First Law of Thermodynamics*.

9.5 The structure of Part III: Modelling Irreversible Behaviour

In Part III, Chapters 14-21 deal with the *irreversible* soil behaviour. Chapter 14 and 15 introduce different *failure criteria* and explore *post-failure plastic flow*, respectively. Chapters 16-19 introduce the *pre-failure plastic yielding*, with such features as *strain-hardening* (Chapter 16), *dilatancy (contraction)* (Chapter 17), *pre-consolidation* (Chapter 18) and *critical state* (Chapter 19). Chapter 20 gives an insight into *irreversibility of cyclic soil behaviour* introducing some *advanced aspects* of the constitutive modelling of soils. Finally, Chapter 21 explores simple rheological models for *rate- and time-dependency*. In all these chapters it will be demonstrated that the models do not violate *the First and the Second Laws of Thermodynamics*.

9.6 Appendices

Appendix A focuses on modelling the *undrained soil behaviour*. Appendix B demonstrates an example of *incorporation of an irreversible constitutive model* into a numerical algorithm for solution of boundary value problems.

Part II

Constitutive Modelling of Reversible Soil Behaviour

Chapter 10

Isotropic Elastic Behaviour

TABLE OF CONTENTS

Chapter 10

Isotropic elastic behaviour

10.1 Introduction

For a slightly overconsolidated clay sample, in the initial part of a triaxial compression stress path, the sample behaviour may be assumed *reversible* for both its deviatoric (Figure 10.1a) and volumetric (Figure 10.1b) components. This means, that before the stress path reaches the yield stress state $Y(p'_y, q_y)$, the unloading stress-strain curves will approximately follow the corresponding loading curves, and no residual strain will appear in a closed loading cycle. The simplest way to model this pre-yielding behaviour is to approximate it as *isotropic linear elastic* (Figure 10.1):

$$\begin{Bmatrix} p' - p'_0 \\ q \end{Bmatrix} = \begin{bmatrix} K & 0 \\ 0 & 3G \end{bmatrix} \begin{Bmatrix} \varepsilon_v - \varepsilon_{v0} \\ \varepsilon_s \end{Bmatrix} \tag{10.1}$$

where K and G are the secant bulk and shear modulus, respectively. It follows, that for the triaxial compression stress path $\Delta q/\Delta p' = 3$, the pre-yielding strain path will also be linear $\Delta\varepsilon_v/\Delta\varepsilon_s = G/K$ (Figure 10.1c).

This Chapter provides a broader theoretical background for the above relationship, including its place within the Theory of Elasticity, its interpretation in the full six-dimensional stress-strain space and its thermomechanical validation.

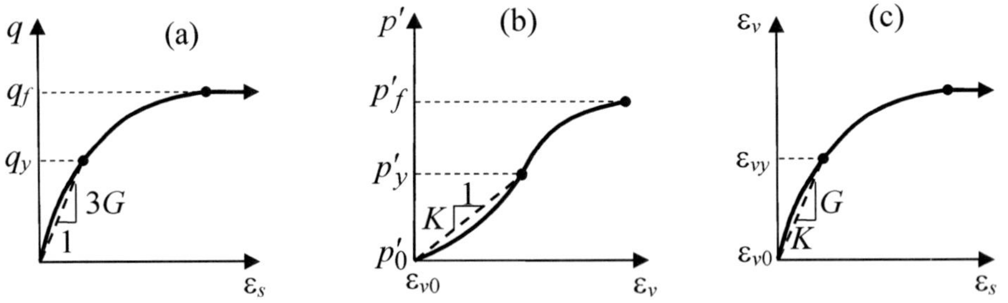

Figure 10.1 Linear approximation of the reversible stress-strain behaviour: (a) deviatoric; (b) volumetric; (c) the strain path.

10.2 Definition of elastic material

Elasticity is the simplest form of the constitutive relationship for a deformable (non-rigid) material. The constitutive behaviour is *elastic*, if the state of stress is a function of the current state of deformation only. In other words, no permanent strain is allowed and the mechanical behaviour is completely reversible: unloading follows the loading curve (Figure 10.2).

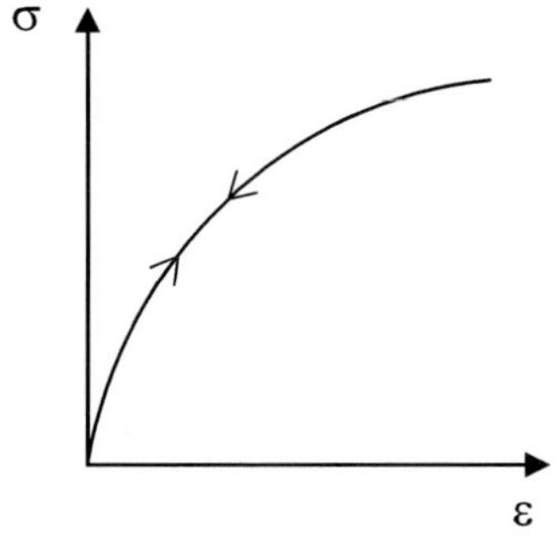

Figure 10.2 Elastic behaviour

Though the term elastic is normally associated with linear elasticity, this is just a particular case of a general non-linear elasticity. A non-linear elastic law can be written in a general form:

$$\sigma_{ij} = f_{ij}\left(\varepsilon_{kl}\right) \tag{10.2}$$

where f_{ij} is a response function.

The Generalized Hooke's *linear elastic* law is given by:

$$\sigma_{ij} = D_{ijkl}\varepsilon_{kl} \tag{10.3}$$

where D_{ijlk} is the fourth order stiffness tensor – the simplest form of a response function – a constant. Because the fourth order stiffness tensor D_{ijlk} is difficult to visualize, it is advantageous to rewrite equation (10.3) in a vector-matrix form:

$$\{\sigma\} = \begin{Bmatrix} \sigma_{11} \\ \sigma_{22} \\ \sigma_{33} \\ \sigma_{12} \\ \sigma_{23} \\ \sigma_{13} \end{Bmatrix} = \begin{bmatrix} D_{11} & D_{12} & D_{13} & D_{14} & D_{15} & D_{16} \\ D_{21} & D_{22} & D_{23} & D_{24} & D_{25} & D_{26} \\ D_{31} & D_{32} & D_{33} & D_{34} & D_{35} & D_{36} \\ D_{41} & D_{42} & D_{43} & D_{44} & D_{45} & D_{46} \\ D_{51} & D_{52} & D_{53} & D_{54} & D_{55} & D_{56} \\ D_{61} & D_{62} & D_{63} & D_{64} & D_{65} & D_{66} \end{bmatrix} \begin{Bmatrix} \varepsilon_{11} \\ \varepsilon_{22} \\ \varepsilon_{33} \\ \varepsilon_{12} \\ \varepsilon_{23} \\ \varepsilon_{13} \end{Bmatrix} = [D]\{\varepsilon\} \tag{10.4}$$

Because the stress σ_{ij} and ε_{ij} strain tensors have each only 6 independent components out of 9, the corresponding vectors $\{\sigma\}$ and $\{\varepsilon\}$ are only 6 dimensional. Therefore, while the fourth order stiffness tensor D_{ijlk} has in general 81 terms, the stiffness matrix $[D]$ has *maximum* 36 independent components. This stiffness matrix will be simplified drastically if we assume that the material is isotropic: it will only have 12 non-zero components and only 2 of them will be independent!

10.3 Isotropic linear elasticity

10.3.1 Isotropy

Material is called *isotropic*, if its mechanical properties are the same in all geometric directions. Constitutive relationships for these isotropic materials should be *invariant* to transformations of the coordinate system. When the constitutive equations are expressed via stress and strain tensor invariants, this invariance (and isotropy) are satisfied automatically. In other cases, it has to be proven. Consider, for example, the following non-linear relationship:

$$\sigma_{ij} = a_0 \delta_{ij} + a_1 \varepsilon_{ij} \tag{10.5}$$

where the coefficients are polynomial functions of strain tensor invariants:

$$a_n = f_n(I_{1\varepsilon}, I_{2\varepsilon}) = f_n(I_1, I_2) \tag{10.6}$$

Relationship (10.5) is by definition elastic. But is it isotropic?

When the coordinate system transforms:

$$\overset{*}{\sigma}_{ab} = T_{ai} T_{bj} \sigma_{ij} \tag{10.7}$$

we substitute equation (10.5) into (10.7):

$$\overset{*}{\sigma}_{ab} = a_0 \delta_{ij} T_{ai} T_{bj} + a_1 T_{ai} T_{bj} \varepsilon_{ij} \tag{10.8}$$

and using the fact that coefficients $a_n = f_n(I_1, I_2)$ are invariants and that

$\overset{*}{\varepsilon}_{ab} = T_{ai} T_{bj} \varepsilon_{ij}$ and $T_{ai} T_{bi} = \delta_{ab}$, we obtain

$$\overset{*}{\sigma}_{ab} = a_0 \delta_{ab} + a_1 \overset{*}{\varepsilon}_{ab} \tag{10.9}$$

which has a functional form identical to equation (10.5), proving its invariance to the coordinate system transformation, and, hence, its isotropy. In this proof, the crucial argument was that the coefficients $a_n = f_n(I_1, I_2)$, including material constants, are invariants. This, in fact, is an alternative definition of an isotropic material. Also very important is the inverse statement: *constitutive relationships can be expressed via stress and strain tensor invariants only for isotropic materials.*

10.3.2 Linearity

Isotropic elastic relationship (10.5) becomes linear, if we adopt:

$$a_0 = \lambda I_1; \quad a_1 = \mu, \tag{10.10}$$

where λ and μ are constants, so that:

$$\sigma_{ij} = \lambda I_1 \delta_{ij} + \mu \varepsilon_{ij} \tag{10.11}$$

Note, that $\sigma_{ij} = 0$ at $\varepsilon_{ij} = 0$.

What is the physical meaning of the parameters λ and μ? In order to interpret parameters of any constitutive relationship, the model should be subjected to some simple mechanical tests, where its behaviour could be compared to the corresponding engineering definitions of mechanical properties. For example, the *secant shear modulus* G can be defined in a *simple shear deformation test* (Figure 10.3a):

$$\varepsilon_{11} = \varepsilon_{22} = \varepsilon_{33} = \varepsilon_{23} = \varepsilon_{13} = 0; \quad \varepsilon_{12} = \varepsilon_{21}, \tag{10.12}$$

from the following expression:

$$G = \frac{\tau_{12}}{\gamma_{12}} = \frac{\sigma_{12}}{2\varepsilon_{12}} \tag{10.13}$$

where $\tau_{12} = \sigma_{12}$ and $\gamma_{12} = 2\varepsilon_{12}$ are the shear stress and strain, respectively. Substituting (10.12) into (10.11) we obtain: $I_1 = \varepsilon_{ii} = 0$ and $\sigma_{12} = \mu\varepsilon_{12}$, so that from (10.13):

$$\mu = 2G \tag{10.14}$$

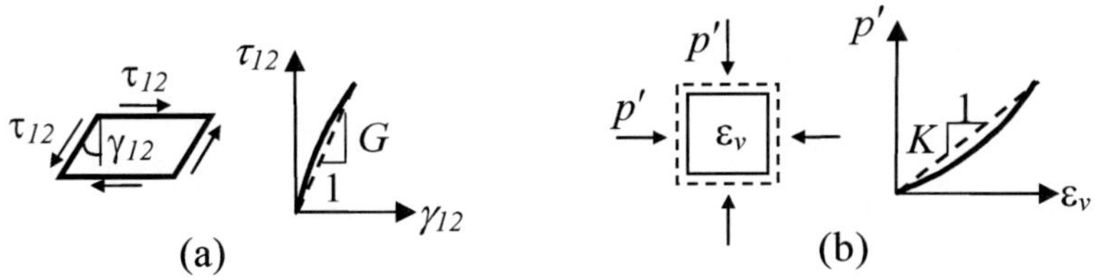

Figure 10.3 Elementary strain controlled tests: a) Simple shear; b) Volumetric contraction (expansion).

Another useful elementary test is the *uniform volumetric contraction test* (Figure 10.3b):

$$\varepsilon_{11}=\varepsilon_{22}=\varepsilon_{33}=\varepsilon; \quad \varepsilon_{12}=\varepsilon_{23}=\varepsilon_{13}=0, \tag{10.15}$$

which is used to define the *secant bulk modulus*:

$$K=\frac{p}{\varepsilon_v}=\frac{\sigma_{ii}/3}{\varepsilon_{jj}}=\frac{J_1}{3I_1} \tag{10.16}$$

where $p=\frac{\sigma_{ii}}{3}=\frac{J_1}{3}$ and $\varepsilon_{vol}=\varepsilon_{ii}=I_1$ are the hydrostatic stress and the volumetric strain, respectively. Substituting (10.15) into (10.11) we obtain: $\varepsilon_{ij}=\frac{I_1}{3}\delta_{ij}$ and $\sigma_{ij}=\lambda I_1\delta_{ij}+2G\frac{I_1}{3}\delta_{ij}$, so that

$$J_1=\sigma_{ii}=\left(\lambda+\frac{2G}{3}\right)I_1\delta_{ii}=3\left(\lambda+\frac{2G}{3}\right)I_1 \tag{10.17}$$

and from (10.16):

$$\lambda=K-\frac{2G}{3} \tag{10.18}$$

As a result, the elastic model (10.11) can be written as:

$$\sigma_{ij}=\lambda I_1\delta_{ij}+\mu\varepsilon_{ij}=KI_1\delta_{ij}+2G\left(\varepsilon_{ij}-\frac{I_1}{3}\delta_{ij}\right) \tag{10.19}$$

which is easily recognizable as *Hooke's linear elastic law*, where λ and μ are Lame's elastic constants.

In the vector-matrix form Hooke's law for isotropic material is given by:

$$\begin{Bmatrix}\sigma_{11}\\ \sigma_{22}\\ \sigma_{33}\\ \sigma_{12}\\ \sigma_{23}\\ \sigma_{13}\end{Bmatrix}=\begin{bmatrix} K+\frac{4G}{3} & K-\frac{2G}{3} & K-\frac{2G}{3} & 0 & 0 & 0\\ K-\frac{2G}{3} & K+\frac{4G}{3} & K-\frac{2G}{3} & 0 & 0 & 0\\ K-\frac{2G}{3} & K-\frac{2G}{3} & K+\frac{4G}{3} & 0 & 0 & 0\\ 0 & 0 & 0 & 2G & 0 & 0\\ 0 & 0 & 0 & 0 & 2G & 0\\ 0 & 0 & 0 & 0 & 0 & 2G\end{bmatrix}\begin{Bmatrix}\varepsilon_{11}\\ \varepsilon_{22}\\ \varepsilon_{33}\\ \varepsilon_{12}\\ \varepsilon_{23}\\ \varepsilon_{13}\end{Bmatrix} \tag{10.20}$$

The inverse linear elastic law for isotropic material is given by:

$$\varepsilon_{ij} = \left(\frac{1}{9K} - \frac{1}{6G}\right) J_1 \delta_{ij} + \frac{1}{2G}\sigma_{ij} \tag{10.21}$$

10.3.3 Decoupling

Important property of the isotropic linear elastic model is that volumetric and deviatoric components of the stress-strain behaviour in this model are completely *uncoupled*. In other words, the hydrostatic stress does not change the shape, while deviatoric stresses do not affect the volume of the material element. In fact, using the definition of the deviatoric strain tensor:

$$e_{ij} = \varepsilon_{ij} - \frac{I_1}{3}\delta_{ij} \tag{10.22}$$

we rewrite equation (10.19):

$$\sigma_{ij} = K I_1 \delta_{ij} + 2G e_{ij} \tag{10.23}$$

It follows that:

$$J_1 = \sigma_{ii} = K I_1 \delta_{ii} = 3K I_1 \tag{10.24}$$

and using the definition of the deviatoric stress tensor:

$$s_{ij} = \sigma_{ij} - \frac{J_1}{3}\delta_{ij} \tag{10.25}$$

we obtain from equations (10.23) and (10.24):

$$s_{ij} = \sigma_{ij} - \frac{J_1}{3}\delta_{ij} = K I_1 \delta_{ij} + 2G e_{ij} - \frac{3K I_1}{3}\delta_{ij} = 2G e_{ij} \tag{10. 26}$$

Equations (10.24) and (10.26) are an alternative presentation of the Hooke's linear elastic law, which demonstrates its uncoupled nature.

This decoupling is also reflected in modelling the *triaxial behaviour* in Figure 10.1. Indeed, recalling definitions of the second invariants of the deviatoric stress and strain tensors and using equation (10.26), we obtain:

$$\sqrt{J_{2D}} = \sqrt{\frac{1}{2} s_{ij} s_{ij}} = 2G\sqrt{\frac{1}{2} e_{ij} e_{ij}} = 2G\sqrt{I_{2D}} \tag{10. 27}$$

And substituting definitions of triaxial stresses and strains from Chapter 9:

$$p' - p'_0 = \frac{J_1}{3}; \quad q = \sqrt{3J_{2D}}; \quad \varepsilon_v - \varepsilon_{v0} = I_1; \quad \varepsilon_s = \frac{2}{3}\sqrt{3I_{2D}} \tag{10.28}$$

into equations (10.24) and (10.27) we obtain equation (10.1):

$$\begin{Bmatrix} p' - p'_0 \\ q \end{Bmatrix} = \begin{bmatrix} K & 0 \\ 0 & 3G \end{bmatrix} \begin{Bmatrix} \varepsilon_v - \varepsilon_{v0} \\ \varepsilon_s \end{Bmatrix}$$

which as we now see, is indeed the triaxial version of the isotropic linear elastic Hooke's law.

10.3.4 Elastic constants

The Hooke law for isotropic material is completely defined by two material constants. In the literature one can find at least 4 different couples of constants describing the same linear elastic model. We have already mentioned the bulk and shear moduli (K, G) and Lame constants (λ, μ). Another two couples of elastic constants are defined below:

<u>*Young's modulus and Poisson's ratio:*</u>
Young's modulus E and Poisson's ratio ν:

$$E = \frac{\sigma_{11}}{\varepsilon_{11}}; \qquad \nu = -\frac{\varepsilon_{22}}{\varepsilon_{11}}, \tag{10.29}$$

are defined from the uniaxial compression test (Figure 10.4a):

$$\sigma_{ij} = \begin{bmatrix} \sigma_{11} & 0 & 0 \\ 0 & 0 & 0 \\ 0 & 0 & 0 \end{bmatrix} \tag{10.30}$$

which results in $J_1 = \sigma_{11}$, so that from equation (10.21):

$$\varepsilon_{11} = \frac{3K+G}{9KG}\sigma_{11}; \qquad \varepsilon_{22} = \varepsilon_{33} = -\frac{3K-2G}{18KG}\sigma_{11}. \tag{10.31}$$

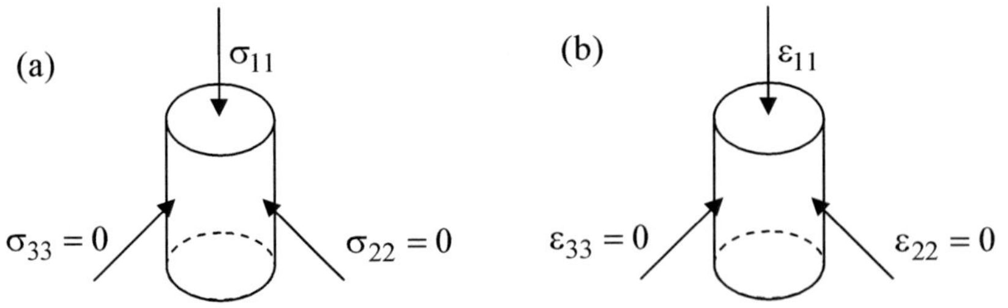

Figure 10.4 Elementary tests: (a) uniaxial compression and (b) oedometer tests.

Substitution of equations (10.31) into (10.29) gives:

$$E = \frac{\sigma_{11}}{\varepsilon_{11}} = \frac{9KG}{3K+G}; \qquad \nu = -\frac{\varepsilon_{22}}{\varepsilon_{11}} = \frac{3K-2G}{6K+2G}. \tag{10.32}$$

Constrained modulus and at rest lateral pressure coefficient:

Constrained modulus M and at rest lateral pressure coefficient K_0:

$$M = \frac{\sigma_{11}}{\varepsilon_{11}}; \qquad K_0 = \frac{\sigma_{22}}{\sigma_{11}} \tag{10.33}$$

are defined from the oedometer test (Figure 10.4b):

$$\varepsilon_{ij} = \begin{bmatrix} \varepsilon_{11} & 0 & 0 \\ 0 & 0 & 0 \\ 0 & 0 & 0 \end{bmatrix} \tag{10.34}$$

which results in $I_1 = \varepsilon_{11}$, so that from equation (10.19):

$$\sigma_{11} = \left(K + \frac{4G}{3}\right)\varepsilon_{11}; \qquad \sigma_{22} = \sigma_{33} = \left(K - \frac{2G}{3}\right)\varepsilon_{11}. \tag{10.35}$$

Substitution of equations (10.35) into (10.33) gives:

$$M = \frac{\sigma_{11}}{\varepsilon_{11}} = K + \frac{4G}{3} = \frac{E(1-\nu)}{(1+\nu)(1-2\nu)} \tag{10.36}$$

$$K_0 = \frac{\sigma_{22}}{\sigma_{11}} = \frac{3K-2G}{3K+4G} = \frac{\nu}{1-\nu} \tag{10.37}$$

Note, that in spite of the fact that the Young's modulus $E = \sigma_{11}/\varepsilon_{11}$ and the constrained modulus $M = \sigma_{11}/\varepsilon_{11}$ are defined from the same stress-strain ratio, these are two different constants, because the ratio will be different for different loading and boundary conditions.

Summary of elastic constants

A summary of the constants describing the isotropic linear elastic model and relationships between these constants are presented in Table 10.1:

Table 10.1 Summary of elastic constants

Constants	K, G	λ, μ	E, ν	M, K_0
K	K	$\lambda + \frac{\mu}{3}$	$\frac{E}{3(1-2\nu)}$	$\frac{M(1+2K_0)}{3}$
G	G	$\frac{\mu}{2}$	$\frac{E}{2(1+\nu)}$	$\frac{M(1-K_0)}{2}$
λ	$K - \frac{2G}{3}$	λ	$\frac{E\nu}{(1+\nu)(1-2\nu)}$	MK_0
μ	$2G$	μ	$\frac{E}{1+\nu}$	$M(1-K_0)$
E	$\frac{9KG}{3K+G}$	$\frac{3\lambda+\mu}{2\lambda+\mu}\mu$	E	$\frac{M(1+2K_0)(1-K_0)}{1+K_0}$
ν	$\frac{3K-2G}{6K+2G}$	$\frac{\lambda}{2\lambda+\mu}$	ν	$\frac{K_0}{1+K_0}$
M	$K + \frac{4G}{3}$	$\lambda + \mu$	$\frac{E(1-\nu)}{(1+\nu)(1-2\nu)}$	M
K_0	$\frac{3K-2G}{3K+4G}$	$\frac{\lambda}{\lambda+\mu}$	$\frac{\nu}{1-\nu}$	K_0

10.4 Thermomechanics of isotropic linear elasticity

Classical elasticity requires a unique relationship between stresses and strains. This implies that the stress-strain behaviour is reversible: in a closed stress cycle no residual strains can occur. From Section 8.4 we know, that a reversible material satisfies the Second Law of Thermodynamics automatically. Is the First Law also satisfied automatically? It is not, unless the stress-strain relationship can be derived from an energy potential function (equation 8.25).

10.4.1 Hyperelasticity

In elasticity it is common to define $U_0(\varepsilon_{ij})$ - the internal strain energy per unit volume, and $U_{c0}(\sigma_{ij}) = \sigma_{ij}\varepsilon_{ij} - U_0(\varepsilon_{ij})$ - the complementary strain energy per unit volume (Figure 10.5). For isothermal conditions, these are equivalent to the Helmholtz free energy $f(\varepsilon_{ij}, \theta)$, and the Gibbs free energy with the negative sign: $-g(\sigma_{ij}, \theta)$.

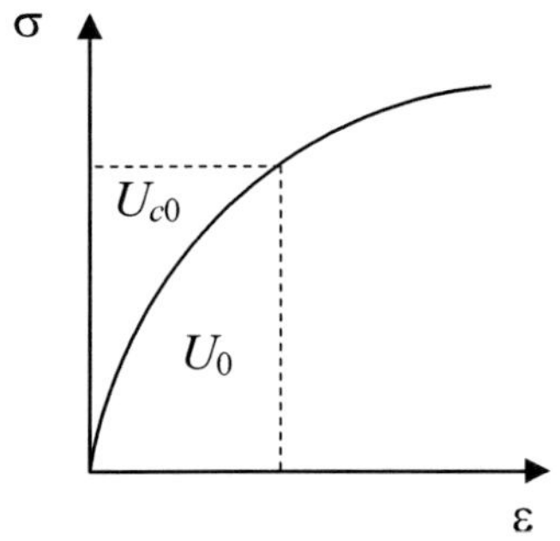

Figure 10.5 Strain energy.

From Table 5.1 it follows, that only the elastic constitutive equations of the form:

$$\sigma_{ij}(\varepsilon_{ij}) = \frac{\partial f}{\partial \varepsilon_{ij}} = \frac{\partial U_0}{\partial \varepsilon_{ij}} \quad \text{and} \quad \varepsilon_{ij}(\sigma_{ij}) = -\frac{\partial g}{\partial \sigma_{ij}} = \frac{\partial U_{c0}}{\partial \sigma_{ij}} \tag{10.38}$$

satisfy the First Law of Thermodynamics. These kind of relations are termed "hyperelastic". "Hyper" in ancient Greek means "more, over, beyond", and Hyperelasticity is more restrictive than classical elasticity, because it requires existence of a strain energy potential function. This restrictiveness becomes clear when we consider isotropic materials.

10.4.2 Isotropy and Thermomechanics

In section 10.3.1 we have proven that the following non-linear elastic relationship:

$$\sigma_{ij} = a_0 \delta_{ij} + a_1 \varepsilon_{ij} \tag{10.39}$$

where the coefficients are polynomial functions of strain tensor invariants:

$$a_n = f_n(I_{1\varepsilon}, I_{2\varepsilon}) = f_n(I_1, I_2) \tag{10.40}$$

is isotropic. But does it automatically satisfy the First Law of Thermodynamics?

For an isotropic material it should be possible to express the strain energy function via strain tensor invariants. Assume for simplicity, that it is only the function of the two first invariants of the strain tensor:

$$U_0 = U_0(I_{1\varepsilon}, I_{2\varepsilon}) = U_0(I_1, I_2) \tag{10.41}$$

Substituting this strain energy function into the first equation (10.38), and using

$$\frac{\partial I_1}{\partial \varepsilon_{ij}} = \delta_{ij}; \qquad \frac{\partial I_2}{\partial \varepsilon_{ij}} = \varepsilon_{ij}$$

we obtain:

$$\sigma_{ij} = \frac{\partial U_0}{\partial I_1}\left(\frac{\partial I_1}{\partial \varepsilon_{ij}}\right) + \frac{\partial U_0}{\partial I_2}\left(\frac{\partial I_2}{\partial \varepsilon_{ij}}\right) = \frac{\partial U_0}{\partial I_1}\delta_{ij} + \frac{\partial U_0}{\partial I_2}\varepsilon_{ij} \tag{10.42}$$

The inverse relation can be obtained from the complementary strain energy potential $U_{c0}(J_1, J_2)$:

$$\varepsilon_{ij} = \frac{\partial U_{c0}}{\partial \sigma_{ij}}(J_1, J_2) = \frac{\partial U_{c0}}{\partial J_1}\delta_{ij} + \frac{\partial U_{c0}}{\partial J_2}\sigma_{ij} \tag{10.43}$$

Clearly, the constitutive relationship (10.42) is nothing else but a particular case of the non-linear isotropic elastic law (10.39), with

$$a_0(I_1, I_2) = \frac{\partial U_0}{\partial I_1}; \qquad a_1(I_1, I_2) = \frac{\partial U_0}{\partial I_2}. \tag{10.44}$$

There is, however, a very important difference: here a_0 and a_1 are not independent, which is clearly demonstrated by the following identity:

$$\frac{\partial a_0}{\partial I_2} = \frac{\partial a_1}{\partial I_1} \tag{10.45}$$

It follows that not any non-linear isotropic elastic law of the type of equation (10.39) automatically satisfies the First Law of Thermodynamics, but only those, whose coefficients satisfy condition (10.45).

10.4.3 Linearity and Thermomechanics

In order to make the isotropic elastic relationship (10.39) linear, in Section 10.3.2 we adopted the following coefficients:

$$a_0 = \lambda I_1; \quad a_1 = \mu, \tag{10.46}$$

where λ and μ are Lame's constants. Clearly, these coefficients automatically satisfy condition (10.45), and, consequently, the First Law of Thermodynamics.

In fact, consider the following quadratic strain energy function:

$$U_0 = \mu I_2 + \frac{\lambda}{2} I_1^2 \tag{10.47}$$

After substitution into equation (10.42), this function produces nothing else but the Hooke law for isotropic linear elastic materials:

$$\sigma_{ij} = \frac{\partial\left(\mu I_2 + \lambda I_1^2\right)}{\partial I_1}\delta_{ij} + \frac{\partial\left(\mu I_2 + \lambda I_1^2\right)}{\partial I_2}\varepsilon_{ij} = \lambda I_1 \delta_{ij} + \mu\varepsilon_{ij} \tag{10.48}$$

10.4.4 Decoupling and Thermomechanics

Using expressions

$$\lambda = K - \frac{2G}{3}; \quad \mu = 2G \quad \text{and} \quad I_{2D} = I_2 - \frac{I_1^2}{6} \tag{10.49}$$

the strain energy function (10.47) can be decoupled into deviatoric and volumetric parts:

$$U_0 = 2G\left(I_2 - \frac{I_1^2}{6}\right) + \frac{K}{2}I_1^2 = 2GI_{2D} + \frac{K}{2}I_1^2 \tag{10.50}$$

For the triaxial case, using $\varepsilon_v - \varepsilon_{v0} = I_1$ and $\varepsilon_s = \frac{2}{3}\sqrt{3I_{2D}}$:

$$U_0(\varepsilon_v, \varepsilon_s) = f(\varepsilon_v, \varepsilon_s) = \frac{3G}{2}\varepsilon_s^2 + \frac{K}{2}\left(\varepsilon_v - \varepsilon_{v0}\right)^2 + p_0'\left(\varepsilon_v - \varepsilon_{v0}\right) \tag{10.51}$$

and the isotropic linear elastic relationships are then derived as:

$$p' = \frac{\partial U_0(\varepsilon_v, \varepsilon_s)}{\partial \varepsilon_v} = p_0' + K(\varepsilon_v - \varepsilon_{v0}) \qquad q = \frac{\partial U_0(\varepsilon_v, \varepsilon_s)}{\partial \varepsilon_s} = 3G\varepsilon_s \tag{10.52}$$

The complimentary strain energy function for the triaxial case is given by:

$$U_{c0}(p', q) = -g(p', q) = \frac{q^2}{6G} + \frac{(p' - p_0')^2}{2K} + (p' - p_0')\varepsilon_{v0} \tag{10.53}$$

Existence of the strain energy (and complimentary strain energy) function is the direct proof that the Hooke's isotropic linear elastic law satisfies the Laws of Thermodynamics.

10.4.5 Restrictions on elastic constants

Because the strain energy U_0 must be positive, for the case of isotropic linear elastic material this results in the following restrictions on elastic components:

$$G \geq 0; \quad K \geq 0; \quad E \geq 0; \quad -1 \leq \nu \leq \frac{1}{2}. \tag{10.54}$$

Chapter 11

Anisotropy and Coupling

TABLE OF CONTENTS

Chapter 11

Anisotropy and Coupling

11.1 Introduction

Isotropic linear elastic behaviour, though very simple to model (Chapter 10), is a strong idealisation of real soil behaviour, which is in most cases anisotropic, non-linear and, sometimes, inelastic (irreversible) even at very small strains. While the non-linearity and small strain irreversibility will be treated later in the book, this chapter will be devoted to modelling *anisotropy*: i.e., the changing material behaviour with geometric direction of loading. Soils deposits are formed in nature under gravity, resulting in different soil properties in vertical and horizontal planes. In triaxial tests, cylindrical samples are normally extracted from a borehole log, with their axis being vertical (Figure 11.1a). If a second sample was taken from the same depth in such a way that its axis was horizontal, comparison between the results of two triaxial compression tests would immediately indicate the anisotropy (Figure 11.1b). (For isotropic material the two stress-strain curves would be identical.)

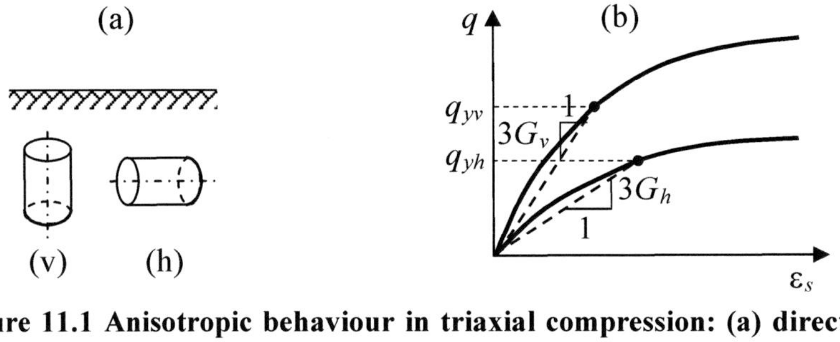

Figure 11.1 Anisotropic behaviour in triaxial compression: (a) direction of the sample axes; (b) deviatoric stress-strain curves.

But if the both samples have been extracted with their axes in vertical direction, is it still possible to spot in a triaxial test that the soil is anisotropic?

It is, if after subjecting the first sample to triaxial compression (Figure 11.2a), the second sample is brought through the same pre-consolidation p'_c to the same initial stress p'_0, but then subjected to a different stress path, e.g., isotropic consolidation (Figure 11.2a). The isotropic consolidation will not produce any deviatoric stress-strain curve (Figure 11.2b), but if material is anisotropic, the secant bulk modulus K_{ic}, obtained in Figure 11.2c from pre-yielding volumetric stress-strain curve $\left(\text{for } p'_0 \le p' \le p'_c\right)$, will be different to that obtained in triaxial compression K_{tc}. Another important indication of anisotropy can be observed in a strain path plot (Figure 11.2d): hydrostatic loading produces shear strains, i.e., for anisotropic materials the volumetric and deviatoric behaviours are *coupled*! As a result, any straight probing stress path emanating from p'_0 in Figure 11.2a will produce its own shear and bulk moduli, as well as a different strain path in Figures 11.2b-d. This phenomenon is called *stress path dependency of stiffness*.

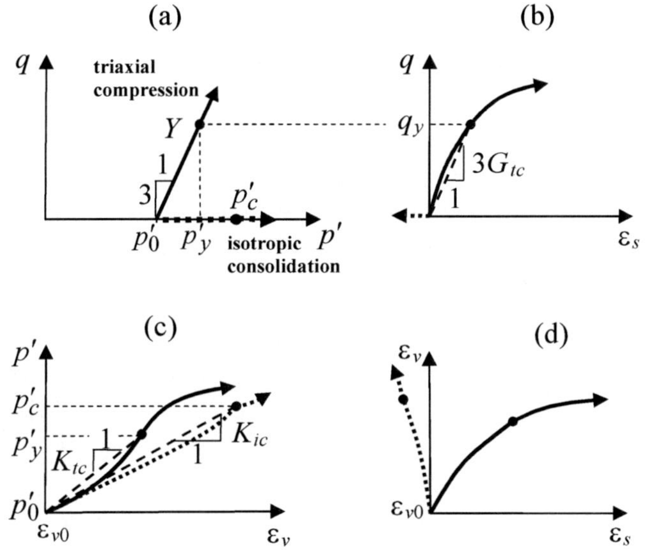

Figure 11.2 Anisotropic behaviour in triaxial compression vs. isotropic consolidation: (a) stress paths; (b) deviatoric stress-strain curves; (c) volumetric stress-strain curves; (b) strain paths.

To summarise the above, soil *anisotropy* leads to the *coupling* between volumetric and deviatoric components of the reversible triaxial behaviour, which in turns leads to the *stress path dependency of stiffness*. All these complex phenomena have to be modelled properly, and the simplest way to do that for a triaxial test is to use the following linear elastic constitutive relationship:

$$\begin{Bmatrix} \varepsilon_v - \varepsilon_{v0} \\ \varepsilon_s \end{Bmatrix} = \begin{bmatrix} 1/K & -1/J \\ -1/J & 1/3G \end{bmatrix} \begin{Bmatrix} p' - p'_0 \\ q \end{Bmatrix} \tag{11.1}$$

where J is the coupling modulus; K is the bulk modulus for the isotropic consolidation ($q = 0$); G is the shear modulus for the pure shear ($p' = p'_0$). All the stiffness parameters can be obtained from the test results in Figure

11.2 as follows:

$$\frac{1}{K}=\frac{1}{K_{ic}};\qquad \frac{1}{J}=\frac{1}{3K_{ic}}-\frac{1}{3K_{tc}};\qquad \frac{1}{G}=\frac{1}{G_{tc}}+\frac{1}{3K_{ic}}-\frac{1}{3K_{tc}}. \tag{11.2}$$

This chapter explores the special type of anisotropy typical for soils – cross-anisotropic behaviour and explains what is required from an anisotropic linear elastic constitutive relationship to satisfy the Laws of Thermodynamics. It also shows how the cross-anisotropic model leads to the coupling modelled by relationship (11.1), and explains why it is intrinsically wrong to express the anisotropic behaviour in the form of such a relationship!

11.2 Cross-anisotropic linear elastic behaviour

11.2.1 Isotropic behaviour

In order to better understand the cross-anisotropic behaviour, let us first have another look at the isotropic behaviour, which was explored in detail in the previous Chapter. One way to express the Hooke law for isotropic materials is in the following vector-matrix form:

$$\begin{Bmatrix}\varepsilon_{xx}\\ \varepsilon_{yy}\\ \varepsilon_{zz}\\ \varepsilon_{xy}\\ \varepsilon_{yz}\\ \varepsilon_{xz}\end{Bmatrix}=\begin{bmatrix}\frac{1}{E} & -\frac{\nu}{E} & -\frac{\nu}{E} & 0 & 0 & 0\\ -\frac{\nu}{E} & \frac{1}{E} & -\frac{\nu}{E} & 0 & 0 & 0\\ -\frac{\nu}{E} & -\frac{\nu}{E} & \frac{1}{E} & 0 & 0 & 0\\ 0 & 0 & 0 & \frac{1+\nu}{E} & 0 & 0\\ 0 & 0 & 0 & 0 & \frac{1+\nu}{E} & 0\\ 0 & 0 & 0 & 0 & 0 & \frac{1+\nu}{E}\end{bmatrix}\begin{Bmatrix}\sigma_{xx}\\ \sigma_{yy}\\ \sigma_{zz}\\ \sigma_{xy}\\ \sigma_{yz}\\ \sigma_{xz}\end{Bmatrix} \tag{11.3}$$

Where E is the Young modulus and ν is the Poisson ratio. Consider now, for example the third component of the strain vector, the normal strain in the direction z:

$$\varepsilon_{zz}=\frac{\sigma_{zz}}{E}-\left(\frac{\sigma_{xx}}{E}\right)\nu-\left(\frac{\sigma_{yy}}{E}\right)\nu \tag{11.4}$$

Each compressive stress σ_{aa} contributes to the contraction σ_{aa}/E in its own direction a, and to the expansion $-\nu\sigma_{aa}/E$ in two other orthogonal directions (Figure 11.3a). Therefore, expression (11.4) is the sum of the strain

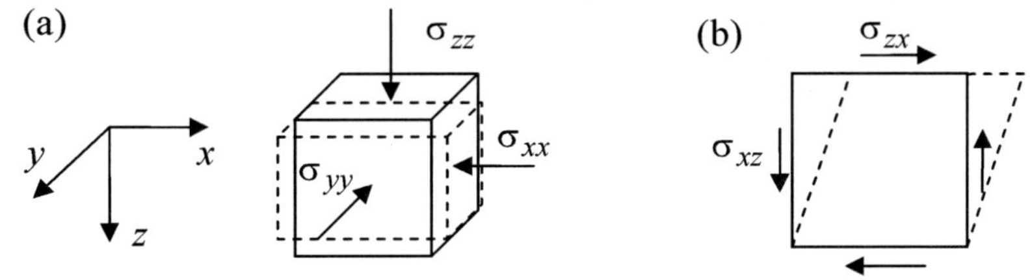

Figure 11.3 Deformations of a soil element: (a) normal; (b) shear.

contributions in the direction z of all the normal stresses. In contrast, shear stresses contribute to the shear strains only in the plane of their action, e.g. (Figure 11.3b):

$$\varepsilon_{xz} = \frac{\sigma_{xz}}{2G} = \frac{1+\nu}{E}\sigma_{xz} \tag{11.5}$$

Because the material is assumed to be isotropic, the same two elastic constants have been used for all the loading directions.

11.2.2 Cross-anisotropic behaviour

Geologic materials are formed in nature under the action of gravity, resulting in different soil properties in vertical and horizontal planes, i.e. in *anisotropy*. In fact, this is a special kind of anisotropy, which follows the axial symmetry with respect to any vertical axis. This means that the mechanical behaviour in all horizontal planes (h) is identical. The mechanical behaviour in all vertical planes (v), passing through the axis of symmetry, is also identical, but different to the one in the horizontal planes. This type of anisotropic behaviour is called *cross-anisotropic*. According to the linear elastic cross-anisotropic model, when the z axis is vertical and to others are horizontal (like in Figure 11.3), the normal strain in the direction z is given by

$$\varepsilon_{zz} = \frac{\sigma_{zz}}{E_v} - \left(\frac{\sigma_{xx}}{E_h}\right)\nu_{hv} - \left(\frac{\sigma_{yy}}{E_h}\right)\nu_{hv} \tag{11.6}$$

This expression reflects the fact, that:

- contribution of the compressive stress σ_{aa} to the contraction in its own direction a, can be different, depending on whether this direction is vertical or horizontal (σ_{vv}/E_v and σ_{hh}/E_h, respectively).
- contribution of the compressive stress σ_{aa} to the expansion in another orthogonal direction (Figure 11.3a), will depend on whether the direction a and this another direction form a vertical plane (producing a strain component $-\nu_{vh}\sigma_{vv}/E_v$ or $-\nu_{hv}\sigma_{hh}/E_h$) or a horizontal plane (producing $-\nu_{hh}\sigma_{hh}/E_h$).

Here, E_v and E_h are the Young moduli in the vertical and horizontal directions, respectively; ν_{ab} is the Poison ratio, where a indicates direction of the stress (vertical v or horizontal h), while where b indicates direction of the strain component (also vertical v or horizontal h) caused by this stress.

Example of a horizontal normal strain (in the direction x) is given by

$$\varepsilon_{xx} = \frac{\sigma_{xx}}{E_h} - \left(\frac{\sigma_{yy}}{E_h}\right)\nu_{hh} - \left(\frac{\sigma_{zz}}{E_v}\right)\nu_{vh} \tag{11.7}$$

A shear stress σ_{ab} will also produce different shear strains, depending on whether both directions a and b are horizontal (producing a shear strain of $\sigma_{ab}/2G_{hh}$) or one of them is vertical (producing $\sigma_{ab}/2G_{vh}$). Here, G_{vh} and G_{hh} are the shear moduli in the vertical and horizontal planes.

In general, the linear elastic cross-anisotropic model is described by

$$\begin{Bmatrix} \varepsilon_{xx} \\ \varepsilon_{yy} \\ \varepsilon_{zz} \\ \varepsilon_{xy} \\ \varepsilon_{yz} \\ \varepsilon_{xz} \end{Bmatrix} = \begin{bmatrix} \frac{1}{E_h} & -\frac{\nu_{hh}}{E_h} & -\frac{\nu_{vh}}{E_v} & 0 & 0 & 0 \\ -\frac{\nu_{hh}}{E_h} & \frac{1}{E_h} & -\frac{\nu_{vh}}{E_v} & 0 & 0 & 0 \\ -\frac{\nu_{hv}}{E_h} & -\frac{\nu_{hv}}{E_h} & \frac{1}{E_v} & 0 & 0 & 0 \\ 0 & 0 & 0 & \frac{1}{2G_{hh}} & 0 & 0 \\ 0 & 0 & 0 & 0 & \frac{1}{2G_{vh}} & 0 \\ 0 & 0 & 0 & 0 & 0 & \frac{1}{2G_{vh}} \end{bmatrix} \begin{Bmatrix} \sigma_{xx} \\ \sigma_{yy} \\ \sigma_{zz} \\ \sigma_{xy} \\ \sigma_{yz} \\ \sigma_{xz} \end{Bmatrix} \tag{11.8}$$

The model has seven constants: E_h, E_v, ν_{hh}, ν_{vh}, ν_{hv}, G_{hh}, G_{vh}. Not all of them are, however, independent. First of all, because the behaviour in a horizontal plane is isotropic, the following relationship from Chapter 10 applies for the horizontal shear stiffness:

$$G_{hh} = \frac{E_h}{2(1+\nu_{hh})} \tag{11.9}$$

Second, if we require that the compliance matrix in equation (11.8) is symmetric (we are going to justify this requirement later), then:

$$\frac{\nu_{hv}}{E_h} = \frac{\nu_{vh}}{E_v}, \text{ so that } \nu_{hv} = \nu_{vh}\frac{E_h}{E_v} \tag{11.10}$$

and we are left with only 5 independent model parameters: E_h, E_v, ν_{hh}, ν_{vh} and G_{vh}.

11.2.3 Transformation of coordinates

The isotropic elastic model has been shown in the previous Chapter to be invariant to coordinate transformations. Is that also true for the cross-anisotropic model (11.8)? For simplicity, instead of the general transformation let as consider here a very simple rotation of the original coordinate system (Figure 11.4a) by 90 degrees anticlockwise, so that the axes of the new coordinate system (Figure 11.4b) are related to the old ones in the following way:

$$x^* = -z\,;\; y^* = y\,;\; z^* = x \tag{11.11}$$

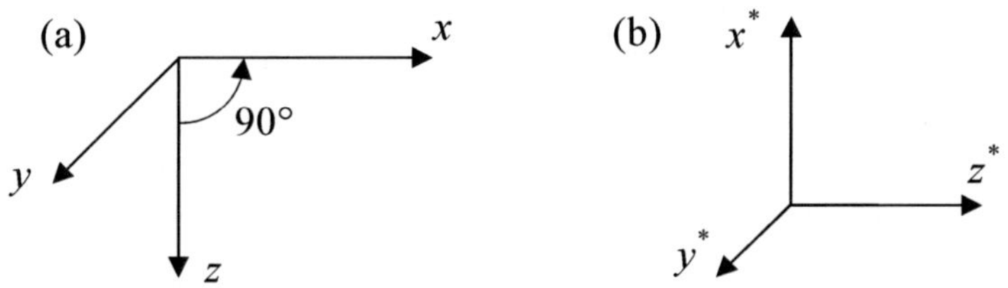

Figure 11.4 Coordinate systems: (a) original; (b) transformed.

To further simplify the transformation, we assume that x, y, z are principle axes, and, therefore the new axes x^*, y^*, z^* are principal as well (the orthogonal directions did not change).

Then the new components of the strain and stress tensors can be easily expressed through the old ones:

$$\begin{aligned} \varepsilon_{xx}^* &= \varepsilon_{zz}\,; \qquad \sigma_{xx}^* = \sigma_{zz}\,; \\ \varepsilon_{yy}^* &= \varepsilon_{yy}\,; \qquad \sigma_{yy}^* = \sigma_{yy}\,; \\ \varepsilon_{zz}^* &= \varepsilon_{xx}\,; \qquad \sigma_{zz}^* = \sigma_{xx}\,. \end{aligned} \tag{11.12}$$

leading to the new formulation of the stress-strain relationship, e.g.,

$$\begin{aligned} \varepsilon_{xx}^* = \varepsilon_{zz} &= \left(-\frac{\nu_{hv}}{E_h}\right)\sigma_{xx} + \left(-\frac{\nu_{hv}}{E_h}\right)\sigma_{yy} + \frac{1}{E_v}\sigma_{zz} \\ &= \frac{1}{E_v}\sigma_{xx}^* + \left(-\frac{\nu_{hv}}{E_h}\right)\sigma_{yy}^* + \left(-\frac{\nu_{hv}}{E_h}\right)\sigma_{zz}^* \end{aligned} \tag{11.13}$$

Comparing this with the original expression:

$$\varepsilon_{xx} = \frac{1}{E_h}\sigma_{xx} + \left(-\frac{\nu_{hh}}{E_h}\right)\sigma_{yy} + \left(-\frac{\nu_{hv}}{E_h}\right)\sigma_{zz} \tag{11.14}$$

we observe that expressions (11.13) and (11.14) are not identical. Although this was a particular case, it is sufficient for concluding that for cross-anisotropic materials the constitutive law is *not invariant*! If the material was isotropic, with $\nu_{vh} = \nu_{hv} = \nu_{hh}$ and $E_v = E_h$, expressions (11.13) and (11.14) would be identical.

This result leads to the very important conclusion: *an anisotropic constitutive law cannot be expressed via stress and strain tensor invariants.* (An exception is made when the formulation includes mixed invariants with the fabric tensor, reflecting material anisotropy).

11.3 Coupling and stress path dependency of stiffness

11.3.1 Cross-anisotropic behaviour in a triaxial test

Let us now investigate what kind of behaviour the cross-anisotropic linear elastic model would produce in a triaxial test. Assuming that x, y, z are principle axes, with the maximum principal stress σ_1 acting in the direction of the vertical axis z, the cross-anisotropic linear elastic model (11.8)-(11.10) for the triaxial stress-strain state can be rewritten in the following way:

$$\begin{Bmatrix} \varepsilon_{xx} \\ \varepsilon_{yy} \\ \varepsilon_{zz} \end{Bmatrix} = \begin{Bmatrix} \varepsilon_3 \\ \varepsilon_3 \\ \varepsilon_1 \end{Bmatrix} = \begin{bmatrix} \frac{1}{E_h} & -\frac{\nu_{hh}}{E_h} & -\frac{\nu_{vh}}{E_v} \\ -\frac{\nu_{hh}}{E_h} & \frac{1}{E_h} & -\frac{\nu_{vh}}{E_v} \\ -\frac{\nu_{hv}}{E_h} & -\frac{\nu_{hv}}{E_h} & \frac{1}{E_v} \end{bmatrix} \begin{Bmatrix} \sigma_3 \\ \sigma_3 \\ \sigma_1 \end{Bmatrix} \tag{11.15}$$

which can be further simplified as

$$\begin{Bmatrix} \varepsilon_3 \\ \varepsilon_1 \end{Bmatrix} = \begin{bmatrix} \frac{1-\nu_{hh}}{E_h} & -\frac{\nu_{vh}}{E_v} \\ -\frac{2\nu_{vh}}{E_v} & \frac{1}{E_v} \end{bmatrix} \begin{Bmatrix} \sigma_3 \\ \sigma_1 \end{Bmatrix} \tag{11.16}$$

Using the definitions of the triaxial strains and stresses:

$$\begin{Bmatrix} \varepsilon_v - \varepsilon_{v0} \\ \varepsilon_s \end{Bmatrix} = \begin{bmatrix} 2 & 1 \\ -2/3 & 2/3 \end{bmatrix} \begin{Bmatrix} \varepsilon_3 \\ \varepsilon_1 \end{Bmatrix}; \qquad \begin{Bmatrix} \sigma_3 \\ \sigma_1 \end{Bmatrix} = \begin{bmatrix} 1 & -1/3 \\ 1 & 2/3 \end{bmatrix} \begin{Bmatrix} p' - p'_0 \\ q \end{Bmatrix}. \tag{11.17}$$

and substituting the second equation (11.17) into equation (11.16) and the result into the first equation (11.17) we obtain:

$$\begin{Bmatrix} \varepsilon_v - \varepsilon_{v0} \\ \varepsilon_s \end{Bmatrix} = \begin{bmatrix} 1/K & -1/J \\ -1/J & 1/3G \end{bmatrix} \begin{Bmatrix} p' - p'_0 \\ q \end{Bmatrix} \tag{11.18}$$

where

$$\frac{1}{K} = 2\frac{1-\nu_{hh}}{E_h} + \frac{1-4\nu_{vh}}{E_v} \tag{11.19}$$

$$\frac{1}{J} = \frac{2}{3}\left(\frac{1-\nu_{hh}}{E_h} - \frac{1-\nu_{vh}}{E_v}\right) \tag{11.20}$$

$$\frac{1}{G} = \frac{2}{3}\left(\frac{1-\nu_{hh}}{E_h} + 2\frac{1+2\nu_{vh}}{E_v}\right) \tag{11.21}$$

Equation (11.18) is identical to the relationship (11.1) suggested in the Introduction to this chapter for simulating the *stress path dependency of stiffness* and *coupling* in triaxial tests.

11.3.2 Stress path dependency of stiffness

Consider a straight probing stress path (Figure 11.5a):

$$q = (p' - p'_0)\tan\theta \tag{11.22}$$

The corresponding shear and bulk moduli (Figures 11.5b,c)

$$K_\theta = \frac{p' - p'_0}{\varepsilon_v - \varepsilon_{v0}}; \qquad 3G_\theta = \frac{q}{\varepsilon_s} \tag{11.23}$$

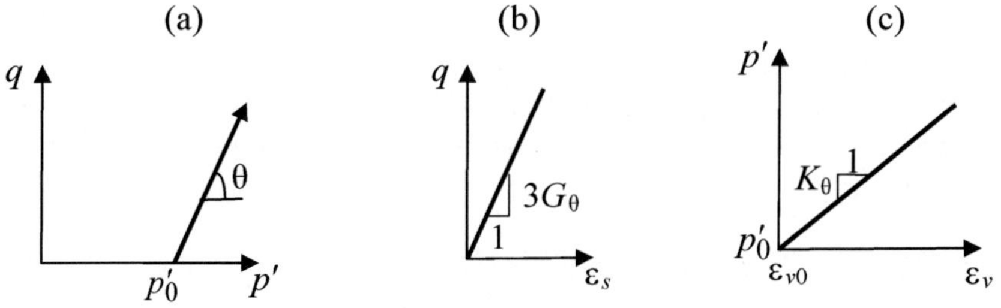

Figure 11.5 Stress path dependency of stiffness in a triaxial test: (a) a probing stress path; (b) deviatoric; and (c) volumetric stress-strain curves.

can be defined using equation (11.18) together with (11.22):

$$K_\theta = \frac{KJ}{J - K\tan\theta}; \qquad G_\theta = \frac{3GJ\tan\theta}{J\tan\theta - 3G}. \tag{11.24}$$

The stiffness in the model (11.18) clearly depends on the direction of the stress path.

11.3.3 Coupling

Two particular stress paths (11.22) are of special interest: isotropic consolidation $\theta = 0\,(q = 0)$ and pure shear $\theta = \pi/2\,(p' = p'_0)$. For isotropic linear elastic material, these two stress paths would produce zero deviatoric and volumetric strains, respectively. In the model (11.18) they result, respectively, in:

$$(\varepsilon_s)_{\theta=0} = -\frac{p' - p'_0}{J}; \qquad (\varepsilon_v - \varepsilon_{v0})_{\theta=\pi/2} = -\frac{q}{J}, \tag{11.25}$$

which clearly indicates that the model (11.18) produces coupling between the deviatoric and volumetric behaviour.

11.3.4 Relationship between the anisotropy and coupling models

This relationship between the cross-anisotropic (11.8)-(11.10) and coupling (11.18) models looks pretty straightforward. In fact, it may be a cause of some serious errors.

First of all, while the cross-anisotropic model (11.8)-(11.10) can always simulate the stress path dependency of stiffness and coupling (11.18) in a triaxial test, the inverse statement is *wrong*. The coupling model (11.18) is correct only if the major principle axis stays vertical. But even in this case, generalisation of the coupling model (11.18) to the general stress-strain space (by substituting instead of the triaxial strains and stresses corresponding invariants) would fail to model cross-anisotropic behaviour. Because only isotropic models can be expressed via tensor invariants (see Section 11.2.3)! Therefore, the model (11.18), though capable of simulating the stress path dependency of stiffness and coupling in a triaxial test, *is not cross-anisotropic*. In order to model the cross-anisotropic behaviour in a boundary value problem, the general cross-anisotropic model (11.8)-(11.10) has to be used!

And here comes the second problem – derivation of parameters. The stiffness parameters for coupling model (11.18) can be easily defined from the two standard triaxial tests (triaxial compression and isotropic consolidation) as shown in Figure 11.2 and equations (11.2). They are sufficient for predicting the triaxial behaviour along arbitrary stress paths. However, when it comes to derivation of parameters for the general cross-anisotropic model (11.8)-(11.10), from equations (11.19)-(11.21) it follows, that only three out of five parameters can be defined via triaxial constants:

$$\frac{1-\nu_{hh}}{E_h} = -\frac{2}{9K} + \frac{7}{3J} - \frac{1}{6G} \tag{11.26}$$

$$\frac{1}{E_v} = -\frac{1}{9K} + \frac{2}{3J} - \frac{1}{3G} \tag{11.27}$$

$$\nu_{vh} = \left(\frac{1}{9K} - \frac{1}{6J} - \frac{1}{6G}\right) \Big/ \left(-\frac{1}{9K} + \frac{2}{3J} - \frac{1}{3G}\right) \tag{11.28}$$

In order to separate between parameters ν_{hh} and E_h, and define the parameter G_{vh}, additional tests (or assumptions) are necessary.

Therefore, triaxial compression and isotropic consolidation tests are not sufficient for derivation of parameters for the general cross-anisotropic model (11.8)-(11.10).

11.4 Thermomechanics of anisotropy and coupling

11.4.1 Anisotropy and Thermomechanics

Does the cross-anisotropic linear elastic model (11.8) satisfy the First Law of Thermodynamics? It does not, unless the compliance matrix in (11.8) is symmetric. In fact, for the generalized Hooke law

$$\varepsilon_{ij} = C_{ijkl}\sigma_{lk} \tag{11.29}$$

where C_{ijkl} is the compliance tensor, to satisfy the First Law of Thermodynamics, equation (11.29) should be derived from a complimentary strain energy function U_{c0} (the Gibbs free energy $-g$):

$$\varepsilon_{ij} = \frac{\partial U_{c0}}{\partial \sigma_{ij}} = -\frac{\partial g}{\partial \sigma_{ij}} \tag{11.30}$$

which is only possible when $g(\sigma_{ij})$ is a quadratic function of stress:

$$g = -\frac{1}{2}C_{ijkl}\sigma_{ij}\sigma_{kl} \tag{11.31}$$

Parameters of the compliance tensor can be then defined from the double differentiation of the Gibbs free energy function:

$$C_{ijkl} = -\frac{\partial g}{\partial \sigma_{ij}\partial \sigma_{kl}} = -\frac{\partial g}{\partial \sigma_{kl}\sigma_{ij}} = C_{ijkl} \tag{11.32}$$

And using the symmetry of the strain tensor:

$$C_{ijkl} = -\frac{\partial g}{\partial \sigma_{ij} \partial \sigma_{kl}} = -\frac{\partial g}{\partial \sigma_{ji} \partial \sigma_{kl}} = C_{jikl}$$

$$= -\frac{\partial g}{\partial \sigma_{ij} \partial \sigma_{lk}} = C_{ijlk} \tag{11.33}$$

$$= -\frac{\partial g}{\partial \sigma_{ji} \partial \sigma_{lk}} = C_{jilk}$$

The above symmetry relations lead to the symmetry of the compliance matrix of the generalized Hooke law.

This important condition can be also derived directly from the vector-matrix form of the generalized Hooke law $\{\varepsilon\} = [C]\{\sigma\}$ (or $\varepsilon_i = C_{ij}\sigma_j$, for $i = 1...6$, $j = 1...6$), where:

$$\varepsilon_i = -\frac{\partial g}{\partial \sigma_i}; \qquad g = -\frac{1}{2} C_{ij}\sigma_i\sigma_j, \tag{11.34}$$

so that

$$C_{ij} = -\frac{\partial g}{\partial \sigma_i \partial \sigma_j} = -\frac{\partial g}{\partial \sigma_j \partial \sigma_i} = C_{ji} \tag{11.35}$$

i.e., the compliance matrix of generalized Hooke's Law is indeed symmetric and can have no more than 21 independent components. The cross-anisotropic linear elastic relationship (11.8) is a particular case of the generalized Hooke law, therefore, in order to satisfy the First Law of Thermodynamics its compliance matrix has to be symmetric, which is only satisfied by the requirement (11.10).

In other words, *condition* $\nu_{hv} = \nu_{vh} E_h / E_v$ *is necessary for the cross-anisotropic linear elastic relationship to satisfy the law of energy conservation.*

11.4.2 Coupling and Thermomechanics

Let us now see if the coupling model (11.18) is also thermomechanically consistent. There is no reason why it should not be, because it is a particular case of the general cross-anisotropic model (11.8)-(11.10), which has been shown above to satisfy the First Law of Thermodynamics. There is, however, a simple direct proof: existence of the strain energy (and complimentary strain energy) functions.

For the isotropic linear elastic triaxial behaviour, the complimentary strain energy (Gibbs free energy) function is given by:

$$U_{c0}(p',q) = -g(p',q) = \frac{q^2}{6G} + \frac{(p'-p'_0)^2}{2K} + (p'-p'_0)\varepsilon_{v0} \qquad (11.36)$$

and the isotropic linear elastic relationships can be then derived as:

$$\varepsilon_v = -\frac{\partial g(p',q)}{\partial p'} = \varepsilon_{v0} + \frac{p'-p'_0}{K}; \qquad \varepsilon_s = -\frac{\partial g(p',q)}{\partial q} = \frac{q}{3G}. \qquad (11.37)$$

For the coupled linear elastic triaxial behaviour, the complimentary strain energy (Gibbs free energy) function has an additional term:

$$U_{c0}(p',q) = -g(p',q) = \frac{q^2}{6G} + \frac{(p'-p'_0)^2}{2K} - \frac{q(p'-p'_0)}{J} + (p'-p'_0)\varepsilon_{v0} \qquad (11.38)$$

and the coupled linear elastic relationships become:

$$\begin{aligned} \varepsilon_v &= -\frac{\partial g(p',q)}{\partial p'} = \varepsilon_{v0} + \frac{p'-p'_0}{K} - \frac{q}{J} \\ \varepsilon_s &= -\frac{\partial g(p',q)}{\partial q} = \frac{q}{3G} - \frac{p'-p'_0}{J} \end{aligned} \qquad (11.39)$$

which can be easily recognised as the coupled model (11.18).

Existence of the complimentary strain energy function is the direct proof that the Hooke's cross-anisotropic linear elastic model (11.18) satisfies the Laws of Thermodynamics.

Chapter 12

Pressure Dependency of Stiffness

TABLE OF CONTENTS

Chapter 12

Pressure Dependency of Stiffness

12.1 Introduction

At this stage we know how to model isotropic and cross-anisotropic linear elastic behaviour. In these models the soil stiffness parameters are constants. In reality, however, the soil stiffness depends on the confining stress. Consider, for example, two triaxial samples brought through the same pre-consolidation p'_c to two different initial stresses p'_{01} and p'_{02} ($p'_{02} > p'_{01}$), and then subjected to triaxial compression (Figure 12.1a). The resulting deviatoric (Figure 12.1b) and volumetric (Figure 12.1c) stress-strain curves, will be different, with the corresponding pre-yielding secant shear and bulk stiffness being larger for the test with the larger initial stress ($G_2 > G_1$ and $K_2 > K_1$). This phenomenon is called *pressure dependency of stiffness*.

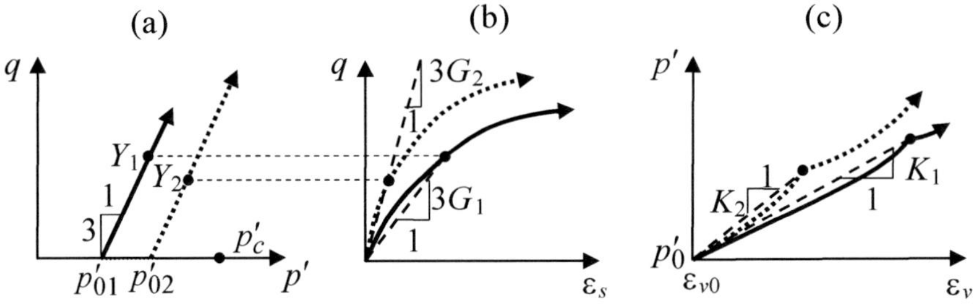

Figure 12.1 Pressure dependency of stiffness in triaxial compression: (a) stress paths; (b) deviatoric; and (c) volumetric stress-strain curves.

Another way this pressure dependency expresses itself, is in the isotropic reconsolidation test with $q = 0$ (Figure 12.2). When typical test results are plotted in a semi-logarithmic plot ($\varepsilon_v - \ln p'$, Figure 12.2a), the reversible swelling-reloading stress-strain curve can be approximated by a straight line:

$$\varepsilon_v = \varepsilon_{v0} + \frac{1}{k} \ln \frac{p'}{p'_0} \tag{12.1}$$

In an arithmetic scale plot ($p' - \varepsilon_v$, Figure 12.2b), this results in a non-linear relationship. The bulk modulus is not constant any more, and the tangent bulk

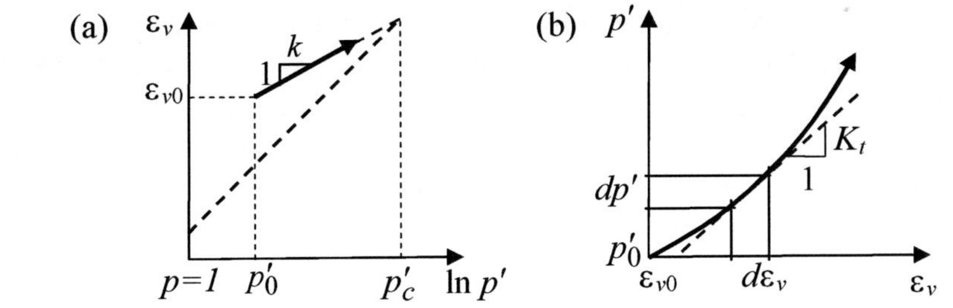

Figure 12.2 Dependency of stiffness on pressure in isotropic re-loading: (a) semi-logarithmic; and (b) arithmetic plots.

modulus K_t can be defined as follows:

$$K_t(p') = \frac{dp'}{d\varepsilon_v} = kp' \tag{12.2}$$

Dependency of the bulk stiffness on the pressure is obvious. The larger is the mean effective stress, the stiffer is the material, which makes sense – the soil becomes denser. The simplest way to model the phenomena illustrated in Figures 12.1 and 12.2, is to use the tangent compliance matrix:

$$\begin{Bmatrix} d\varepsilon_v \\ d\varepsilon_s \end{Bmatrix} = \begin{bmatrix} 1/K_t(p') & 0 \\ 0 & 1/(3G_t(p')) \end{bmatrix} \begin{Bmatrix} dp' \\ dq \end{Bmatrix} \tag{12.3}$$

This Chapter explains how the pressure dependency of stiffness can be modelled without violating the Laws of Thermodynamics both in triaxial and general stress spaces. In particular, it will be shown why it is intrinsically wrong to express the pressure dependency of stiffness in the form of relationship (12.3)!

12.2 Elasticity with pressure dependent stiffness

12.2.1 Hypoelasticity

It has been already mentioned that the real pre-yielding soil behaviour is neither linear, nor reversible. Linear elasticity, whether isotropic or anisotropic, cannot model these phenomena. But it is such a convenient form of describing material behaviour that it has been difficult just to let it go. This resulted in the generalization of the theory of elasticity by adopting the generalized incremental law:

$$d\varepsilon_{ij} = C_{ijkl}(\sigma_{mn})d\sigma_{kl} \tag{12.4}$$

where the components of the *tangent* compliance matrix $C_{ijkl}(\sigma_{mn})$ are the functions of stresses.

This generalization is called "Hypoelasticity". "Hypo" in ancient Greek means "less", and Hypoelasticity is less restrictive than elasticity – it allows for inelastic (irreversible) behaviour. Indeed, because the stiffness is a function of stress or strain and relationship (12.4) is given in the incremental form, it may not always be integrable, and can produce some residual permanent strains over a closed stress cycle (Figures 12.3).

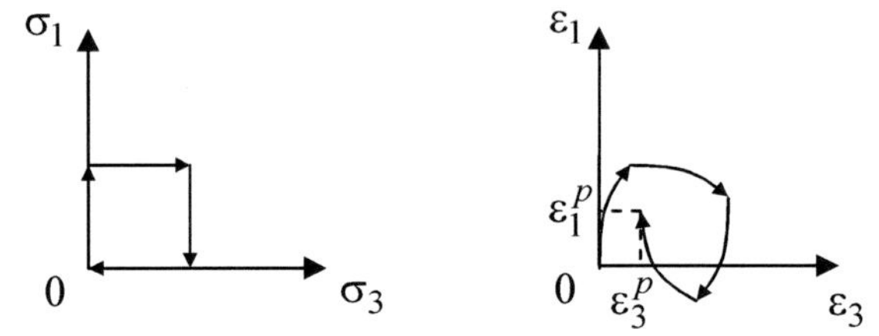

Figure 12.3 Inelastic behaviour in hypoelasticity.

12.2.2 Pressure dependency of stiffness

Hypoelasticity looks like a correct framework to describe pressure dependency of soil stiffness:

$$d\varepsilon_{ij} = C_{ijkl}(\sigma_{nn})d\sigma_{kl} \tag{12.5}$$

and using analogy with Chapter 10, for isotropic material this results in:

$$d\varepsilon_{ij} = \frac{1}{9K_t(\sigma_{nn})}d\sigma_{nn}\delta_{ij} + \frac{1}{2G_t(\sigma_{nn})}ds_{ij} \tag{12.6}$$

For triaxial stress space, the above model produces relationship (12.3):

$$\begin{Bmatrix} d\varepsilon_v \\ d\varepsilon_s \end{Bmatrix} = \begin{bmatrix} 1/K_t(p') & 0 \\ 0 & 1/(3G_t(p')) \end{bmatrix} \begin{Bmatrix} dp' \\ dq \end{Bmatrix}$$

which we were planning to use for modelling the pressure dependency of stiffness.

At this point it would be possible to finish this chapter, if not for a rather annoying obstacle: relationship (12.3), in general, violates the First Law of Thermodynamics!

12.3 Thermomechanics of the pressure dependent stiffness

12.3.1 Hypoelasticity and Thermomechanics

In order to satisfy the First Law of Thermodynamics, a reversible behaviour should be hyperelastic. *Hyperelasticity* requires existence of the strain energy potential $U_0(\varepsilon_{ij})$ for stresses and/or complimentary energy potential $U_{c0}(\sigma_{ij})$ for strains. For isothermal conditions, typical for most of the geotechnical problems, these two potentials correspond to the specific Helmholtz free energy $f(\varepsilon_{ij})$ and the specific Gibbs free energy $g(\sigma_{ij})$, respectively:

$$\sigma_{ij} = \frac{\partial U_0}{\partial \varepsilon_{ij}} = \frac{\partial f}{\partial \varepsilon_{ij}}; \quad \varepsilon_{ij} = \frac{\partial U_{c0}}{\partial \sigma_{ij}} = -\frac{\partial g}{\partial \sigma_{ij}}, \tag{12.7}$$

where σ_{ij} and ε_{ij} are the effective stress and elastic strain tensors, respectively. These two potentials are related through the following Legendre Transformation: $g(\sigma_{ij}) = f(\varepsilon_{ij}) - \sigma_{ij}\varepsilon_{ij}$. The choice between the Helmholtz and Gibbs free energy formulation is based on whether the proposed model is strain or stress-space based.

Incremental form of the second relationships (12.7) is given by

$$d\sigma_{ij} = D_{ijkl}(\varepsilon_{mn})d\varepsilon_{kl}; \quad d\varepsilon_{ij} = C_{ijkl}(\sigma_{mn})d\sigma_{kl}, \tag{12.8}$$

where

$$D_{ijkl}(\varepsilon_{mn}) = \frac{\partial^2 f(\varepsilon_{mn})}{\partial \varepsilon_{ij}\partial \varepsilon_{kl}}; \quad C_{ijkl}(\sigma_{mn}) = -\frac{\partial^2 g(\sigma_{mn})}{\partial \sigma_{ij}\partial \sigma_{kl}} \tag{12.9}$$

are the *tangent* stiffness and compliance tensors, respectively.

Although the second expression (12.8) looks identical to the hypoelastic incremental law (12.4), there is one very important difference: restriction imposed by the second equation (12.9). When the stress dependency of the tangent compliance tensor $C_{ijkl}(\sigma_{mn})$ in equation (12.4) is specified arbitrary (without satisfying condition 12.9), the resulting hypoelastic incremental law may violate the First Law of Thermodynamics.

12.3.2 Isotropic hypoelasticity and Thermomechanics

The best illustration of the above statement can be given by considering isotropic materials in triaxial space. The Gibbs free energy function for linear elastic isotropic material is given by

$$U_{c0}(p',q) = -g(p',q) = \frac{q^2}{6G} + \frac{(p'-p'_0)^2}{2K} + (p'-p'_0)\varepsilon_{v0} \tag{12.10}$$

and the stress-strain relationships can be then derived as:

$$\varepsilon_v = -\frac{\partial g(p',q)}{\partial p'} = \varepsilon_{v0} + \frac{p'-p'_0}{K}; \qquad \varepsilon_s = -\frac{\partial g(p',q)}{\partial q} = \frac{q}{3G} \tag{12.11}$$

producing the *secant* compliance matrix:

$$\begin{Bmatrix} \varepsilon_v - \varepsilon_{v0} \\ \varepsilon_s \end{Bmatrix} = \begin{bmatrix} 1/K & 0 \\ 0 & 1/(3G) \end{bmatrix} \begin{Bmatrix} p' - p'_0 \\ q \end{Bmatrix} \tag{12.12}$$

The *tangent* compliance matrix is obtained by the double differentiation of the Gibbs free energy function (equation 12.9):

$$\begin{Bmatrix} d\varepsilon_v \\ d\varepsilon_s \end{Bmatrix} = -\begin{bmatrix} \dfrac{\partial^2 g}{\partial p'^2} & \dfrac{\partial^2 g}{\partial p' \partial q} \\ \dfrac{\partial^2 g}{\partial q \partial p'} & \dfrac{\partial^2 g}{\partial q^2} \end{bmatrix} \begin{Bmatrix} dp' \\ dq \end{Bmatrix} = \begin{bmatrix} 1/K & 0 \\ 0 & 1/(3G) \end{bmatrix} \begin{Bmatrix} dp' \\ dq \end{Bmatrix} \tag{12.13}$$

Note, that for linear elastic materials, with constant bulk and shear moduli, the secant and tangent compliance matrices are identical and satisfy the First Law of Thermodynamics. Clearly, at this point it is very tempting to introduce experimentally calibrated pressure dependency of stiffness in the form of the relationship (12.3):

$$\begin{Bmatrix} d\varepsilon_v \\ d\varepsilon_s \end{Bmatrix} = \begin{bmatrix} 1/K_t(p') & 0 \\ 0 & 1/(3G_t(p')) \end{bmatrix} \begin{Bmatrix} dp' \\ dq \end{Bmatrix}$$

The problem is, that if we introduce this experimentally calibrated stress dependency of stiffness into the Gibbs free energy function

$$U_{c0}(p',q) = -g(p',q) = \frac{q^2}{6G(p')} + \frac{(p' - p'_0)^2}{2K(p')} + (p' - p'_0)\varepsilon_{v0} \tag{12.14}$$

its double differentiation (12.13) would never produce the tangent compliance matrix in the form of the relationship (12.3)! First of all, non-diagonal terms would not be zero, but even diagonal terms would not be the same as in (12.3).

This is bad news, but not yet the end of the world. Can it be that there exists another functional form of the Gibbs free energy function, which could produce the tangent compliance matrix in the form of the relationship (12.3)?

12.3.3 Stress induced coupling

The unfortunate reality is that as long as the shear modulus G is a function of p', no Gibbs free energy function can produce the tangent compliance matrix (12.3). Indeed, if its second derivative with respect to q is a function of p':

$$\frac{\partial^2 g}{\partial q^2} = -\frac{1}{3G(p')} \tag{12.15}$$

the mixed derivative cannot be zero:

$$\frac{\partial^2 g}{\partial p' \partial q} = \frac{1}{J(p',q)} \neq 0 \tag{12.16}$$

This means, that in order to satisfy the First Law of Thermodynamics, the stress dependent material behaviour has to be *coupled*! In Chapter 11 we explored the volumetric-deviatoric coupling caused by the structural anisotropy of the material. Here the coupling occurs for structurally isotropic materials – it is induced by the changing stress state, and is, therefore, called the *stress-induced coupling*. It should be, however, by no means mixed with *stress-induced anisotropy*! Even with coupling, the triaxial material model stays isotropic – simply due to the fact that it is expressed via stress and strain invariants. Stress-induced anisotropy appears only in the general stress space formulation $d\varepsilon_{ij} = C_{ijkl}(\sigma_{mn})d\sigma_{kl}$.

This thermomechanically compulsory stress-induced coupling also means that for the stress dependent shear modulus the hypoelastic relationship (12.3) will always violate the First Law of Thermodynamics. Is there a way of modelling the stiffness dependency on pressure in a thermomechanically consistent manner?

12.4 Thermomechanically consistent models

To make sure that the proposed models are thermomechanically consistent we derive them directly from the energy function.

12.4.1 The constant shear modulus model

Consider the following Gibbs free energy function, with the constant shear modulus G:

$$U_{c0}(p',q) = -g(p',q) = \frac{q^2}{6G} + \frac{p'}{k}\left(\ln\left(\frac{p'}{p'_0}\right) - 1\right) + (p' - p'_0)\varepsilon_{v0} \tag{12.17}$$

The incremental relationship is obtained by the double differentiation of the Gibbs free energy function (equation 12.13):

$$\begin{Bmatrix} d\varepsilon_v \\ d\varepsilon_s \end{Bmatrix} = -\begin{bmatrix} \dfrac{\partial^2 g}{\partial p'^2} & \dfrac{\partial^2 g}{\partial p' \partial q} \\ \dfrac{\partial^2 g}{\partial q \partial p'} & \dfrac{\partial^2 g}{\partial q^2} \end{bmatrix} \begin{Bmatrix} dp' \\ dq \end{Bmatrix} = \begin{bmatrix} \dfrac{1}{kp'} & 0 \\ 0 & \dfrac{1}{3G} \end{bmatrix} \begin{Bmatrix} dp' \\ dq \end{Bmatrix} \tag{12.18}$$

In this model, there is no volumetric-deviatoric coupling, the shear modulus G is constant, the tangent bulk modulus $K_t(p') = kp'$ follows the experimental relationship (12.1).

12.4.2 The linear model

Consider now the Gibbs free energy function, with the shear modulus G linearly depending on the mean effective stress $G(p')=g_s p'$:

$$U_{c0}(p',q)=-g(p',q)=\frac{q^2}{6g_s p'}+\frac{p'}{k}\left(\ln\left(\frac{p'}{p'_0}\right)-1\right)+(p'-p'_0)\varepsilon_{v0} \tag{12.19}$$

The incremental relationship is obtained again by the double differentiation of the Gibbs free energy function (equation 12.13):

$$\begin{Bmatrix} d\varepsilon_v \\ d\varepsilon_s \end{Bmatrix}=\begin{bmatrix} \frac{1}{kp'}\left(1+\frac{kq^2}{3g_s p'^2}\right) & -\frac{q}{3g_s p'^2} \\ -\frac{q}{3g_s p'^2} & \frac{1}{3g_s p'} \end{bmatrix}\begin{Bmatrix} dp' \\ dq \end{Bmatrix} \tag{12.20}$$

In this model, there is a volumetric-deviatoric coupling, but on the isotropic axis $q=0$ the model degenerates into a simple form

$$\begin{Bmatrix} d\varepsilon_v \\ d\varepsilon_s \end{Bmatrix}=\begin{bmatrix} \frac{1}{kp'} & 0 \\ 0 & \frac{1}{3g_s p'} \end{bmatrix}\begin{Bmatrix} dp' \\ dq \end{Bmatrix} \tag{12.21}$$

with no coupling and linear dependency of both tangent bulk and shear moduli on the mean effective stress:

$$K_t(p')=kp'; \quad G(p')=g_s p', \tag{12.22}$$

which is a reasonable approximation of the behaviour of clay during isotropic swelling and reloading.

12.4.3 The power law model

Finally, consider the following Gibbs free energy function:

$$U_{c0}(p',q)=-g(p',q)=\frac{p_s^{2-n}}{p_r^{1-n}k(1-n)(2-n)}+\frac{p'}{k(1-n)}+(p'-p'_0)\varepsilon_{v0} \tag{12.23}$$

where p_r is a reference stress; n is a parameter;

$$p_s^2=p'^2+\frac{k(1-n)q^2}{3g_s} \tag{12.24}$$

The corresponding incremental relationship is given by:

$$\begin{Bmatrix} d\varepsilon_v \\ d\varepsilon_s \end{Bmatrix} = \begin{bmatrix} \dfrac{1}{k(1-n)p_r^{1-n}p_s^n}\left(1-\dfrac{np'^2}{p_s^2}\right) & -\dfrac{np'q}{3g_s p_r^{1-n}p_s^{n+2}} \\ -\dfrac{np'q}{3g_s p_r^{1-n}p_s^{n+2}} & \dfrac{1}{3g_s p_r^{1-n}p_s^n}\left(1-\dfrac{nk(1-n)q^2}{3g_s p_s^2}\right) \end{bmatrix} \begin{Bmatrix} dp' \\ dq \end{Bmatrix}$$

(12.25)

In this model, there is also a volumetric-deviatoric coupling, but on the isotropic axis $q = 0$ the model degenerates into a simple form

$$\begin{Bmatrix} d\varepsilon_v \\ d\varepsilon_s \end{Bmatrix} = \begin{bmatrix} \dfrac{1}{kp_r^{1-n}p'^n} & 0 \\ 0 & \dfrac{1}{3g_s p_r^{1-n}p'^n} \end{bmatrix} \begin{Bmatrix} dp' \\ dq \end{Bmatrix} \qquad (12.26)$$

with no coupling and the power dependency of both tangent shear and bulk moduli on the mean effective stress:

$$\frac{K_t(p')}{p_r} = k\left(\frac{p'}{p_r}\right)^n; \qquad \frac{G_t(p')}{p_r} = g_s\left(\frac{p'}{p_r}\right)^n, \qquad (12.27)$$

which is a reasonable approximation of the behaviour of sand during isotropic swelling and reloading.

12.4.4 The general stress formulation

In the general stress formulation, the compliance tensor tor the power law model is given by the second equation (12.9):

$$C_{ijkl} = \frac{1}{p_r}\left(\frac{p_r}{p_s}\right)^n\left(\left(\frac{1}{k}+\frac{n\sigma'_{mn}\sigma'_{mn}}{2g_s p_s^2}\right)\frac{\delta_{ij}\delta_{kl}}{9} - \frac{n\sigma_{mm}}{18g_s p_s^2}\left(\sigma'_{ij}\delta_{kl}+\delta_{ij}\sigma'_{kl}\right) + \frac{1}{2g_s}\left(\delta_{ik}\delta_{jl}-\frac{1}{3}\delta_{kl}\delta_{ij}\right) - \frac{nk(1-n)}{4g_s^2 p_s^2}\sigma'_{ij}\sigma'_{kl}\right) \qquad (12.28)$$

where $p_s^2 = \dfrac{\sigma_{mm}\sigma_{nn}}{9} + \dfrac{k(1-n)\sigma'_{mn}\sigma'_{mn}}{2g_s}$. (12.29)

This can also be applied for the compliance tensor of the linear law model, by setting $n = 1$. This formulation is capable of modelling not just the stress-induced coupling but also the *stress-induced anisotropy*.

Chapter 13

Small Strain Nonlinearity

TABLE OF CONTENTS

Chapter 13

Small Strain Nonlinearity

13.1 Introduction

In the previous chapter, in the models with pressure dependent stiffness, the shear modulus G was either constant or increased with the mean effective stress. These models used to describe the pre-yielding deviatoric behaviour of soils in standard triaxial compression tests reasonably well. The problem was that the corresponding numerical models considerably over-predicted displacements in many boundary value problems. In late 1980s this discrepancy lead researchers at Imperial College, London (e.g., Burland, 1989), to an idea to conduct laboratory tests with strains measured locally on the sample, as opposed to the external strain measurements in standard triaxial tests (Figure 13.1a). When plotted in a wide strain range (Figure 13.1b), the deviatoric stress-strain curves of the two tests do not differ that much. The curve for the local strain test (dashed line) goes slightly higher, but this does not affect the pre-yielding secant shear modulus G significantly. A different picture is observed when we zoom (Figure 13.1c) into the area of very small strains (up to 0.01-0.1%). While the externally measured stress-strain test curve (solid line) is almost linear, the locally measured stress-strain curve proves to be highly non-linear, with the initial tangent shear modulus G_0 almost an order of magnitude higher than the pre-yielding secant shear modulus G. Another important discovery was that even at the very small strains the stress-strain behaviour is not entirely reversible, i.e. it exhibits some very small permanent strains in a closed loading cycle (Figure 13.1c).

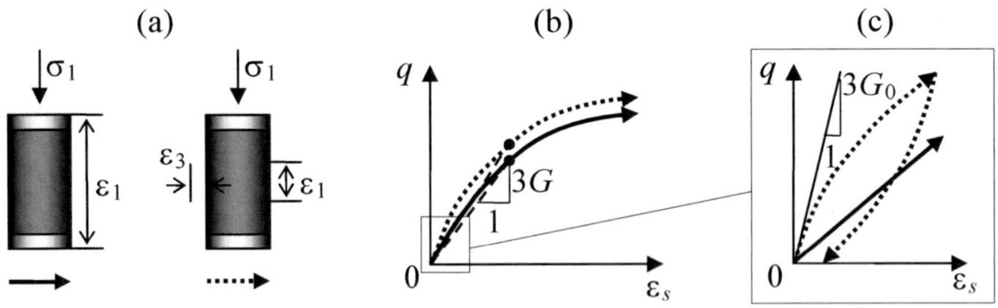

Figure 13.1 External vs. local deformation measurements in a triaxial test: (a) schematic setup; (b) deviatoric stress-strain curves; (c) zoom into small strains.

Why is this high stiffness small strain behaviour so important? The reality is that in many practical geotechnical problems only a relatively small volume of soil experiences large deformations. The strains in the remaining part are very small. But to calculate displacements, these small strains are integrated over the large area and their overall contribution can be quite significant. Therefore, underestimation of the soil stiffness at small strains in standard tests results (for the same stress level) in overestimating the strains and displacements in geotechnical boundary value problems.

The new understanding of the highly non-linear nature of the small strain behaviour of soils obtained from the local deformation measurement tests allowed for this problem to be solved. Provided, of course, that this non-linear behaviour is properly modelled.

13.2 Non-linear models of the small-strain behaviour

In this section we are going to consider a number of analytical functions for curve-fitting the deviatoric stress-strain behaviour of soils at small strains. What are the requirements this function should satisfy?

First of all it should be simple and contain a minimum number of parameters. These parameters should have a clear physical meaning and be easy to derive from standard lab tests. Finally, most importantly, the function should provide a good fit to the experimental deviatoric stress-strain curves. And this is exactly the catch: when our "elegant" function is not sufficiently accurate, we "modify" it by adding parameters, they loose their physical meaning, become difficult to derive, etc.. And this may still not be sufficient, if the intrinsic shape of the function is not the proper one. The best known "victim" of such modifications is the hyperbolic function, described below.

There are also some mathematical conditions the curve-fitting function should satisfy, but in order to formulate them, we first introduce a normalized deviatoric stress-strain space.

13.2.1 Normalized pre-yielding behaviour

The curve-fitting function does not necessarily have to fit the deviatoric stress-strain curve all the way up to the yield stress q_y, because the high non-linearity occurs normally at rather small strains. After a certain limiting deviator stress q_L and up to the yield stress q_y, the stress strain curve can be often well approximated by a straight line with inclination $3G_L$, corresponding to the tangent shear modulus G_L at the limiting stress q_L (Figure 13.2a). In this case an analytical function has to simulate the soil behaviour up to limiting deviator stress q_L (corresponding to the limiting shear strain ε_{sL}) only. The following normalisation of the curve within the above limits was found to be very convenient (Figure 13.2b):

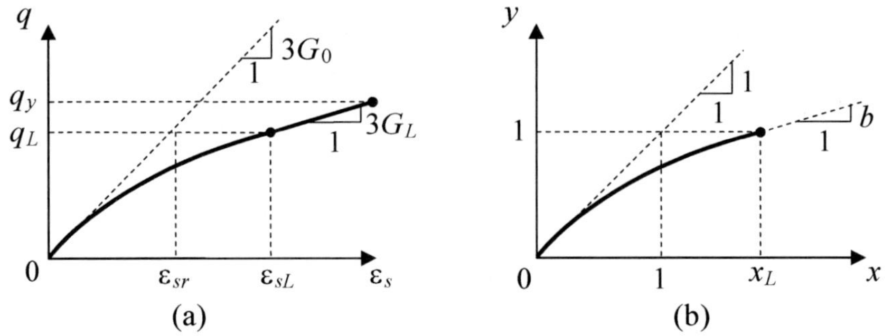

Figure 13.2 Deviatoric stress-strain curve: (a) true; (b) normalised.

$$y = \frac{q}{q_L}; \qquad x = \frac{\varepsilon_s}{\varepsilon_{sr}}; \qquad \varepsilon_{sr} = \frac{q_L}{3G_0}; \qquad x_L = \frac{\varepsilon_{sL}}{\varepsilon_{sr}}, \tag{13.1}$$

where y and x are the normalized deviator stress and strain, respectively; G_0 is the initial tangent shear modulus; x_L is the normalized limiting shear strain.

It follows, that

$$\frac{dy}{dx} = \frac{1}{3G_0}\frac{dq}{d\varepsilon_s} = \frac{G_t(x)}{G_0} \tag{13.2}$$

i.e., the tangent shear modulus in the normalized space is equal to the true tangent shear modulus G_t normalized by the initial modulus G_0. As a result, the normalized analytical curve-fitting function has to satisfy the following conditions at the end points of the fitting interval (Figure 13.2b):

$$x = 0; \qquad y = 0; \qquad \frac{dy}{dx} = 1, \tag{13.3}$$

$$x = x_L; \quad y = 1; \qquad \frac{dy}{dx} = b = \frac{G_L}{G_0}. \tag{13.4}$$

Let us now consider some popular normalized curve-fitting functions.

13.2.2 The hyperbolic function

The hyperbolic model was proposed for modelling soil behaviour by Konder (1963). In terms of simplicity it is the absolute champion – in the normalised form it does not have any parameters:

$$y = \frac{x}{1+x} \tag{13.5}$$

Of course, there is a price to pay – while conditions (13.3) are fully satisfied, conditions (13.4) are not. Furthermore $y = 1$ is reached only at the infinite strain $x = \infty$, where $dy/dx = 0$, i.e. the material is slowly approaching failure (Figure 13.3a). This means, that this function can only be applied in the full range of strains with $q_L = q_f$, i.e. the limiting deviator stress is equal to the failure stress (shear strength). In order to be able to apply this function for modelling pre-yielding behaviour, it has to be slightly modified:

$$y = \frac{x}{1+\alpha x}; \qquad \frac{dy}{dx} = \frac{1}{(1+\alpha x)^2}, \tag{13.6}$$

where an additional parameter

$$\alpha = \frac{x_L - 1}{x_L} \tag{13.7}$$

is introduced to satisfy the first condition (13.4). Example of this modified hyperbolic curve for $x_L = 5$ is given in Figure 13.3b by the dotted line. Further modifications are needed in order to also satisfy the second condition (13.4), but these are already too cumbersome and it is better to use other functions, which are designed to satisfy all these conditions, e.g. the Ramberg-Osgood function.

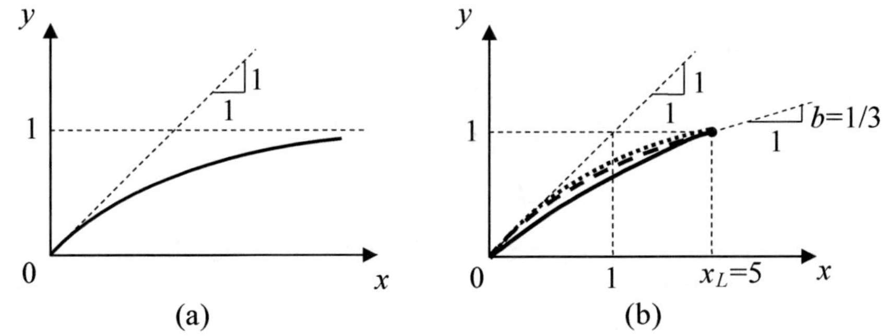

Figure 13.3 Normalized curve fitting functions: (a) original hyperbolic; (b) the modified hyperbolic, Ramberg-Osgood and logarithmic for $x_L = 5$, $b = 1/3$.

13.2.3 The Ramberg-Osgood function

The Ramberg-Osgood function was proposed by Ramberg and Osgood (1943):

$$x = y + \alpha y^R; \qquad \frac{dy}{dx} = \frac{1}{1 + \alpha R y^{R-1}}, \tag{13.8}$$

where parameters

$$\alpha = x_L - 1; \qquad R = \frac{1-b}{(x_L - 1)b} \tag{13.9}$$

are chosen to satisfy both conditions (13.3) and (13.4) in full. Example of this modified hyperbolic curve for $x_L = 5$ and $b = 1/3$ is given in Figure 13.3b by the dashed line.

The Ramberg-Osgood function was originally designed to model cyclic behaviour of soils, because parameter R could be alternatively found from the damping ratio of soil at the strain amplitude x_L. This will be explored more in detail in Section 13.3. For modelling the small strain behaviour of soils, however, the shape of the Ramberg-Osgood function is not as well fitting as that of some other functions, e.g. of the logarithmic function.

13.2.4 The logarithmic function

The logarithmic function was proposed by Puzrin and Burland (1996):

$$y = x - \alpha x[\ln(1+x)]^R \tag{13.10}$$

with parameters

$$R = \frac{(1+x_L)\ln(1+x_L)}{(x_L - 1)}\left(\frac{1}{x_L} - b\right); \qquad \alpha = \frac{x_L - 1}{x_L[\ln(1+x_L)]^R} \tag{13.11}$$

chosen to satisfy both conditions (13.3) and (13.4) in full. On top of that, however, this function has a correct shape - for realistic (for soils) values of x_L, it satisfies the following condition:

$$\left.\frac{d(y/x)}{dx}\right|_{x=0} = -\infty \tag{13.12}$$

which causes the curve to have a very high level of non-linearity at very small strains. This useful property proved to be instrumental in providing a very good fit to a large range of the small strain experimental data (Puzrin and Burland, 1996; 1998). Example of this modified hyperbolic curve for $x_L = 5$ and $b = 1/3$ is given in Figure 13.3b by the solid line.

13.3 Modelling irreversible small strain behaviour

Another important feature of the small strain behaviour of soils discovered in local deformation measurement tests is its irreversibility. The unloading and reloading stress-strain curves do not follow the initial loading curve (Figure 13.4a). Elastic behaviour, in principle, does not allow for plastic (permanent) strains, and the non-linear models presented above are elastic. This is less important when the soil element only experiences loading. But when certain stress reversals take place (e.g., in cyclic loading), modelling of the inelastic behaviour becomes an issue. In order to address this issue properly, the soil behaviour has to be considered within an elastic-plastic framework, as will be described in Chapter 20. Here we are going to present a simplified approach (sometimes called *deformational plasticity*) in which the small strain non-linear models are extended by assigning different loading, unloading and reloading shear moduli (G_L, G_U and G_R in Figure 13.4a, respectively). This can be done by adopting the so called Masing rules.

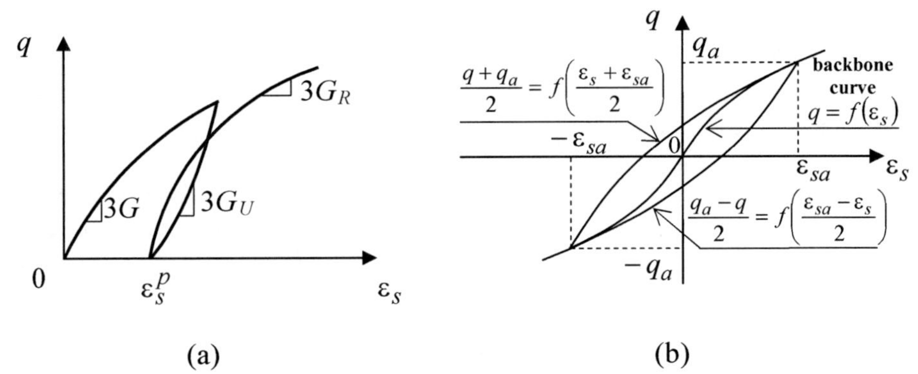

Figure 13.4 Irreversible stress-strain behaviour: (a) loading, unloading and reloading curves; (b) the Masing rules.

13.3.1 Masing rules

The Masing rules are a set of two original (Figure 13.4b) and two extended (e.g., Pyke, 1979) rules describing the one dimensional non-linear cyclic soil behaviour. Assume that the initial loading in a cyclic triaxial test can be described by the skeleton (or backbone) stress-strain curve (Figure 13.4b):

$$q = f(\varepsilon_s); \qquad 3G_t(\varepsilon_s) = \partial f(\varepsilon_s)/\partial \varepsilon_s \,. \tag{13.13}$$

This initial loading takes place up to the point *A* with deviator stress q_a, corresponding to the shear strain ε_{sa} ($q_a = f(\varepsilon_{sa})$), after which the sample was unloaded to $-q_a$ and reloaded back to q_a (Figure 13.4b). The Masing rules can be then formulated as follows (Figure 13.4b):

1. The reloading curve can be obtained by scaling the backbone curve by a factor of 2 both along the q and ε axes. The unloading curve has the same shape as the reloading one, rotated by 180 degrees:

$$\frac{q_R + q_a}{2} = f\left(\frac{\varepsilon_s + \varepsilon_{sa}}{2}\right); \qquad \frac{q_a - q_U}{2} = f\left(\frac{\varepsilon_{sa} - \varepsilon_s}{2}\right). \tag{13.14}$$

   It follows, that the second stress reversal point B is located at $(-\varepsilon_a, -q_a)$, and the reloading curve meets the backbone curve again at the point $A(\varepsilon_a, q_a)$.
2. The initial tangent shear modulus G_{t0} after all the stress reversals is the same and is equal to the initial tangent shear modulus G_0 of the backbone curve (which is in fact the consequence of the Rule 1).

The first two rules (Masing, 1926) are sufficient to describe the regular cyclic loading (with the constant amplitude) only. For the general loading, two additional rules are necessary:

3. After the reloading/unloading curve meets the backbone curve again, further reloading/unloading continues along this backbone curve.
4. In general, every time a stress-strain curve meets a curve from a previous cycle, it will follow this previous curve.

The Masing rules appear to model the experimental irreversible small strain behaviour rather well (e.g., Puzrin and Shiran, 2000), provided the backbone curve $q = f(\varepsilon_s)$ is chosen properly.

13.3.2 Modelling cyclic and dynamic small strain behaviour

Non-linear stress-strain functions, in combination with the Masing rules, allow for a simplified equivalent-linear analysis of shear wave propagation in soils, which is an important component of many seismic and dynamic geotechnical problems. Soil parameters required by this equivalent-linear analysis for each value of the strain amplitude ε_{sa} are the secant shear modulus G_s and the damping ratio ξ, which can be expressed both in true and normalized stresses and strains:

$$G_s(\varepsilon_{sa}) = \frac{q_a}{3\varepsilon_{sa}} = \frac{f(\varepsilon_{sa})}{3\varepsilon_{sa}} \quad \text{or} \quad G_s(x_a) = \frac{y(x_a)}{x_a} G_0 \tag{13.15}$$

$$\xi(\varepsilon_{sa})=\frac{2}{\pi}\left(\frac{2\int_0^{\varepsilon_{sa}} f(\varepsilon_s)d\varepsilon_s}{f(\varepsilon_{sa})\varepsilon_{sa}}-1\right) \quad \text{or} \quad \xi(x_a)=\frac{2}{\pi}\left(\frac{2\int_0^{x_a} y(x)dx}{y(x_a)x_a}-1\right) \tag{13.16}$$

The damping ratio is defined as the ratio between the W_D - energy dissipated in the closed cycle (equal to the area of the hysteresis loop in Figure 13.4b) and $4\pi W_s$, where $W_s = q_a \varepsilon_{sa}/2$ is the maximum strain energy of the equivalent elastic material.

For example, for the modified hyperbolic function (13.6) it gives:

$$G_s(x_a)=\frac{G_0}{1+\alpha x_a}; \qquad \xi(x_a)=\frac{4}{\pi}\left(1-\frac{\ln(1+\alpha x_a)}{\alpha x_a}\right)\left(1+\frac{1}{\alpha x_a}\right)-\frac{2}{\pi}. \tag{13.17}$$

For the Ramberg-Osgood function (13.8), which is more suitable for a stress controlled cyclic loading with the amplitude y_a, we obtain:

$$G_s(y_a)=\frac{G_0}{1+\alpha y_a^{R-1}}; \qquad \xi(y_a)=\frac{2}{\pi}\frac{\alpha y_a^{R-1}}{1+\alpha y_a^{R-1}}\frac{R-1}{R+1}. \tag{13.18}$$

Finally, for the logarithmic function (13.10):

$$G_s(x_a)=G_0\left(1-\alpha[\ln(1+x_a)]^R\right)$$

$$\xi(x_a)=\frac{4}{\pi}\frac{\int_0^{x_a}\left(1-\alpha[\ln(1+x)]^R\right)x\,dx}{\left(1-\alpha[\ln(1+x_a)]^R\right)x_a^2}-\frac{2}{\pi} \tag{13.19}$$

The integral in the numerator of the last expression can be elaborated to a closed form using an exponential integral function. It is preferable, however, to calculate this integral numerically, except for some particular cases, when parameter R is an integer number.

13.3.3 Comparison between the non-linear models

Hyperbolic functions (13.5) and (13.6) are not really suitable for modelling dynamic behaviour of soils, because they cannot be fitted to the experimental damping ratio data $\xi(x_a)$, and they tend to overestimate the damping ratio at the medium to larger strains quite considerably.

An advantage of the Ramberg-Osgood function (13.8) is that one of its parameters, *R*, instead of being fitted using inclination *b* of the stress-strain curve at the normalised limiting strain x_L, can be calculated using equation (13.18) to provide the prescribed damping ratio $\xi_L = \xi(x_L)$ at this strain:

$$R = \frac{2(x_L - 1) - \pi x_L \xi_L}{2(x_L - 1) + \pi x_L \xi_L} \tag{13.20}$$

Nevertheless, at very small strains the Ramberg-Osgood model tends to under-predict the damping ratios, while at the medium strains to overpredict them (Puzrin and Shiran, 2000).

Finally, for the logarithmic function (13.10), its parameter R can also be defined via $\xi_L = \xi(x_L)$. However, generally, there is no need in that: the correct shape of the logarithmic function automatically guaranties a good fit to the experimental damping ratios both at small and medium strains (Puzrin and Shiran, 2000).

13.3.4 The proper theoretical framework

The proper way to model both the non-linearity and the irreversibility of the small strain behaviour is within the framework of the multiple surfaces kinematic hardening plasticity (e.g., Puzrin and Burland, 2000). In this case, the normalised non-linear deviatoric stress-strain curve is used to calibrate the hardening rule. Within this framework, all the four Masing rules for the irreversible cyclic behaviour are satisfied automatically! Chapter 20 of this book gives additional insight into this type of models.

13.4 Thermomechanics of the small strain non-linearity

Although the small strain non-linearity introduces a certain level of complexity into the constitutive modelling of soil, following some simple rules can prevent the corresponding models from violating the Laws of Thermodynamics.

13.4.1 Reversible non-linear behaviour and Thermomechanics

If the deviatoric behaviour is considered to be non-linear but reversible, it will always satisfy the First Law of Thermodynamics, provided the model parameters such as initial shear modulus G_0 and limiting stress q_L are both pressure independent. The bulk modulus $K(p')$ can still be pressure dependent. In this case, the Helmholz free energy function

$$f(\varepsilon_v, \varepsilon_s) = \frac{p_0'}{k} e^{k(\varepsilon_v - \varepsilon_{v0})} + \frac{q_L^2}{G_0} \int_0^{\varepsilon_s G_0 / q_L} y(x) dx \tag{13.21}$$

will produce the following decoupled volumetric and deviatoric behaviour:

$$p' = \frac{\partial f}{\partial \varepsilon_v} = p_0' e^{k(\varepsilon_v - \varepsilon_{v0})}; \qquad q = \frac{\partial f}{\partial \varepsilon_s} = q_L y\left(\frac{\varepsilon_s G_0}{q_L}\right), \tag{13.22}$$

where the first expression corresponds to the linear pressure dependency of the bulk stiffness $K = \partial^2 f / \partial \varepsilon_v^2 = kp'$; while the second expression reproduces the chosen deviatoric non-linear stress-strain curve in the normalised form $y = y(x)$.

For the stress controlled behaviour, the corresponding Gibbs free energy function

$$g(p', q) = -\frac{p'}{k}\left(\ln\left(\frac{p'}{p'_0}\right) - 1\right) - (p' - p'_0)\varepsilon_{v0} - \frac{q_L^2}{G_0}\int_0^{q/q_L} x(y)dy \tag{13.23}$$

will produce the same (but inverse) decoupled volumetric and deviatoric behaviour:

$$\varepsilon_v = -\frac{\partial g}{\partial p'} = \varepsilon_{v0} + \frac{1}{k}\ln\frac{p'}{p'_0}; \qquad \varepsilon_s = -\frac{\partial g}{\partial q} = \frac{q_L}{G_0} x\left(\frac{q}{q_L}\right) \tag{13.24}$$

For example, for the modified hyperbolic function (13.6), we obtain:

$$f(\varepsilon_v, \varepsilon_s) = \frac{p'_0}{k} e^{k(\varepsilon_v - \varepsilon_{v0})} + \frac{q_L^2}{G_0 \alpha^2}\left(\alpha\varepsilon_s \frac{G_0}{q_L} - \ln\left(1 + \alpha\varepsilon_s \frac{G_0}{q_L}\right)\right) \tag{13.25}$$

For the Ramberg-Osgood function (13.8), which is more suitable for a stress controlled loading, the Gibbs free energy is:

$$\begin{aligned} g(p', q) = & -\frac{p'}{k}\left(\ln\left(\frac{p'}{p'_0}\right) - 1\right) - (p' - p'_0)\varepsilon_{v0} \\ & - \frac{q_L^2}{G_0}\left(\frac{1}{2}\left(\frac{q}{q_L}\right)^2 + \frac{\alpha}{R+1}\left(\frac{q}{q_L}\right)^{R+1}\right) \end{aligned} \tag{13.26}$$

Finally, for the logarithmic function (13.10), the Helmholz free energy is:

$$f(\varepsilon_v, \varepsilon_s) = \frac{p'_0}{k} e^{k(\varepsilon_v - \varepsilon_{v0})} + \frac{q_L^2}{G_0}\int_0^{\varepsilon_s G_0 / q_L} \left(1 - \alpha[\ln(1 + x)]^R\right) x dx \tag{13.27}$$

The integral in the last expression can again be elaborated to a closed form using an exponential integral function.

In principle, also the reversible behaviour with shear stiffness dependent on pressure can be incorporated in a thermomechanically consistent way. For example, when the initial shear modulus G_0 and limiting stress q_L are both linearly dependent on the pressure:

$$G_0(p') = g_s p'; \qquad q_L = q_l p', \tag{13.28}$$

the following Gibbs free energy function:

$$g(p',q) = -\frac{p'}{k}\left(\ln\left(\frac{p'}{p'_0}\right) - 1\right) - (p' - p'_0)\varepsilon_{v0} - \frac{q_l^2 p'}{g_s}\int_0^{q/q_l p'} x(y)dy \tag{13.29}$$

will produce the corresponding volumetric and deviatoric behaviour:

$$\varepsilon_v = -\frac{\partial g}{\partial p'} = \varepsilon_{v0} + \frac{1}{k}\ln\frac{p'}{p'_0} + \frac{q_l^2}{g_s}\int_0^{q/q_l p'} x(y)dy - \frac{q_l}{g_s p'} x\left(\frac{q}{q_l p'}\right) \tag{13.30}$$

$$\varepsilon_s = -\frac{\partial g}{\partial q} = \frac{q_l}{g_s} x\left(\frac{q}{q_l p'}\right) \tag{13.31}$$

In this model, there is a *volumetric-deviatoric coupling*, but on the isotropic axis $q = 0$ the model degenerates into a simple form with the linear pressure dependency of the bulk stiffness $K = kp'$, while for the pure shear $p' = p'_0$, the model reproduces the chosen deviatoric non-linear stress-strain curve in the normalised form $x = x(y)$.

13.4.2 Irreversible non-linear behaviour and Thermomechanics

As mentioned above, the proper way to model both the non-linearity and the irreversibility of the small strain behaviour is within the framework of the multiple surfaces kinematic hardening plasticity. This framework can be made thermomechanically consistent by applying to it the principles of *Continuous Hyperplasticity* (e.g., Puzrin et al., 2001; Puzrin and Houlsby, 2006). In this case, the resulting models will satisfy both the First and the Second Laws of Thermodynamics. The following Part III of this book is in general devoted to modelling of irreversible soil behaviour and its thermomechanical consistency.

Part III

Constitutive Modelling of Irreversible Soil Behaviour

Chapter 14

Failure

TABLE OF CONTENTS

Chapter 14

Failure

14.1 Introduction

In a typical drained triaxial compression test on slightly overconsolidated clay, after the sample is brought to certain small overconsolidation ratio $OCR = p'_c / p'_0 \leq 2$, it is sheared along the straight drained stress path with the inclination $\Delta q / \Delta p' = 3$ (Figure 14.1a).

In the initial part of this stress path, the deviatoric (Figure 14.1b) behaviour of the sample is *reversible* until the stress path reaches the yield stress state $Y(p'_y, q_y)$. Starting from this stress state the material is yielding plastically and its behaviour becomes *irreversible*. The plastic yielding continues until the soil sample reaches the *failure* stress state $F(p'_f, q_f)$. At failure, the plastic flow continues at constant stresses.

Important task in modelling the soil behaviour is to define this failure stress state. If we perform another triaxial compression test, with a different initial mean effective stress p'_0, the sample will fail at a different stress state. All possible failure stress states will form a *failure surface* $F(p'_f, q_f) = 0$. For many types of soils, their failure surfaces in the effective triaxial stress space can be described by the straight line:

$$q_f = \mathrm{M} p'_f + d_0 \tag{14.1}$$

where M and d_0 are the inclination and the intercept of the failure surface, respectively.

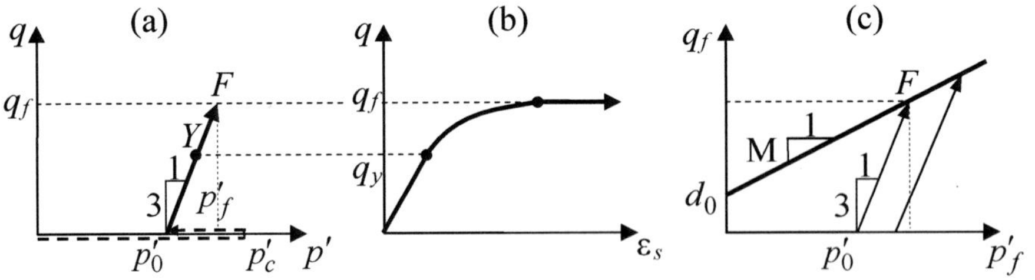

Figure 14.1: Failure in a triaxial compression test on a slightly overconsolidated clay: (a) the stress path; (b) deviatoric behaviour; (d) the failure envelope.

This chapter provides a broader theoretical background for the above relationship, including its interpretations in the principal stress-strain space, its relation to the standard soil strength parameters, and its thermomechanical consistency.

14.2 Visualisation of failure surfaces

In the principal stress space, a failure surface is given by the equation

$$F(\sigma_1, \sigma_2, \sigma_3) = 0 \tag{14.2}$$

or, if the soil is isotropic, via the stress tensor invariants

$$F(J_1, J_2, J_3) = 0 \tag{14.3}$$

The 3D visualisation of different failure surfaces (also called *failure criteria*) is most conveniently done in the Haigh-Westergaard stress space, which provides a clear geometric relationship between expressions (14.2) and (14.3).

14.2.1 The Haigh-Westergaard stress space

The stress state P in a principal stress space can be expressed in the Cartesian $P(\sigma_1, \sigma_2, \sigma_3)$ (no particular order of principal stresses has been chosen) or cylindrical $P(z, r, \theta)$ coordinates (Figure 14.2), also called the Haigh-Westergaard stress space. Advantages of the latter presentation become clear after we express the cylindrical coordinates via the principal stresses.

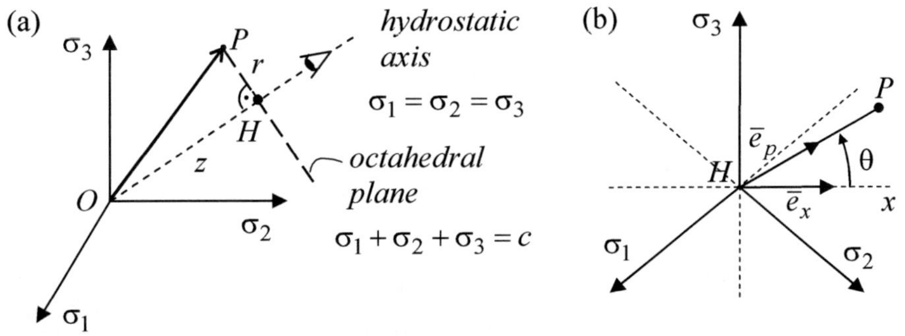

Figure 14.2: The principal stress space: (a) general; (b) octahedral plane.

First of all consider the *hydrostatic axis* (Figure 14.2a) given by the equation

$$\sigma_1 = \sigma_2 = \sigma_3 \tag{14.4}$$

and imagine a plane passing through the stress state $P(\sigma_1, \sigma_2, \sigma_3)$ perpendicular to this axis, intersecting it at the point H. Such a plane is called the octahedral plane and is given by the equation:

$$\sigma_1+\sigma_2+\sigma_3=c \quad \text{or} \quad J_1=c \tag{14.5}$$

Then, the coordinates of the intersection point H are found by solving the two equations (14.4) and (14.5) together:

$$\overline{OH}=\left\{\frac{c}{3};\frac{c}{3};\frac{c}{3}\right\}=\left\{\frac{J_1}{3};\frac{J_1}{3};\frac{J_1}{3}\right\} \tag{14.6}$$

so that

$$\overline{HP}=\left\{\sigma_1-\frac{J_1}{3};\sigma_2-\frac{J_1}{3};\sigma_3-\frac{J_1}{3}\right\} \tag{14.7}$$

where s_1, s_2 and s_3 are principal deviator stresses. In this case, the cylindrical coordinates z and r can be defined as:

$$z=\left|\overline{OH}\right|=\sqrt{\left(\frac{J_1}{3}\right)^2+\left(\frac{J_1}{3}\right)^2+\left(\frac{J_1}{3}\right)^2}=\frac{J_1}{\sqrt{3}} \tag{14.8}$$

$$r=\left|\overline{HP}\right|=\sqrt{\left(\sigma_1-\frac{J_1}{3}\right)^2+\left(\sigma_2-\frac{J_1}{3}\right)^2+\left(\sigma_3-\frac{J_1}{3}\right)^2}=\sqrt{2J_{2D}} \tag{14.9}$$

Advantages of the above two cylindrical coordinates immediately become obvious: they are proportional to the stress tensor invariants, representing generalized hydrostatic and shear behaviour, respectively. The third cylindrical coordinate, the angle θ, is measured between the rays HP and Hx in the anticlockwise direction (Figure 14.2b). Axis Hx belongs to the octahedral plane and is perpendicular to the direction σ_3. The direction of Hx is then defined by the unit vector

$$\bar{e}_x=\left\{-\frac{1}{\sqrt{2}};\frac{1}{\sqrt{2}};0\right\} \tag{14.10}$$

The scalar product of two unit vectors:

$$\cos\theta=\bar{e}_p\cdot\bar{e}_x \tag{14.11}$$

where

$$\bar{e}_p=\frac{\overline{HP}}{\left|\overline{HP}\right|}=\frac{1}{\sqrt{2J_{2D}}}\left\{\sigma_1-\frac{J_1}{3};\sigma_2-\frac{J_1}{3};\sigma_3-\frac{J_1}{3}\right\} \tag{14.12}$$

produces the following expressions for the angle θ:

$$\cos\theta = \frac{\sigma_2 - \sigma_1}{2\sqrt{J_{2D}}}; \qquad \sin\theta = \sqrt{1-\cos^2\theta} = \frac{2\sigma_3 - \sigma_1 - \sigma_2}{2\sqrt{3J_{2D}}}; \quad (14.13)$$

$$\tan\theta = -\frac{\mu}{\sqrt{3}} \quad (14.14)$$

where

$$\mu = \frac{2\sigma_3 - \sigma_1 - \sigma_2}{\sigma_1 - \sigma_2} \quad (14.15)$$

is the so called *Lode parameter*, in general defined as

$$\mu = \frac{2\sigma_{\text{int}} - \sigma_{\min} - \sigma_{\max}}{\sigma_{\min} - \sigma_{\max}} \quad (14.16)$$

where $\sigma_{\max} \geq \sigma_{\text{int}} \geq \sigma_{\min}$ are the maximal, intermediate and minimal principal stresses, respectively. Equation (14.15) is a particular case of (14.16), for the $\sigma_2 \geq \sigma_3 \geq \sigma_1$ sector of the octahedral plane (Figure 14.3), where we assumed our stress state P to be located (see also Figure 14.2b).

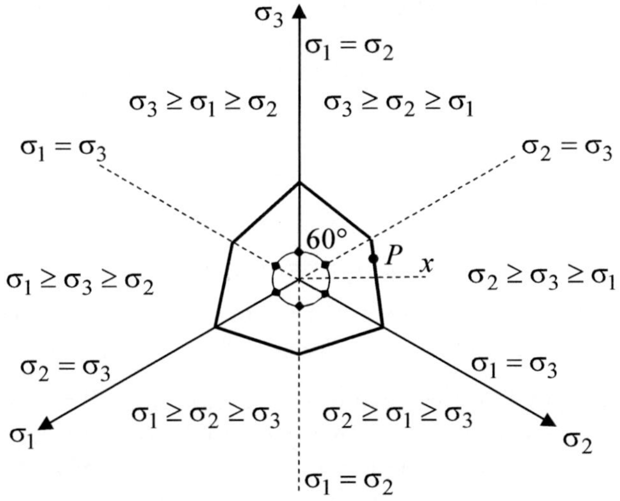

Figure 14.3: The six sectors of the octahedral plane.

This sector is 60° wide and is one of the six sectors, dividing the octahedral plane into equal parts, each having its own relationship, $\sigma_{\max} \geq \sigma_{\text{int}} \geq \sigma_{\min}$, and, consequently, its own definition (14.16) of the Lode parameter μ.

In each sector, however, the Lode parameter μ varies between $\mu = 1$ (for $\sigma_{max} \geq \sigma_{int} = \sigma_{min}$) and $\mu = -1$ (for $\sigma_{max} = \sigma_{int} \geq \sigma_{min}$), from equation (14.14) corresponding to $-30^\circ \leq \theta \leq 30^\circ$. This means that any isotropic failure surface in each sector will be self-similar (Figure 14.3). Defining μ and θ, however, differently for each sector is rather cumbersome, and it would be extremely advantageous to be able to define θ via stress invariants, which are independent of the order of the principal stresses. Luckily, this appears to be possible, and after certain mathematical elaboration we obtain:

$$\theta = -\frac{1}{3}\sin^{-1}\left(\frac{3\sqrt{3}}{2}\frac{J_{3D}}{J_{2D}^{3/2}}\right), \quad \text{for } -30^\circ \leq \theta \leq 30^\circ \tag{14.17}$$

where

$$J_{2D} = \frac{1}{6}\left[(\sigma_{max} - \sigma_{int})^2 + (\sigma_{int} - \sigma_{min})^2 + (\sigma_{max} - \sigma_{min})^2\right] \tag{14.18}$$

$$J_{3D} = \left(\sigma_{max} - \frac{J_1}{3}\right)\left(\sigma_{int} - \frac{J_1}{3}\right)\left(\sigma_{min} - \frac{J_1}{3}\right) \tag{14.19}$$

are the second and the third invariants of the deviatoric stress tensor, respectively.

Together with equations (14.8) and (14.9), equation (14.17) completes expressing of the Haigh-Westergaard cylindrical coordinates via the stress tensor invariants. This allows for an easy mathematical description, geometric visualisation and mechanical interpretation of different isotropic failure surfaces.

14.2.2 Interpretation of various shear tests in the octahedral plane

Using definition (14.16) of the Lode parameter μ, the triaxial compression and extension (Figure 14.4), correspond to $\mu = 1$ and $\mu = -1$, respectively. From equation (14.14), this results in the values of $\theta = -30^\circ$ and $\theta = 30^\circ$ (measured from the *Hx* direction in Figure 14.5) for the triaxial compression and extension, respectively. It has to be mentioned, however, stress path vectors do not belong to one octahedral plane, because the mean effective stress $p' = J_1/3$, defining the position of this plane, is changing during the test.

Another important direction on the octahedral plane is *Hx* ($\theta = 0^\circ$), bisecting the sector between the compression and extension lines, and corresponding to $\mu = 0$, i.e. $\sigma_{int} = (\sigma_{min} + \sigma_{max})/2$. Together with the condition of the constant mean effective stress during the test $p' = J_1/3 = \sigma$, this produces the following relationship between the principal stresses:

$$\sigma_{\max} = \sigma + \tau; \qquad \sigma_{\text{int}} = \sigma; \qquad \sigma_{\min} = \sigma - \tau, \tag{14.20}$$

corresponding to the pure shear test conditions (Figure 14.6). The stress path of this test entirely belongs to the octahedral plane.

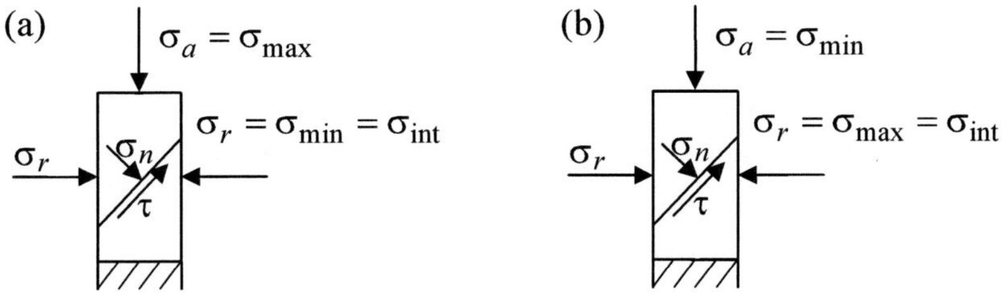

Figure 14.4: Triaxial (a) compression and (b) extensions tests.

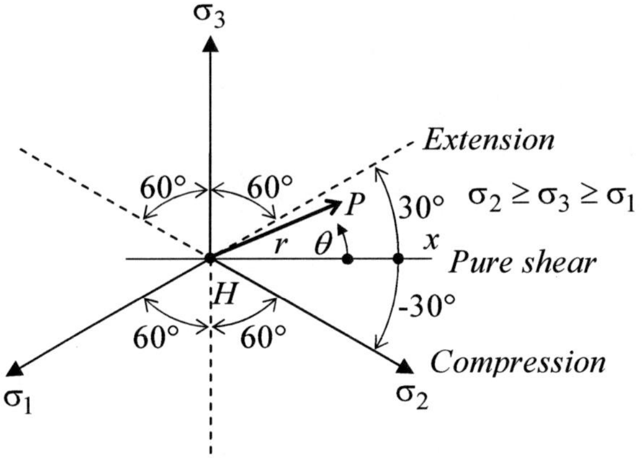

Figure 14.5: Shear tests in the octahedral plane.

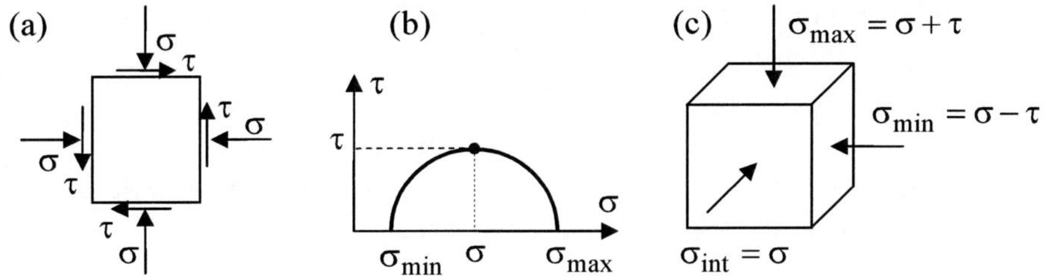

Figure 14.6: Pure shear test (a) the soil element, (b) the Mohr circle, (c) in principal stresses.

14.3 Failure criteria

14.3.1 Mohr-Coulomb failure criterion

In a drained triaxial compression test (Figure 14.7) the failure takes place when the Mohr circle touches the failure envelope, approximated by the following straight line:

$$\tau_f = c' + \sigma_{nf} \tan \varphi' \tag{14.21}$$

where τ_f and σ_{nf} are the shear and effective normal stresses on the failure plane, respectively; φ' is the effective angle of internal friction; c' is the cohesion. Equation (14.21) represents a so-called the Mohr-Coulomb failure envelope and in principal stresses can be expressed as

$$\frac{\sigma_{max} - \sigma_{min}}{2} = \frac{\sigma_{max} + \sigma_{min}}{2} \sin \varphi' + c' \cos \varphi' \tag{14.22}$$

with no dependency on the intermediate principal stress.

In the principal stress space, equation (14.22) represents a plane, which is parallel to the intermediate principal stress axis and produces a straight line at the intersection with an octahedral plane (Figure 14.8). The combination of

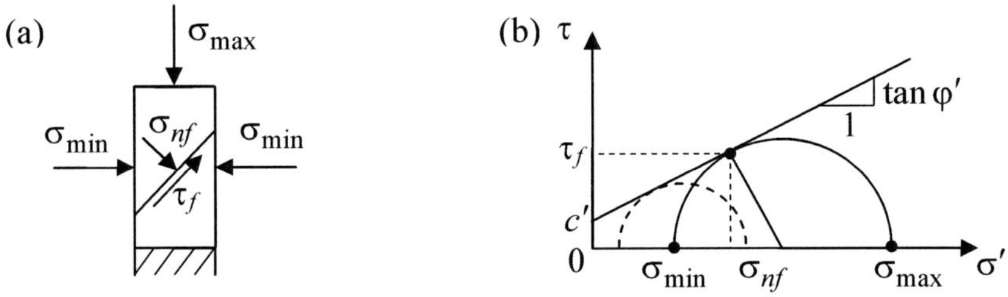

Figure 14.7: Failure in a triaxial compression test (a) the stress-state, (b) the Mohr-Coulomb failure envelope.

the six planes (each one for its own sector of the octahedral plane), forms a non-equilateral hexagonal pyramid. Rewriting expressions (14.13):

$$\frac{\sigma_{max} - \sigma_{min}}{2} = \sqrt{J_{2D}} \cos \theta ; \qquad \frac{\sigma_{max} + \sigma_{min}}{2} = \frac{J_1}{3} - \sqrt{\frac{J_{2D}}{3}} \sin \theta \tag{14.23}$$

and substituting them into (14.22), we obtain the Mohr-Coulomb failure surface expressed via the stress tensor invariants:

$$F = \frac{J_1}{3} \sin \varphi' - \sqrt{J_{2D}} \cos \theta - \frac{\sqrt{J_{2D}}}{\sqrt{3}} \sin \varphi' \sin \theta + c' \cos \varphi' = 0 \tag{14.24}$$

where $\theta = -\frac{1}{3}\sin^{-1}\left(\frac{3\sqrt{3}}{2}\frac{J_{3D}}{J_{2D}^{3/2}}\right)$, for $-30^\circ \le \theta \le 30^\circ$.

Let us explore the properties of this surface. Using definition of the mean effective stress $p' = J_1/3$, we can reformulate expression (14.24):

$$\sqrt{J_{2D}} = \sqrt{3}\,\frac{p'\sin\varphi' + c'\cos\varphi'}{\sqrt{3}\cos\theta + \sin\varphi'\sin\theta} \tag{14.25}$$

It follows, that the hexagonal pyramid has an origin on the hydrostatic axis with $p' = -c'\cot\varphi'$ and the size of its section (proportional to material strength) increases linearly with the mean effective stress.

The diameter of the largest Mohr circle at failure $(\sigma_{max} - \sigma_{min})$ is a reasonable generalized measure of the shear strength of the material. This diameter can be calculated for different testing conditions by substituting equation (14.25) into the first equation (14.23):

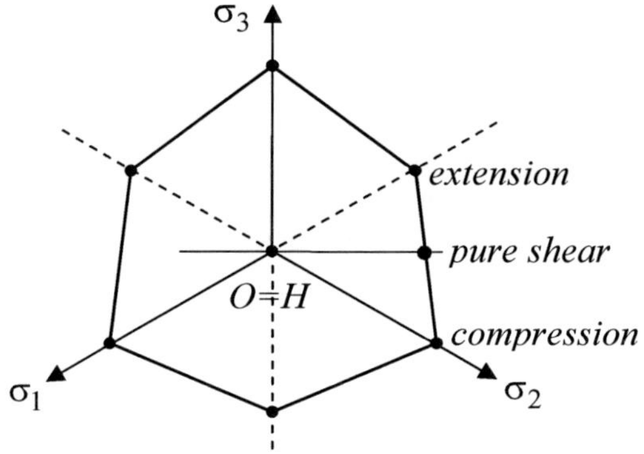

Figure 14.8: A cross-section of the Mohr-Coulomb failure surface by an octahedral plane.

- triaxial compression ($\theta = -30^\circ$):

$$\sigma_{max} - \sigma_{min} = 2\cos\theta\sqrt{J_{2D}} = \frac{6\sin\varphi'}{3-\sin\varphi'}p' + \frac{6c'\cos\varphi'}{3-\sin\varphi'} \tag{14.26}$$

- triaxial extension ($\theta = 30^\circ$):

$$\sigma_{max} - \sigma_{min} = 2\cos\theta\sqrt{J_{2D}} = \frac{6\sin\varphi'}{3+\sin\varphi'}p' + \frac{6c'\cos\varphi'}{3+\sin\varphi'} \tag{14.27}$$

- pure shear test ($\theta = 0^\circ$):

$$\sigma_{max} - \sigma_{min} = 2\cos\theta\sqrt{J_{2D}} = 2\sin\varphi' p' + 2c'\cos\varphi' \tag{14.28}$$

Note that according to the Mohr-Coulomb failure criterion, the highest strength achieved in the triaxial compression, the lowest in extension and the pure shear produces the intermediate value of shear strength. This is in a good agreement with the experimental data on soils.

For triaxial tests, expressions (14.26) and (14.27) can be generalized:

$$q_f = \mathrm{M}p'_f + d_0 \tag{14.29}$$

where

$$\mathrm{M} = \frac{6\sin\varphi'}{3\mp\sin\varphi'}; \qquad d_0 = \frac{6c'\cos\varphi'}{3\mp\sin\varphi'} \tag{14.30}$$

with the upper sign in the denominator standing for compression. Equation (14.29) is identical to the failure envelope (14.1), while equations (14.30) relate its parameters to the strength parameters of soil.

14.3.2 Drucker-Prager failure criterion

The Mohr-Coulomb failure surface is the direct extension of the Coulomb law of dry friction into the 3D principal stress space. It has been shown to provide a good fit to the experimental data in triaxial compression and extension. And yet, in spite of these significant advantages, it is not the most convenient model to use, either for analytical or numerical analysis. Analytical derivations are complicated by the extremely cumbersome expression (14.24), which this surface has in the stress tensor invariant space. Numerical analysis may run into difficulties, because the surface is not smooth – it has "corners". Both these limitations can be eliminated by using the Drucker-Prager failure criterion:

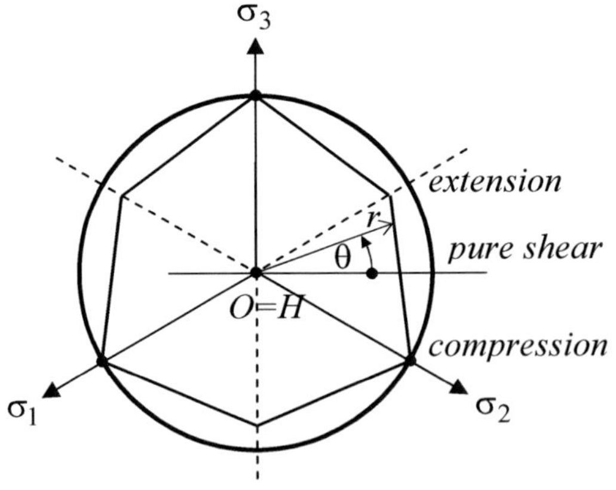

Figure 14.9: A cross-section of the Drucker-Prager failure surface by an octahedral plane.

$$F = \sqrt{J_{2D}} - \alpha J_1 - k = 0 \tag{14.31}$$

where α and k are constants.

Let us explore the properties of this surface. Using definition of the mean effective stress $p' = J_1/3$, we can reformulate expression (14.0:

$$\sqrt{J_{2D}} = 3\alpha p' + k \tag{14.32}$$

It follows, that the failure surface is a cone with its cross-section by an octahedral plane being a circle (Figure 14.9). The cone has an origin on the hydrostatic axis with $p' = -k/3\alpha$ and its diameter (proportional to material strength) increases linearly with the mean effective stress. This criterion has an extremely simple analytical expression and the surface is smooth. The limitations are that the predicted triaxial shear strengths in compression and extension ($\theta = \mp 30^\circ$) are identical:

$$\sigma_{max} - \sigma_{min} = 2\cos\theta\sqrt{J_{2D}} = 3\sqrt{3}\alpha p' + \sqrt{3}k \tag{14.33}$$

and the shear strength in pure shear ($\theta = 0^\circ$) is even larger than these two:

$$\sigma_{max} - \sigma_{min} = 2\cos\theta\sqrt{J_{2D}} = 6\alpha p' + 2k \tag{14.34}$$

which does not fit well the experimental data.

For triaxial tests, expression (14.33) can also be written in the form of the failure envelope (14.1)

$$q'_f = \mathrm{M}p'_f + d_0 \tag{14.35}$$

where

$$\mathrm{M} = 3\sqrt{3}\alpha; \qquad d_0 = \sqrt{3}k \tag{14.36}$$

which after comparison with equations (14.30) allow for the Drucker-Prager parameters to be related to the strength parameters of soil:

$$\alpha = \frac{2\sin\varphi'}{\sqrt{3}(3 \mp \sin\varphi')} \qquad k = \frac{6c'\cos\varphi'}{\sqrt{3}(3 \mp \sin\varphi')} \tag{14.37}$$

with the sign in the denominator depending on whether the cone fits the Mohr-Coulomb pyramid in compression or in extension. Normally, it is chosen to fit in compression (Figure 14.9), therefore, the upper sign should be used in the denominator.

14.3.3 Failure criteria for undrained total stress analysis

When a soil sample is tested in undrained triaxial compression, it will fail when its total stress Mohr circle touches the strength envelope, which can be approximated by a straight line, parallel to the total normal stress axis (Figure

14.10a):

$$\tau = s_u \tag{14.38}$$

where $\tau = s_u$ is the undrained shear strength.

This failure envelope can be considered as a particular case of the Mohr-Coulomb failure envelope (14.22) with $\varphi = 0$ and $c = s_u$, leading to the so called *Tresca failure criterion*:

$$\sigma_{\max} - \sigma_{\min} = 2s_u \tag{14.39}$$

which produces in the principal stress space an equilateral hexagonal prism (Figure 14.10b). Consequently, using the first equation (14.23), the Tresca failure surface can be expressed via the stress tensor invariants:

$$F = \sqrt{J_{2D}}\cos\theta - s_u = 0 \tag{14.40}$$

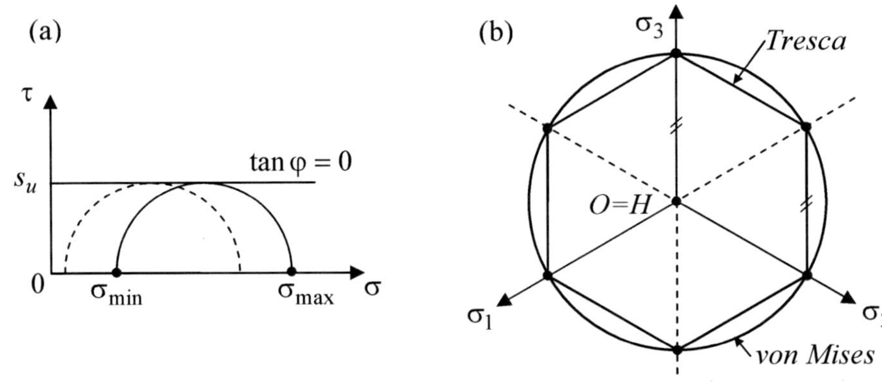

Figure 14.10: Failure in an undrained triaxial compression test: (a) failure envelope; (b) Tresca and von Mises failure criteria in an octahedral plane.

where $\theta = -\frac{1}{3}\sin^{-1}\left(\frac{3\sqrt{3}}{2}\frac{J_{3D}}{J_{2D}^{3/2}}\right)$, for $-30^\circ \le \theta \le 30^\circ$.

An interesting property of the Tresca failure criterion, is that the predicted triaxial shear strengths in compression and extension ($\theta = \mp 30^\circ$) and the shear strength in pure shear ($\theta = 0^\circ$) are all identical:

$$\sigma_{\max} - \sigma_{\min} = 2s_u \tag{14.41}$$

Alternatively, the strength envelope (14.38) can be considered as a particular case of the Drucker-Prager failure envelope (14.31) with $\alpha = 0$, in which case it produces the *von Mises failure criterion*:

$$F = \sqrt{J_{2D}} - k = 0 \tag{14.42}$$

which produces in the principal stress space a cylinder (Figure 14.10b). Parameter k is selected, using the first equation (14.23), to satisfy the condition (14.41) for triaxial compression and extension ($\theta = \mp 30^\circ$):

$$k = \sqrt{J_{2D}} = \frac{\sigma_{max} - \sigma_{min}}{2\cos\theta} = \frac{2}{\sqrt{3}} s_u \tag{14.43}$$

In this case, however, the shear strength in pure shear ($\theta = 0^\circ$) will be considerably overestimated:

$$\sigma_{max} - \sigma_{min} = 2\cos\theta\sqrt{J_{2D}} = \frac{4}{\sqrt{3}} s_u > 2 s_u \tag{14.44}$$

The Tresca and von Mises failure criteria are used in undrained total stress analysis of geotechnical boundary value problems involving rapid loading of saturated clays.

14.3.4 Generalized failure criteria

Using definition (14.18), the von Mises failure criterion (14.42)-(14.43) can be expressed in principal stresses:

$$\frac{(\sigma_{max} - \sigma_{int})^2}{s_u^2} + \frac{(\sigma_{int} - \sigma_{min})^2}{s_u^2} + \frac{(\sigma_{max} - \sigma_{min})^2}{s_u^2} = 8 \tag{14.45}$$

which emphasizes its generalized nature with respect to the Tresca failure criterion (14.39):

$$\frac{(\sigma_{max} - \sigma_{min})^2}{s_u^2} = 4 \tag{14.46}$$

For the Mohr-Coulomb failure criterion (14.22), which for $c' = 0$ can be expressed as:

$$\frac{(\sigma_{max} - \sigma_{min})^2}{\mu^2 \sigma_{max}\sigma_{min}} = 4; \quad \mu = \tan\varphi', \tag{14.47}$$

a similar generalization leads to the following expression:

$$\frac{(\sigma_{max} - \sigma_{int})^2}{\mu^2 \sigma_{max}\sigma_{int}} + \frac{(\sigma_{int} - \sigma_{min})^2}{\mu^2 \sigma_{int}\sigma_{min}} + \frac{(\sigma_{max} - \sigma_{min})^2}{\mu^2 \sigma_{max}\sigma_{min}} = 8 \tag{14.48}$$

which is known in the literature as the *Matsuoka-Nakai failure criterion*. Matsuoka and Nakai (1974) developed this criterion based on their theory of “spatially mobilized planes”. Advantage of this criterion is that it smoothes the corners of the Mohr-Coulomb criterion (Figure 14.11) and fits better the experimental data. Like other criteria it can be expressed in stress tensor invariants:

$$F = 9J_{3D} - 2J_1J_{2D} + 8\mu^2\left(J_{3D} - \frac{1}{3}J_1J_{2D} + \frac{1}{27}J_1^3\right) = 0 \qquad (14.49)$$

or in a more compact form via the coefficients of characteristic equation for stresses:

$$F = 9I_{3\sigma} - I_{1\sigma}I_{2\sigma} + 8\mu^2 I_{3\sigma} = 0 \qquad (14.50)$$

where

$$I_{1\sigma} = \sigma_1 + \sigma_2 + \sigma_3; \quad I_{2\sigma} = \sigma_1\sigma_2 + \sigma_2\sigma_3 + \sigma_1\sigma_3; \quad I_{3\sigma} = \sigma_1\sigma_2\sigma_3. \qquad (14.51)$$

Further generalisation of the von Mises and the Matsuoka-Nakai failure criteria was proposed by Houlsby (1986). After expressing the Mohr-Coulomb failure criterion (14.22) for $c' > 0$ as:

$$\frac{(\sigma_{max} - \sigma_{min})^2}{(c' + \mu\sigma_{max})(c' + \mu\sigma_{min})} = 4; \quad \mu = \tan\varphi', \qquad (14.52)$$

the *Houlsby failure criterion* is obtained by analogy with expression (14.48):

$$\frac{(\sigma_{max} - \sigma_{int})^2}{(c' + \mu\sigma_{max})(c' + \mu\sigma_{int})} + \frac{(\sigma_{int} - \sigma_{min})^2}{(c' + \mu\sigma_{int})(c' + \mu\sigma_{min})} + \frac{(\sigma_{max} - \sigma_{min})^2}{(c' + \mu\sigma_{max})(c' + \mu\sigma_{min})} = 8 \qquad (14.53)$$

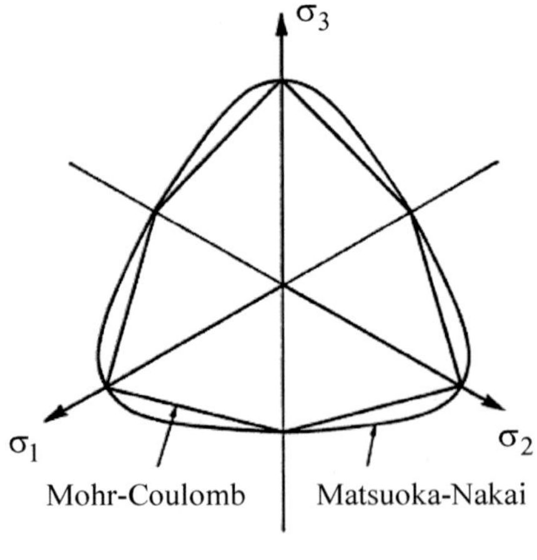

Figure 14.11: Mohr-Coulomb and Matsuoka-Nakai failure criteria in an octahedral plane.

and can also be expressed via the coefficients of characteristic equation for stresses:

$$F = 6c\left(I_{2\sigma} - I_{1\sigma}^2/3\right) + \left(9I_{3\sigma} - I_{1\sigma}I_{2\sigma}\right) + 8\left(c^3 + c^2\mu I_{1\sigma} + c\mu^2 I_{2\sigma} + \mu^3 I_{3\sigma}\right) = 0 \tag{14.54}$$

Note that for $c' = 0$ expression (14.53) degenerates to the Matsuoka-Nakai failure criterion (14.48), while for $\mu = \tan\varphi' = 0$ and $c' = s_u$ it degenerates to the von Mises failure criterion (14.45).

14.4 Thermomechanics of failure

14.4.1 Hyperplasticity

Thermomechanical consistency of any constitutive model simulating irreversible dissipative material behaviour can be validated within the framework of *Hyperplasticity*, described by Houlsby and Puzrin (2006). Main assumptions of this framework were formulated in Chapter 8 of this book. For a particular case of modelling isothermal, rate independent, stress controlled material behaviour in a triaxial stress-strain space, the *sufficient* (but not necessary) conditions for the model to satisfy the First and the Second Laws of Thermodynamics are summarized below:

1. Existence of a Gibbs free energy potential $g(p', q, \alpha_v, \alpha_s)$ for strains:

$$\varepsilon_v = -\frac{\partial g}{\partial p'}; \qquad \varepsilon_s = -\frac{\partial g}{\partial q}, \tag{14.55}$$

   where α_v and α_s are kinematic internal variables, normally associated with irreversible volumetric and deviatoric strains, respectively.

2. Existence of a dissipation function $d(p', q, \alpha_v, \alpha_s, \dot{\alpha}_v, \dot{\alpha}_s)$, non-negative and first-order homogeneous in rates of internal variables:

$$d = \frac{\partial d}{\partial \dot{\alpha}_v}\dot{\alpha}_v + \frac{\partial d}{\partial \dot{\alpha}_s}\dot{\alpha}_s \geq 0 \tag{14.56}$$

3. Functions *g* and *d* satisfy Ziegler's orthogonality condition:

$$-\frac{\partial g}{\partial \alpha_v} = \frac{\partial d}{\partial \dot{\alpha}_v}; \qquad -\frac{\partial g}{\partial \alpha_s} = \frac{\partial d}{\partial \dot{\alpha}_s}. \tag{14.57}$$

14.4.2 Hyperplastic failure surface

For a rigid plastic material, which reaches failure without elastic strains, the Gibbs free energy potential can be expressed in a very simple form:

$$g(p',q,\alpha_v,\alpha_s) = -p'\alpha_v - q\alpha_s \tag{14.58}$$

so that

$$\varepsilon_v = -\frac{\partial g}{\partial p'} = \alpha_v; \qquad \varepsilon_s = -\frac{\partial g}{\partial q} = \alpha_s \tag{14.59}$$

and the total strains are indeed fully irreversible.

Derivation of a dissipation function d for a particular failure surface F is a more complex procedure. In principle, in Hyperplasticity these two functions are related via a degenerate Legendre transformation:

$$\lambda F(p',q,\chi_p,\chi_q) = \chi_p\dot{\alpha}_v + \chi_q\dot{\alpha}_s - d(p',q,\dot{\alpha}_v,\dot{\alpha}_s) = 0 \tag{14.60}$$

where λ is a Lagrangian multiplier, and

$$\chi_p = \frac{\partial d}{\partial \dot{\alpha}_v}; \qquad \chi_q = \frac{\partial d}{\partial \dot{\alpha}_s} \tag{14.61}$$

are generalized dissipative stresses. Equation (14.60) indicates some interesting properties of a hyperplastic failure surface:

- it can be determined only to within an arbitrary multiplicative constant;
- it is expressed as a function of generalized stresses, not the true stresses;
- generalized and true stresses can be related via Ziegler's condition (14.57), which for the rigid plastic Gibbs free energy (14.58) give:

$$\chi_p = \frac{\partial d}{\partial \dot{\alpha}_v} = -\frac{\partial g}{\partial \alpha_v} = p'; \qquad \chi_q = \frac{\partial d}{\partial \dot{\alpha}_s} = -\frac{\partial g}{\partial \alpha_s} = q\,. \tag{14.62}$$

For above reasons, equation (14.60) is not very helpful in deriving a dissipation function in the true stress space for a particular form of the failure surface. It is more convenient to derive a failure surface from a dissipation function, which sometimes has to be complemented by some kind of a kinematic constraint:

$$c(\dot{\alpha}_v,\dot{\alpha}_s) = 0 \tag{14.63}$$

as will be demonstrated in the following section for the failure criteria introduced in this chapter.

14.4.3 Dissipation functions for various failure criteria

<u>*Tresca and von Mises failure criteria*</u>

In the triaxial stress space both the Tresca and von Mises failure criteria are given by:

$$F = q - 2s_u = 0 \tag{14.64}$$

For $\dot{\alpha}_s \geq 0$, this failure surface corresponds to the following dissipation function:

$$d = 2s_u \dot{\alpha}_s \geq 0 \tag{14.65}$$

Indeed, complemented by a constant volume constraint:

$$c(\dot{\alpha}_v, \dot{\alpha}_s) = \dot{\alpha}_v = 0 \tag{14.66}$$

the modified dissipation function becomes:

$$d' = d + \Lambda c = 2s_u \dot{\alpha}_s + \Lambda \dot{\alpha}_v \tag{14.67}$$

with Λ as a Lagrangian multiplier. The generalized stresses are given by

$$\chi_p = \frac{\partial d'}{\partial \dot{\alpha}_v} = \Lambda; \qquad \chi_q = \frac{\partial d'}{\partial \dot{\alpha}_s} = 2s_u, \tag{14.68}$$

which, after being related to the true stresses through Ziegler's condition (14.62), produce

$$p' = \Lambda; \qquad F = q - 2s_u = 0, \tag{14.69}$$

where the first expression indicates indeterminacy of the hydrostatic stress, while the second can be recognized as the Tresca and von Mises failure criteria in the triaxial stress space (14.64).

<u>*Mohr-Coulomb and Drucker-Prager failure criteria*</u>

In the triaxial stress space both the Mohr-Coulomb and Drucker-Prager failure criteria are given by:

$$F = q - Mp' - d_0 = 0 \tag{14.70}$$

For $\dot{\alpha}_s \geq 0$, this failure surface corresponds to the following dissipation function:

$$d = \left[\left(M - M^* \right) p' + d_0 \right] \dot{\alpha}_s \geq 0 \tag{14.71}$$

Indeed, complemented by a constant rate of dilation constraint:

$$c(\dot{\alpha}_v, \dot{\alpha}_s) = \dot{\alpha}_v + M^* \dot{\alpha}_s = 0 \tag{14.72}$$

the modified dissipation function becomes:

$$d' = d + \Lambda c = \left[\left(M - M^*\right)p' + d_0 + M^*\Lambda\right]\dot{\alpha}_s + \Lambda\dot{\alpha}_v \qquad (14.73)$$

with Λ as a Lagrangian multiplier. The generalized stresses are given by

$$\chi_p = \frac{\partial d'}{\partial \dot{\alpha}_v} = \Lambda; \qquad \chi_q = \frac{\partial d'}{\partial \dot{\alpha}_s} = \left(M - M^*\right)p' + d_0 + M^*\Lambda\,. \qquad (14.74)$$

Substitution of the first expression (14.62) into the second one produces the failure surface in the generalized stresses:

$$F = \chi_q - \left(M - M^*\right)p' - d_0 - M^*\chi_p = 0 \qquad (14.75)$$

which, after substituting Ziegler's condition (14.62), produces expression (14.70) for the Mohr-Coulomb and Drucker-Prager failure criteria in the triaxial stress space.

Matsuoka-Nakai and Houlsby failure criteria

In the principal stresses space, the Houlsby failure criterion (of which Matsuoka-Nakai is a particular case) is given by:

$$F = \frac{\left(\sigma_1 - \sigma_2\right)^2}{\left(c' + \mu\sigma_1\right)\left(c' + \mu\sigma_2\right)} + \frac{\left(\sigma_2 - \sigma_3\right)^2}{\left(c' + \mu\sigma_2\right)\left(c' + \mu\sigma_3\right)} + \frac{\left(\sigma_1 - \sigma_3\right)^2}{\left(c' + \mu\sigma_1\right)\left(c' + \mu\sigma_3\right)} - 8 = 0 \qquad (14.76)$$

It can be shown (Houlsby and Puzrin, 2006), that this failure surface corresponds to the following dissipation function:

$$d = \sqrt{\frac{8}{9}\left[A_1 + A_2 + A_3\right]} \qquad (14.77)$$

where

$$\begin{aligned} A_1 &= \left(c' + \mu\sigma_1\right)\left(c' + \mu\sigma_2\right)\left(\dot{\alpha}_1 - \dot{\alpha}_2\right)^2 \\ A_2 &= \left(c' + \mu\sigma_2\right)\left(c' + \mu\sigma_3\right)\left(\dot{\alpha}_2 - \dot{\alpha}_3\right)^2 \\ A_3 &= \left(c' + \mu\sigma_1\right)\left(c' + \mu\sigma_3\right)\left(\dot{\alpha}_1 - \dot{\alpha}_3\right)^2 \end{aligned} \qquad (14.78)$$

complemented by the constant volume constraint:

$$c\left(\dot{\alpha}_v, \dot{\alpha}_s\right) = \dot{\alpha}_1 + \dot{\alpha}_2 + \dot{\alpha}_3 = 0 \qquad (14.79)$$

14.4.4 Discussion

Existence for each of the failure criteria of a non-negative dissipation function, which together with the Gibbs free energy potential for a rigid plastic material satisfies the Ziegler's condition, is a proof that all the failure criteria introduced in this chapter do not violate the First and the Second Laws of Thermodynamics.

Certain assumptions in the above derivations have been made for simplicity and are not in fact necessary. Firstly, the proof is also valid for negative values of the deviatoric irreversible strain rates $\dot{\alpha}_s < 0$, in which case they enter dissipation functions as absolute values $|\dot{\alpha}_s|$, keeping dissipation non-negative and resulting in failure surfaces, which are symmetric with respect to p' axis. Secondly, the assumption of the rigid plastic material can also be alleviated. For an isotropic linear elastic-plastic material, the Gibbs free energy potential:

$$g(p', q, \alpha_v, \alpha_s) = -\frac{p'^2}{2K} - \frac{q^2}{6G} - p'\alpha_v - q\alpha_s \tag{14.80}$$

gives different expressions for strains:

$$\varepsilon_v = -\frac{\partial g}{\partial p'} = \frac{p'}{K} + \alpha_v; \qquad \varepsilon_s = -\frac{\partial g}{\partial q} = \frac{q}{3G} + \alpha_s, \tag{14.81}$$

but produces the same expressions of Ziegler's condition (14.62) as in the rigid case, resulting, therefore, in the same expressions of the failure surfaces and dissipation functions. Finally, although presented in the triaxial stress space, the proof is also valid for the failure surfaces expressed in stress tensor invariants and full stress tensors.

Chapter 15

Plastic Flow

TABLE OF CONTENTS

Chapter 15

Plastic Flow

15.1 Introduction

All possible failure stress states form a *failure surface* $F(\sigma_{ij})=0$. For many types of soils, their failure surfaces in the effective triaxial stress space can be described by the straight line (Figure 15.1a):

$$F(p',q)=q-\mathrm{M}p'-d=0 \tag{15.1}$$

where M and d are the inclination and the intercept of the failure surface, respectively. As the stress reaches the failure surface at $F(p_f,q_f)$, the failure takes place, and the strains keep changing at constant stresses (Figure 15.1b). This type of deformation is called the *plastic flow*. In the triaxial test, the shear strain keeps increasing (Figure 15.1c), while the volumetric strain either stays constant (e.g., for slightly overconsolidated clays) or decreases proportionally to the shear strain (e.g., for dense sands):

$$d\varepsilon_v=-\mathrm{M}^* d\varepsilon_s;\quad 0<\mathrm{M}^*<\mathrm{M}. \tag{15.2}$$

The latter phenomenon is called *dilation*.

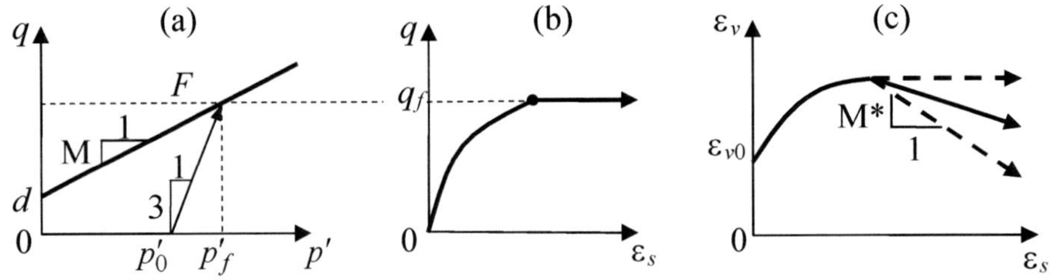

Figure 15.1: Failure in a triaxial compression test: (a) the failure envelope; (b) deviatoric behaviour; (c) the strain path.

An important task in modelling the soil behaviour is to be able to predict accurately this plastic flow. But why would one be interested in the post-failure behaviour? For a triaxial test itself this may be of a minor importance: once the sample fails, it can accommodate almost any strain increment, provided its components satisfy condition (15.2). In a boundary value problem, however, the soil does not fail simultaneously everywhere. The failure is initially confined to certain failure zones, e.g. under the foundation

A.M. Puzrin, *Constitutive Modelling in Geomechanics*,

corners (Figure 15.2a). These zones have to be in equilibrium with the surrounding soil and their deformations have to be compatible. The zones grow as the loading proceeds and contribute to the non-linear behaviour of the foundation (Figure 15.2b). At the failure load P_f, the zones connect to each other and the foundation fails.

In the process of loading, both the stress and strain states in the failure zone may change. It is important to make sure that these stress and strain increments are consistent with each other.

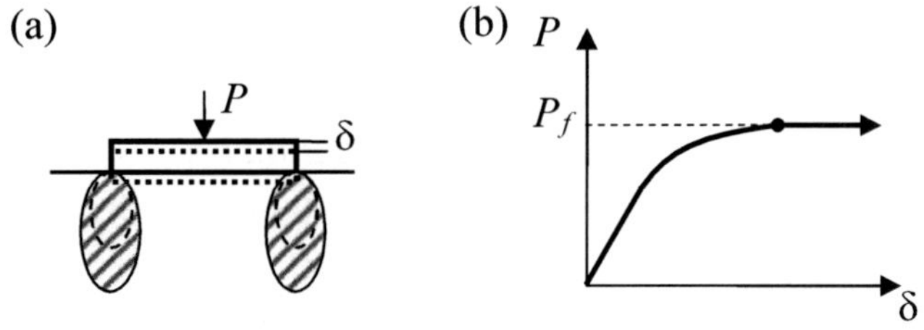

Figure 15.2: Foundation loading test: (a) the failure zones; (b) the load-displacement curve.

This chapter introduces the basic assumptions of the theory of plasticity, such as plastic strains, plastic loading, the associated flow rule, etc., and derives an incremental stress-strain response at failure both in triaxial and general stress spaces. It also explores the thermomechanical aspects of the plastic flow.

15.2 Plastic flow rule in the triaxial stress space

15.2.1 Plastic strains

If the soil behaviour at failure was elastic, the strains would not change, because the stresses at failure stay constant. Therefore, all the strain changes at failure take place due to the contribution of *plastic strains*. It is normally assumed that the total strains can be decomposed into elastic and plastic components (Figure 15.1a):

$$\varepsilon_v = \varepsilon_v^e + \varepsilon_v^p; \qquad \varepsilon_s = \varepsilon_s^e + \varepsilon_s^p, \tag{15.3}$$

where the elastic components are reversible unique functions of stresses. At this stage it is sufficient to identify plastic strains with the irrecoverable component of the total strain (Figure 15.1b), although in reality it is not always correct and we shall return to this definition later, when we talk about yielding.

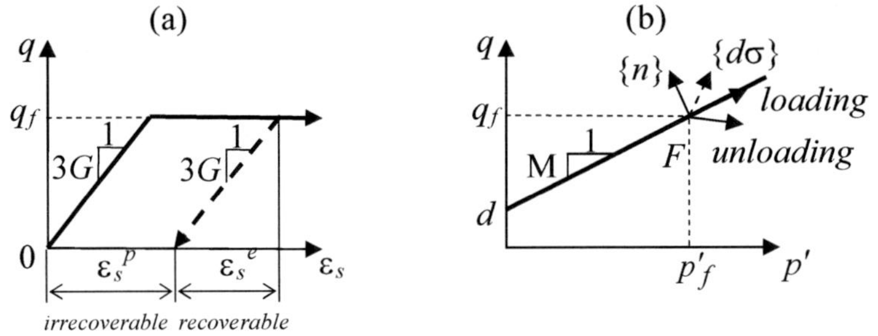

Figure 15.3: Elastic-perfectly plastic model of a triaxial compression test: (a) decomposition of strains; (b) plastic loading.

15.2.2 Plastic loading and consistency condition

Once the stress path reaches the failure envelope, the plastic flow starts at constant stresses. But what happens if after that the stress state experiences a change? The answer depends on the direction of the incremental stress vector $\{d\sigma\} = \{dp', dq\}^T$ (Figure 15.3b). First of all, no stress state outside the failure surface is statically admissible. Second, if the incremental stress vector is directed into the failure surface interior, the soil element experiences an unloading, governed by the elastic law and plastic strains experience no change. Therefore, the plastic flow continues only if the stress state after the increment stays on the failure surface. This kind of the stress increment, which causes plastic strains, is called plastic loading.

Mathematically, the plastic loading condition can be expressed in the following way. Because, both stress states before and after plastic loading should stay on the failure surface, one can write:

$$F(p', q) = F(p' + dp', q + dq) = 0 \tag{15.4}$$

or

$$dF = F(p' + dp', q + dq) - F(p', q) = 0 \tag{15.5}$$

also known as a *consistency condition*. It follows that:

$$dF = \frac{\partial F}{\partial p'} dp' + \frac{\partial F}{\partial q} dq = \{n\}^T \cdot \{d\sigma\} = 0 \tag{15.6}$$

where

$$\{n\} = \left\{ \frac{\partial F}{\partial p'}, \frac{\partial F}{\partial q} \right\}^T \tag{15.7}$$

is the outward normal to the failure surface at the current failure stress state. In other words, the plastic loading occurs if the incremental stress vector is perpendicular to the outward normal to the failure surface (Figure 15.3b). It follows that the *unloading* can be defined by the condition:

$$dF = \frac{\partial F}{\partial p'} dp' + \frac{\partial F}{\partial q} dq = \{n\}^T \cdot \{d\sigma\} < 0 \tag{15.8}$$

For the failure surface (15.1), the outward normal is given by:

$$\{n\} = \{-\mathrm{M}, 1\}^T \tag{15.9}$$

and the plastic loading occurs when

$$dF = -\mathrm{M}dp' + dq = 0 \tag{15.10}$$

i.e., $dq/dp' = \mathrm{M}$ and the plastic loading takes place when the incremental stress vector is directed along the failure envelope (Figure 15.3b).

15.2.3 Plastic work and Drucker's stability postulate

Consider a uniaxial compression test, where the sample reaches failure. In order to increase the strain by the total strain increment $d\varepsilon$, it has to be subjected to the total mechanical work increment:

$$dW = \sigma d\varepsilon = \sigma\left(d\varepsilon^e + d\varepsilon^p\right) = dU + dW^p \tag{15.11}$$

one part of which (the strain energy dU) is stored and can be later recovered, while another part - the *plastic work* increment (Figure 15.4a):

$$dW^p = \sigma d\varepsilon^p \tag{15.12}$$

is dissipated into heat.

If in response to the additional plastic strain $d\varepsilon^p$ the sample experiences a stress increment $d\sigma$ (Figure 15.4a), this plastic work increment can be calculated more accurately:

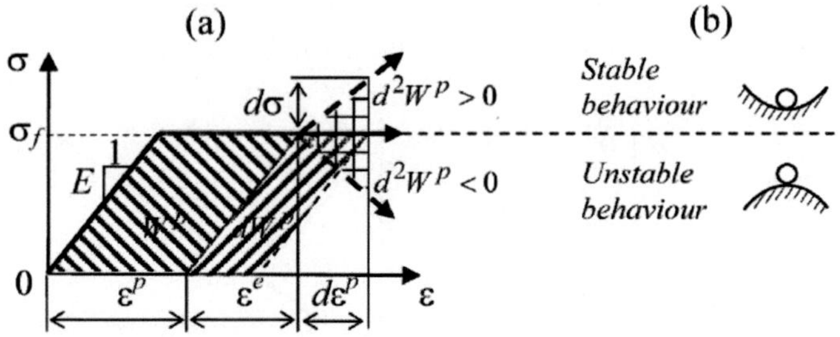

Figure 15.4: Drucker's stability postulate: (a) second order plastic work; (b) analogy with the equilibrium of a ball in a bowl.

$$dW^p = \sigma d\varepsilon^p + \frac{1}{2} d\sigma d\varepsilon^p \tag{15.13}$$

where the additional term

$$d^2W^p = \frac{1}{2} d\sigma d\varepsilon^p \tag{15.14}$$

called the *second order plastic work* term, is an important indicator of the mechanical stability of the post-failure behaviour. In fact, if $d^2W^p > 0$ the material absorbs energy and the behaviour is stable: further deformation requires additional external work and the failure is "progressive". In contrast, if $d^2W^p < 0$ the material releases energy and the behaviour is unstable: further deformation does not require additional external work and the failure is "catastrophic". These two different states are similar to the stable and unstable equilibrium of a ball at the bottom of a bowl and on the top of an overturned bowl, respectively (Figure 15.4b).

The *Drucker Stability Postulate* requires that the material behaviour at failure is stable, e.g., the second order plastic work is non-negative. In triaxial stress space this means:

$$\{d\sigma\}^T \cdot \{d\varepsilon^p\} = dp' d\varepsilon_v^p + dq d\varepsilon_s^p \geq 0 \tag{15.15}$$

In fact, there is no particular reason why the material behaviour should be stable, and many materials exhibit unstable behaviour at failure. Therefore, Drucker's postulate does not have a power of a Thermodynamic Law and simply defines a certain restricted type of material behaviour. This type, however, as we shall see below, has certain mathematical advantages.

15.2.4 Convexity, normality and the associated flow rule

The failure surface is *convex*, if it is located entirely on one side of any plane tangent to it (Figure 15.5a). For a convex failure surface, in order to satisfy both the Drucker postulate (15.15) and the consistency condition (15.6), the incremental plastic strain vector should be *parallel to the normal*:

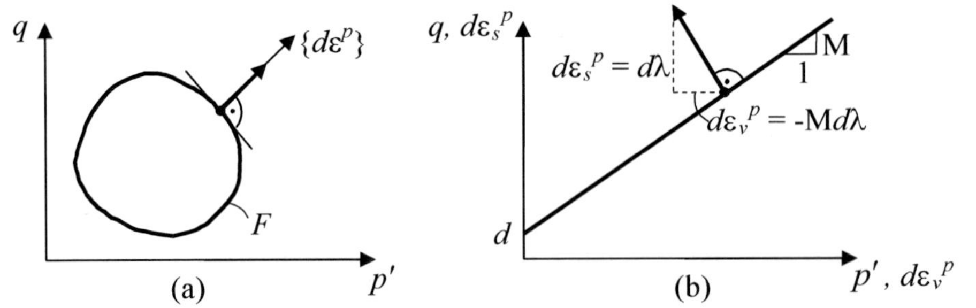

Figure 15.5: Associated flow rule: (a) convexity and normality; (b) the triaxial space example.

$$\{d\varepsilon^p\} = d\lambda\{n\} = d\lambda\left\{\frac{\partial F}{\partial p'}, \frac{\partial F}{\partial q}\right\}^T \qquad (15.16)$$

where $d\lambda$ is an infinitesimal multiplier. For the failure surface (15.1), this gives (Figure 15.5b):

$$\begin{Bmatrix} d\varepsilon_v^p \\ d\varepsilon_s^p \end{Bmatrix} = d\lambda \begin{Bmatrix} -\mathrm{M} \\ 1 \end{Bmatrix} \qquad (15.17)$$

Expression (15.16) represents the so called *associated flow rule*, which defines the direction (but not the length) of the incremental plastic strain vector in the triaxial stress space.

The associated flow rule implies that the principal directions of the incremental plastic strain stress tensor $d\varepsilon_{ij}^p$ coincide with the principal directions of the stress tensor σ_{ij}. This assumption goes against our experience with linear elasticity, where the principal directions of the incremental elastic strain stress tensor $d\varepsilon_{ij}^e$ always coincide with the principal directions of the *incremental* stress tensor $d\sigma_{ij}$.

For the failure, however, the former assumption makes more sense. Consider, for instance, a uniaxial stress state just before the failure (Figure 15.6a). Axial and radial stresses and strains are principal. If we subject the element to a tiny shear stress $\Delta\tau$ (Figure 15.6b), it will cause, of course, some small shear strain $\Delta\gamma$ (Figure 15.6c). But more importantly, it will cause a collapse of the sample, which will result in very large axial plastic strains (and proportionally large plastic radial strains), so that the very small shear strain $\Delta\gamma$ will not have any effect on the principal strain directions (Figure 15.6c). In other words, the principal directions of the incremental strain tensor $\Delta\varepsilon_{ij}^p$ will coincide with those of the stress tensor σ_{ij}.

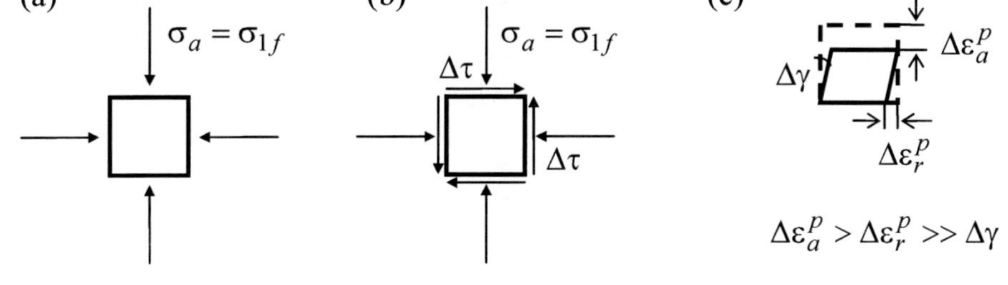

Figure 15.6: Principal directions at failure : (a) uniaxial stress state; (b) the small shear stress increment; (c) deformation of the element.

15.2.5 Incremental elastic-plastic stress-strain response

The *associated flow rule* defines only the direction of the incremental plastic strain vector. In order to find its length, we can use the following procedure. Considering the elastic part of the stress-strain relationship, using strain decomposition (15.3) and substituting plastic strain increments from the flow rule (15.16) we obtain the following expression for the incremental stresses:

$$\begin{Bmatrix} dp' \\ dq \end{Bmatrix} = \begin{bmatrix} K & 0 \\ 0 & 3G \end{bmatrix} \begin{Bmatrix} d\varepsilon_v^e \\ d\varepsilon_s^e \end{Bmatrix} = \begin{bmatrix} K & 0 \\ 0 & 3G \end{bmatrix} \begin{Bmatrix} d\varepsilon_v - d\varepsilon_v^p \\ d\varepsilon_s - d\varepsilon_s^p \end{Bmatrix} = \begin{bmatrix} K & 0 \\ 0 & 3G \end{bmatrix} \begin{Bmatrix} d\varepsilon_v - d\lambda\,\partial F/\partial p' \\ d\varepsilon_s - d\lambda\,\partial F/\partial q \end{Bmatrix} \tag{15.18}$$

Substitution of these stress increments into the consistency (plastic loading) condition (15.6) gives:

$$dF = \frac{\partial F}{\partial p'}dp' + \frac{\partial F}{\partial q}dq = Kd\varepsilon_v \frac{\partial F}{\partial p'} + 3Gd\varepsilon_s \frac{\partial F}{\partial q} - d\lambda\left(K\left(\frac{\partial F}{\partial p'}\right)^2 + 3G\left(\frac{\partial F}{\partial q}\right)^2 \right) = 0 \tag{15.19}$$

which can be resolved with respect to the multiplier $d\lambda$

$$d\lambda = \frac{Kd\varepsilon_v \dfrac{\partial F}{\partial p'} + 3Gd\varepsilon_s \dfrac{\partial F}{\partial q}}{K\left(\dfrac{\partial F}{\partial p'}\right)^2 + 3G\left(\dfrac{\partial F}{\partial q}\right)^2} \tag{15.20}$$

producing the following relationships for the incremental plastic strain components:

$$d\varepsilon_v^p = d\lambda \frac{\partial F}{\partial p'} = \frac{Kd\varepsilon_v\left(\dfrac{\partial F}{\partial p'}\right)^2 + 3Gd\varepsilon_s \dfrac{\partial F}{\partial q}\dfrac{\partial F}{\partial p'}}{K\left(\dfrac{\partial F}{\partial p'}\right)^2 + 3G\left(\dfrac{\partial F}{\partial q}\right)^2} \tag{15.21}$$

$$d\varepsilon_s^p = d\lambda \frac{\partial F}{\partial q} = \frac{Kd\varepsilon_v \dfrac{\partial F}{\partial p'}\dfrac{\partial F}{\partial q} + 3Gd\varepsilon_s\left(\dfrac{\partial F}{\partial q}\right)^2}{K\left(\dfrac{\partial F}{\partial p'}\right)^2 + 3G\left(\dfrac{\partial F}{\partial q}\right)^2} \tag{15.22}$$

Now the plastic flow is fully defined. For the case of the failure surface (15.1) we obtain:

$$d\varepsilon_v^p = \frac{K\mathrm{M}^2 d\varepsilon_v - 3G\mathrm{M} d\varepsilon_s}{K\mathrm{M}^2 + 3G}; \qquad d\varepsilon_s^p = \frac{-\mathrm{M}K d\varepsilon_v + 3G d\varepsilon_s}{K\mathrm{M}^2 + 3G}. \tag{15.23}$$

Note, that plastic strain increments do not depend on stress increments and can only be calculated when total strain increments are known. This makes sense, because at failure the plastic strains can increase at constant stresses.

Substituting expressions (15.21) and (15.22) into (15.18) we calculate *the incremental stress-strain response* at failure:

$$\begin{Bmatrix} dp' \\ dq \end{Bmatrix} = \frac{3GK}{K\left(\dfrac{\partial F}{\partial p'}\right)^2 + 3G\left(\dfrac{\partial F}{\partial q}\right)^2} \begin{bmatrix} \left(\dfrac{\partial F}{\partial q}\right)^2 & -\dfrac{\partial F}{\partial p'}\dfrac{\partial F}{\partial q} \\ -\dfrac{\partial F}{\partial p'}\dfrac{\partial F}{\partial q} & \left(\dfrac{\partial F}{\partial p'}\right)^2 \end{bmatrix} \begin{Bmatrix} d\varepsilon_v \\ d\varepsilon_s \end{Bmatrix} \tag{15.24}$$

Note, that zero stress increments do not lead to zero strain increments, as long as the strain increments satisfy the following condition:

$$\frac{\partial F}{\partial p'} d\varepsilon_s = \frac{\partial F}{\partial q} d\varepsilon_v \tag{15.25}$$

Also note, that the determinant of the elastic-plastic stiffness matrix in (15.24) is equal to zero, i.e. the incremental stress-strain relationship cannot be inverted. This is because the incremental stresses are not independent: they are related via the consistency (plastic loading) condition (15.6). The strain increments are also restricted by the following condition:

$$K d\varepsilon_v \frac{\partial F}{\partial p'} + 3G d\varepsilon_s \frac{\partial F}{\partial q} \geq 0 \tag{15.26}$$

which via equation (15.20) guarantees that $d\lambda \geq 0$, i.e. the incremental plastic strain vector is directed outwards from the failure surface.

For the case of the failure surface (15.1), the incremental stress-strain relationship (15.24) becomes:

$$\begin{Bmatrix} dp' \\ dq \end{Bmatrix} = \frac{3GK}{K\mathrm{M}^2 + 3G} \begin{bmatrix} 1 & \mathrm{M} \\ \mathrm{M} & \mathrm{M}^2 \end{bmatrix} \begin{Bmatrix} d\varepsilon_v \\ d\varepsilon_s \end{Bmatrix} \tag{15.27}$$

It follows that if failure takes place at constant stresses $(dp' = dq = 0)$, the inclination of the strain path is given by (Figure 15.1c):

$$d\varepsilon_v / d\varepsilon_s = -\mathrm{M} \tag{15.28}$$

In general, the irreversible constitutive behaviour of soils is best described via the incremental stress-strain response, where the strains are obtained by integrating strain increments along the stress path. In the case of a failure, however, the stress path is obtained by integrating the stress increments along the particular strain path.

15.3 Incremental response in the general stress space

15.3.1 Assumptions

Derivation of the incremental response at failure in the general space follows the same steps as in the previous section. The main assumptions are:

- decomposition of the strain increment:

$$d\varepsilon_{ij} = d\varepsilon_{ij}^{e} + d\varepsilon_{ij}^{p} \tag{15.29}$$

- elastic relationship:

$$d\sigma_{ij} = D_{ijkl} d\varepsilon_{kl}^{e} \tag{15.30}$$

- the failure surface:

$$F(\sigma_{ij}) = 0 \tag{15.31}$$

- consistency (plastic loading) condition:

$$dF = \frac{\partial F}{\partial \sigma_{ij}} d\sigma_{ij} = 0 \tag{15.32}$$

- and the associated flow rule:

$$d\varepsilon_{ij}^{p} = d\lambda \frac{\partial F}{\partial \sigma_{ij}} \tag{15.33}$$

15.3.2 Derivation

Considering the elastic part (15.29) of the stress-strain relationship, using strain decomposition (15.30) and substituting plastic strain increments from the flow rule (15.33) we obtain the following expression for the incremental stresses:

$$d\sigma_{ij} = D_{ijkl} d\varepsilon_{kl}^{e} = D_{ijkl}\left(d\varepsilon_{kl} - d\varepsilon_{kl}^{p}\right) = D_{ijkl}\left(d\varepsilon_{kl} - d\lambda \frac{\partial F}{\partial \sigma_{kl}}\right) \tag{15.34}$$

Substitution of these stress increments into the consistency (plastic loading) condition (15.32) gives:

$$dF = \frac{\partial F}{\partial \sigma_{ij}} d\sigma_{ij} = \frac{\partial F}{\partial \sigma_{ij}} D_{ijkl} d\varepsilon_{kl} - \frac{\partial F}{\partial \sigma_{ij}} D_{ijkl} \frac{\partial F}{\partial \sigma_{kl}} d\lambda = 0 \tag{15.35}$$

which can be resolved with respect to the multiplier $d\lambda$

$$d\lambda = \frac{\dfrac{\partial F}{\partial \sigma_{ij}} D_{ijkl} d\varepsilon_{kl}}{\dfrac{\partial F}{\partial \sigma_{ij}} D_{ijkl} \dfrac{\partial F}{\partial \sigma_{kl}}} \tag{15.36}$$

Substituting expression (15.36) into (15.34) we calculate *the incremental stress-strain response* at failure:

$$d\sigma_{ij} = D_{ijkl}^{ep} d\varepsilon_{kl} = \left(D_{ijkl}^{e} - D_{ijkl}^{p}\right) d\varepsilon_{kl} \tag{15.37}$$

where the elastic-plastic stiffness matrix can be decomposed into the elastic and plastic components:

$$D_{ijkl}^{e} = D_{ijmn}; \qquad D_{ijkl}^{p} = \frac{D_{ijmn} \dfrac{\partial F}{\partial \sigma_{mn}} \dfrac{\partial F}{\partial \sigma_{pr}} D_{prkl}}{\dfrac{\partial F}{\partial \sigma_{mn}} D_{mnpr} \dfrac{\partial F}{\partial \sigma_{pr}}}. \tag{15.38}$$

15.4 Thermomechanics of the associated plastic flow

Thermomechanical consistency of the associated plastic flow can be easily validated within the framework of Hyperplasticity, introduced in Section 14.4. For an isotropic linear elastic-plastic material, the Gibbs free energy potential:

$$g(p', q, \alpha_v, \alpha_s) = -\frac{p'^2}{2K} - \frac{q^2}{6G} - p'\alpha_v - q\alpha_s \tag{15.39}$$

leads to the *decomposition of strains*:

$$\varepsilon_v = -\frac{\partial g}{\partial p'} = \frac{p'}{K} + \alpha_v; \qquad \varepsilon_s = -\frac{\partial g}{\partial q} = \frac{q}{3G} + \alpha_s, \tag{15.40}$$

which after comparing with equations (15.3) establishes a link between the kinematic variables and plastic strains:

$$\alpha_v = \varepsilon_v^p; \qquad \alpha_s = \varepsilon_s^p. \tag{15.41}$$

In Hyperplasticity the failure surface is related to the dissipation function via a degenerate Legendre transformation:

$$\lambda F(p',q,\chi_p,\chi_q)=\chi_p\dot{\alpha}_v+\chi_q\dot{\alpha}_s-d(p',q,\dot{\alpha}_v,\dot{\alpha}_s)=0 \tag{15.42}$$

where λ is a Lagrangian multiplier, and

$$\chi_p=\frac{\partial d}{\partial\dot{\alpha}_v};\qquad \chi_q=\frac{\partial d}{\partial\dot{\alpha}_s} \tag{15.43}$$

are generalized dissipative stresses. Differentiating expression (15.42) gives:

$$\dot{\alpha}_v=\lambda\frac{\partial F(p',q,\chi_p,\chi_q)}{\partial\chi_p};\qquad \dot{\alpha}_s=\lambda\frac{\partial F(p',q,\chi_p,\chi_q)}{\partial\chi_p}, \tag{15.44}$$

which can be recognized as the *associated flow rule*, but with the *normality* satisfied in the generalized stress space and not in the true stress space.

For an isotropic linear elastic-plastic material, the Gibbs free energy potential (15.39) produces the following Ziegler's condition:

$$\chi_p=\frac{\partial d}{\partial\dot{\alpha}_v}=-\frac{\partial g}{\partial\alpha_v}=p';\qquad \chi_q=\frac{\partial d}{\partial\dot{\alpha}_s}=-\frac{\partial g}{\partial\alpha_s}=q \tag{15.45}$$

and from expressions (15.44) it follows, that *the associated flow rule in the true stress space always satisfies the Laws of Thermodynamics only if the failure surface does not depend on true stresses*:

$$F(p',q,\chi_p,\chi_q)=F(\chi_p,\chi_q)=0 \tag{15.46}$$

which, for example, is the case for the Tresca and von Mises failure criteria (14.68):

$$F=\chi_q-2s_u=0 \tag{15.47}$$

For Mohr-Coulomb and Drucker-Prager failure criteria (14.75), however, condition (15.46) is not satisfied:

$$F=\chi_q-(M-M^*)p'-d_0-M^*\chi_p=0 \tag{15.48}$$

and equations (15.44) combined with (15.41) and (15.45) produce a flow rule

$$d\varepsilon_v^p=d\alpha_v=d\lambda\frac{\partial F}{\partial\chi_p}=-M^*;\qquad d\varepsilon_s^p=d\alpha_s=d\lambda\frac{\partial F}{\partial\chi_q}=1 \tag{15.49}$$

which differs from the associated flow rule (15.17):

$$\begin{Bmatrix} d\varepsilon_v^p \\ d\varepsilon_s^p \end{Bmatrix}=d\lambda\begin{Bmatrix} -\mathrm{M} \\ 1 \end{Bmatrix} \tag{15.50}$$

Thermomechanical inconsistency of the associated flow rule for frictional materials will be discussed in more detail in the following chapter.

Chapter 16

Dilatancy

TABLE OF CONTENTS

Chapter 16

Dilatancy

16.1 Introduction

What kind of a strain path would predict an elastic-plastic model in triaxial compression? Consider the failure surface (Figure 16.1a):

$$F(p',q) = q - \mathrm{M}p' - d = 0 \tag{16.1}$$

and the corresponding associated flow rule:

$$d\varepsilon_v^p = d\lambda \frac{\partial F}{\partial p'} = -d\lambda \mathrm{M}; \qquad d\varepsilon_s^p = d\lambda \frac{\partial F}{\partial q} = d\lambda \tag{16.2}$$

introduced in the previous chapters. Before the stress state reaches the failure surface, the behaviour is linear elastic (Figure 16.1b), and the inclination of the total strain path (Figure 16.1c) can be calculated from the incremental isotropic linear elastic law:

$$\frac{d\varepsilon_v}{d\varepsilon_s} = \frac{d\varepsilon_v^e}{d\varepsilon_s^e} = \frac{dp'/K}{dq/3G} = \frac{G}{K} \tag{16.3}$$

using the fact that the inclination of the effective stress path $dq/dp' = 3$.

At failure, the strain increments are purely plastic, because the stresses are constant, and the inclination of t he total strain path (Figure 16.1c) can be calculated from the associated flow rule (16.2):

$$\frac{d\varepsilon_v}{d\varepsilon_s} = \frac{d\varepsilon_v^p}{d\varepsilon_s^p} = -\mathrm{M} \tag{16.4}$$

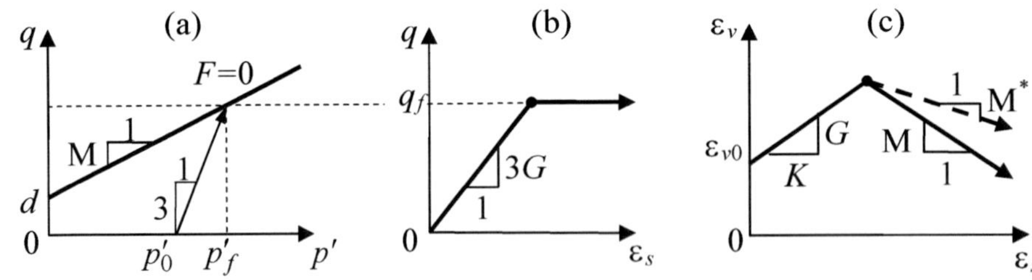

Figure 16.1: Elastic-plastic model of a triaxial compression test: (a) the failure envelope; (b) deviatoric behaviour; (c) the strain path.

That is, the shear strain keeps increasing, while the volumetric decreases proportionally to the shear strain. This increase of volume during shearing is called *dilatancy* and is indeed observed at failure (e.g., in dense sands):

$$d\varepsilon_v = -\mathrm{M}^* d\varepsilon_s \tag{16.5}$$

However, in reality, the experimentally obtained proportionality coefficient M^* is normally smaller than M, or can even be equal to zero (e.g., in slightly overconsolidated clays). I.e., the associated flow rule overestimates dilatancy, which in kinematically constrained problems may lead to the overestimation of the confining pressure, and consequently, of the strength.

How can the plastic flow framework introduced in the previous chapter be modified in order to incorporate the more realistic dilatancy? This is the topic of the present chapter, which introduces the non-associated flow rule, and derives an incremental stress-strain response at failure both in triaxial and general stress spaces. It also explores the thermomechanical aspects of the non-associated plastic flow.

16.2 Non-associated plastic flow in the triaxial stress space

16.2.1 Plastic potential and non-associated flow rule

The associated flow rule postulates the normality of the incremental plastic strain vector to the failure surface. This normality follows from the convexity of the failure surface, the Drucker stability postulate and the consistency condition. However, this normality overestimates dilatancy. Accurate modelling of dilatancy, therefore, requires that the incremental plastic strain vector is not normal to the failure surface. The best way to describe this mathematically is to introduce an additional family of surfaces, which is called the *plastic potential*. The plastic potential family $P(p', q, x)$ has the following properties (Figure 16.2a):

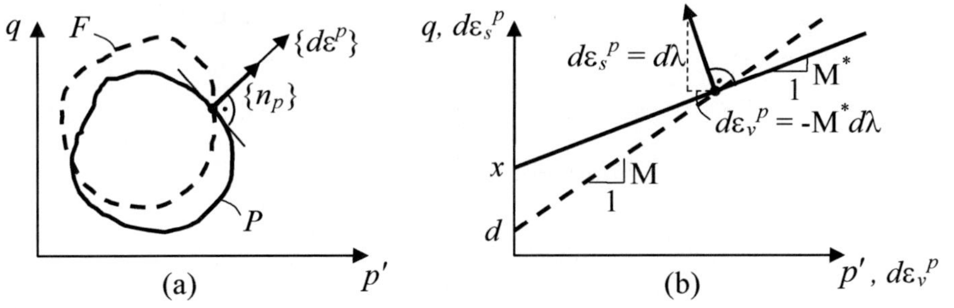

Figure 16.2: Non-associated flow rule: (a) the failure surface and the plastic potential; (b) the triaxial space example.

(a) through each point of the failure surface passes one of the surfaces of the plastic potential family (this allows to define the parameter x);

(b) the incremental plastic strain vector at this point is normal to the plastic potential and not to the failure surface.

Mathematically this can be described as

$$\{d\varepsilon^P\} = d\lambda\{n_p\} = d\lambda\left\{\frac{\partial P}{\partial p'}, \frac{\partial P}{\partial q}\right\}^T \tag{16.6}$$

where $d\lambda$ is an infinitesimal multiplier; $\{n_p\}$ is the normal to the plastic potential surface. Expression (16.6) represents the so called *non-associated flow rule*, which again defines the direction (but not the length) of the incremental plastic strain vector in the triaxial stress space.

For the failure surface (16.1), we can define the plastic potential also as a straight line (Figure 16.2b):

$$P(p', q) = q - \mathrm{M}^* p' - x = 0 \tag{16.7}$$

For each point (p'_f, q_f) on the failure surface F (16.1), the parameter x is defined to ensure that the plastic potential passes through this point:

$$x = \left(\mathrm{M} - \mathrm{M}^*\right)p'_f + d \tag{16.8}$$

The corresponding non-associated flow rule is given by:

$$\begin{Bmatrix} d\varepsilon_v^P \\ d\varepsilon_s^P \end{Bmatrix} = d\lambda \begin{Bmatrix} -\mathrm{M}^* \\ 1 \end{Bmatrix} \tag{16.9}$$

16.2.2 Drucker's stability postulate

The *Drucker Stability Postulate* requires that the material behaviour at failure is stable, e.g., the second order plastic work is non-negative. In triaxial stress space this means:

$$\{d\sigma\}^T \cdot \{d\varepsilon^P\} = dp' d\varepsilon_v^P + dq d\varepsilon_s^P \geq 0 \tag{16.10}$$

The *plastic loading condition* requires that:

$$dF = \frac{\partial F}{\partial p'} dp' + \frac{\partial F}{\partial q} dq = \{n\}^T \cdot \{d\sigma\} = 0 \tag{16.11}$$

where

$$\{n\} = \left\{ \frac{\partial F}{\partial p'}, \frac{\partial F}{\partial q} \right\}^T \tag{16.12}$$

is the outward normal to the failure surface at the current failure stress state. In the associated flow rule incremental plastic strain vector should be parallel to this normal:

$$\{d\varepsilon^p\} = d\lambda\{n\} = d\lambda \left\{ \frac{\partial F}{\partial p'}, \frac{\partial F}{\partial q} \right\}^T \tag{16.13}$$

so that, according to the plastic loading condition:

$$\{d\sigma\}^T \cdot \{d\varepsilon^p\} = \{d\sigma\}^T \cdot d\lambda\{n\} = 0 \tag{16.14}$$

i.e., the Drucker stability postulate is always satisfied.

For the non-associated flow rule, however, the second order plastic work $\{d\sigma\}^T \cdot \{d\varepsilon^p\}$ can become negative. For example, for the failure surface (16.1), the incremental stress vector directed downwards along the failure envelope:

$$\begin{Bmatrix} dp' \\ dq \end{Bmatrix} = d\lambda \begin{Bmatrix} -1 \\ -\mathrm{M} \end{Bmatrix} \tag{16.15}$$

satisfies the plastic loading condition

$$dF = -\mathrm{M}dp' + dq = 0 \tag{16.16}$$

but when substituted into the Drucker stability postulate together with the non-associated flow rule:

$$\begin{Bmatrix} d\varepsilon_v^p \\ d\varepsilon_s^p \end{Bmatrix} = d\lambda \begin{Bmatrix} -\mathrm{M}^* \\ 1 \end{Bmatrix} \tag{16.17}$$

it produces a negative second order plastic work

$$\{d\sigma\}^T \cdot \{d\varepsilon^p\} = dp' d\varepsilon_v^p + dq d\varepsilon_s^p = (d\lambda)^2 \left(\mathrm{M}^* - \mathrm{M}\right) < 0 \tag{16.18}$$

because $\mathrm{M}^* < \mathrm{M}$. That is, Drucker's postulate is violated and the material behaviour is unstable. However, as mentioned before, this postulate does not have a power of a Thermodynamic Law and simply defines a certain restricted type of material behaviour. Later in this Chapter it will be shown that exactly because of the fact that this postulate is violated, the Laws of the Thermodynamics are satisfied!

16.2.3 Incremental elastic-plastic stress-strain response

The *non-associated flow rule* also defines only the direction of the incremental plastic strain vector. In order to find its length, we use the procedure similar to that in the previous Chapter. Considering the elastic part of the stress-strain relationship, using strain decomposition

$$\varepsilon_v = \varepsilon_v^e + \varepsilon_v^p ; \qquad \varepsilon_s = \varepsilon_s^e + \varepsilon_s^p \tag{16.19}$$

and substituting plastic strain increments from the flow rule (16.6) we obtain the following expression for the incremental stresses:

$$\begin{Bmatrix} dp' \\ dq \end{Bmatrix} = \begin{bmatrix} K & 0 \\ 0 & 3G \end{bmatrix} \begin{Bmatrix} d\varepsilon_v^e \\ d\varepsilon_s^e \end{Bmatrix} = \begin{bmatrix} K & 0 \\ 0 & 3G \end{bmatrix} \begin{Bmatrix} d\varepsilon_v - d\varepsilon_v^p \\ d\varepsilon_s - d\varepsilon_s^p \end{Bmatrix}$$
$$= \begin{bmatrix} K & 0 \\ 0 & 3G \end{bmatrix} \begin{Bmatrix} d\varepsilon_v - d\lambda\, \partial P/\partial p' \\ d\varepsilon_s - d\lambda\, \partial P/\partial q \end{Bmatrix} \tag{16.20}$$

Substitution of these stress increments into the consistency (plastic loading) condition (16.11) gives:

$$dF = \frac{\partial F}{\partial p'} dp' + \frac{\partial F}{\partial q} dq$$
$$= K d\varepsilon_v \frac{\partial F}{\partial p'} + 3G d\varepsilon_s \frac{\partial F}{\partial q} - d\lambda \left(K \frac{\partial F}{\partial p'} \frac{\partial P}{\partial p'} + 3G \frac{\partial F}{\partial q} \frac{\partial P}{\partial q} \right) = 0 \tag{16.21}$$

which can be resolved with respect to the multiplier $d\lambda$

$$d\lambda = \frac{K d\varepsilon_v \dfrac{\partial F}{\partial p'} + 3G d\varepsilon_s \dfrac{\partial F}{\partial q}}{K \dfrac{\partial F}{\partial p'} \dfrac{\partial P}{\partial p'} + 3G \dfrac{\partial F}{\partial q} \dfrac{\partial P}{\partial q}} \tag{16.22}$$

producing the following relationships for the incremental plastic strain components:

$$d\varepsilon_v^p = d\lambda \frac{\partial P}{\partial p'} = \frac{K d\varepsilon_v \dfrac{\partial F}{\partial p'} \dfrac{\partial P}{\partial p'} + 3G d\varepsilon_s \dfrac{\partial F}{\partial q} \dfrac{\partial P}{\partial p'}}{K \dfrac{\partial F}{\partial p'} \dfrac{\partial P}{\partial p'} + 3G \dfrac{\partial F}{\partial q} \dfrac{\partial P}{\partial q}} \tag{16.23}$$

$$d\varepsilon_s^p = d\lambda \frac{\partial P}{\partial q} = \frac{K d\varepsilon_v \dfrac{\partial F}{\partial p'} \dfrac{\partial P}{\partial q} + 3G d\varepsilon_s \dfrac{\partial F}{\partial q} \dfrac{\partial P}{\partial q}}{K \dfrac{\partial F}{\partial p'} \dfrac{\partial P}{\partial p'} + 3G \dfrac{\partial F}{\partial q} \dfrac{\partial P}{\partial q}} \tag{16.24}$$

The plastic flow is fully defined. For the case of the failure surface (16.1) and plastic potential (16.7) we obtain:

$$d\varepsilon_v^p = \frac{KMM^* d\varepsilon_v - 3GM^* d\varepsilon_s}{KMM^* + 3G} \qquad d\varepsilon_s^p = \frac{-MKd\varepsilon_v + 3Gd\varepsilon_s}{KMM^* + 3G} \tag{16.25}$$

Note, that plastic strain increments do not depend on stress increments and can only be calculated when total strain increments are known. This makes sense, because at failure the plastic strains can increase at constant stresses.

Substituting expressions (16.23) and (16.24) into (16.20) we calculate *the incremental stress-strain response* at failure:

$$\begin{Bmatrix} dp' \\ dq \end{Bmatrix} = \frac{3GK}{K\dfrac{\partial F}{\partial p'}\dfrac{\partial P}{\partial p'} + 3G\dfrac{\partial F}{\partial q}\dfrac{\partial P}{\partial q}} \begin{bmatrix} \dfrac{\partial F}{\partial q}\dfrac{\partial P}{\partial q} & -\dfrac{\partial P}{\partial p'}\dfrac{\partial F}{\partial q} \\ -\dfrac{\partial F}{\partial p'}\dfrac{\partial P}{\partial q} & \dfrac{\partial F}{\partial p'}\dfrac{\partial P}{\partial p'} \end{bmatrix} \begin{Bmatrix} d\varepsilon_v \\ d\varepsilon_s \end{Bmatrix} \tag{16.26}$$

Note, that zero stress increments do not lead to zero strains increments, as long as the strain increments satisfy the following condition:

$$\frac{\partial P}{\partial p'} d\varepsilon_s = \frac{\partial P}{\partial q} d\varepsilon_v \tag{16.27}$$

which is different to that in the associated flow rule. Also note, that like in the associated flow rule, the determinant of the elastic-plastic stiffness matrix in (16.26) is equal to zero, i.e. the incremental stress-strain relationship cannot be inverted. This is because the incremental stresses are not independent: they are related via the consistency (plastic loading) condition (16.11). The strain increments are also restricted by the condition:

$$d\lambda \geq 0 \tag{16.28}$$

which via equation (16.22) guarantees that the incremental plastic strain vector is directed outwards from the plastic potential surface.

For the case of the failure surface (16.1) and plastic potential (16.7), the incremental stress-strain relationship (16.26) becomes:

$$\begin{Bmatrix} dp' \\ dq \end{Bmatrix} = \frac{3GK}{KMM^* + 3G} \begin{bmatrix} 1 & M^* \\ M & MM^* \end{bmatrix} \begin{Bmatrix} d\varepsilon_v \\ d\varepsilon_s \end{Bmatrix} \tag{16.29}$$

It follows that if the failure takes place at constant stresses $(dp' = dq = 0)$, the inclination of the strain path is given by:

$$d\varepsilon_v / d\varepsilon_s = -M^* \tag{16.30}$$

meaning that the dilatancy measured in the experiments can be now modelled accurately (Figure 16.1c).

16.2.4 Angle of dilation and asymmetry of the stiffness matrix

For triaxial compression, parameter M is defined:

$$M = \frac{6\sin\varphi'}{3-\sin\varphi'} \tag{16.31}$$

where φ' is the effective angle of internal friction. By analogy, we define

$$M^* = \frac{6\sin\psi}{3-\sin\psi} \tag{16.32}$$

where ψ is the *angle of dilation* ($0 \le \psi < \varphi'$). The physical meaning of the both angles follows from the direct shear test (Figure 16.3).

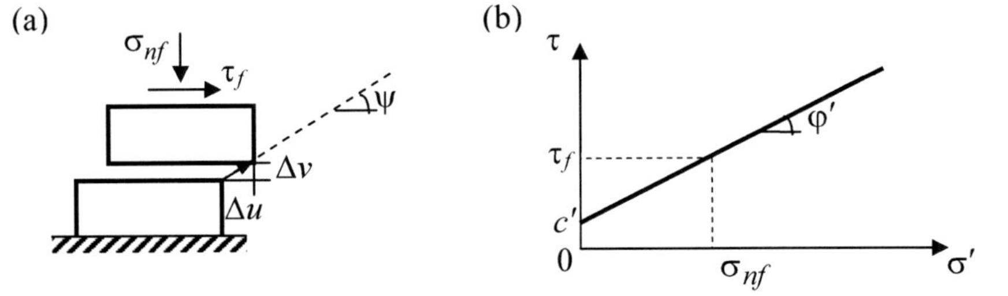

Figure 16.3: Direct shear test: (a) the angle of dilation; (b) the angle of friction.

Note that because $0 \le \psi < \varphi'$, unlike in the associated flow rule, the stiffness matrices in the incremental stress-strain relationships (16.26) and (16.29) are not symmetric, which may lead to certain computational difficulties in the FEM analysis.

16.3 Incremental response in the general stress space

16.3.1 Assumptions

Derivation of the incremental response at failure in the general space follows the same steps as in the previous section. The main assumptions are:

- decomposition of the strain increment:

$$d\varepsilon_{ij} = d\varepsilon_{ij}^e + d\varepsilon_{ij}^p \tag{16. 33}$$

- elastic relationship:

$$d\sigma_{ij} = D_{ijkl} d\varepsilon_{kl}^e \tag{16.34}$$

- the failure surface:

$$F(\sigma_{ij}) = 0 \tag{16.35}$$

- consistency (plastic loading) condition:

$$dF = \frac{\partial F}{\partial \sigma_{ij}} d\sigma_{ij} = 0 \tag{16.36}$$

- the plastic potential:

$$P(\sigma_{ij}, x) = 0 \tag{16.37}$$

- and the associated flow rule:

$$d\varepsilon_{ij}^{p} = d\lambda \frac{\partial P}{\partial \sigma_{ij}} \tag{16.38}$$

16.3.2 Derivation

Considering the elastic part (16.34) of the stress-strain relationship, using strain decomposition (16.33) and substituting plastic strain increments from the flow rule (16.38) we obtain the following expression for the incremental stresses:

$$d\sigma_{ij} = D_{ijkl} d\varepsilon_{kl}^{e} = D_{ijkl}\left(d\varepsilon_{kl} - d\varepsilon_{kl}^{p}\right) = D_{ijkl}\left(d\varepsilon_{kl} - d\lambda \frac{\partial P}{\partial \sigma_{kl}}\right) \tag{16.39}$$

Substitution of these stress increments into the consistency (plastic loading) condition (16.36) gives:

$$dF = \frac{\partial F}{\partial \sigma_{ij}} d\sigma_{ij} = \frac{\partial F}{\partial \sigma_{ij}} D_{ijkl} d\varepsilon_{kl} - \frac{\partial F}{\partial \sigma_{ij}} D_{ijkl} \frac{\partial P}{\partial \sigma_{kl}} d\lambda = 0 \tag{16.40}$$

which can be resolved with respect to the multiplier $d\lambda$

$$d\lambda = \frac{\dfrac{\partial F}{\partial \sigma_{ij}} D_{ijkl} d\varepsilon_{kl}}{\dfrac{\partial F}{\partial \sigma_{ij}} D_{ijkl} \dfrac{\partial P}{\partial \sigma_{kl}}} \tag{16.41}$$

Substituting expression (16.41) into (16.39) we calculate *the incremental stress-strain response* at failure:

$$d\sigma_{ij} = D_{ijkl}^{ep} d\varepsilon_{kl} = \left(D_{ijkl}^{e} - D_{ijkl}^{p}\right) d\varepsilon_{kl} \tag{16.42}$$

where the elastic-plastic stiffness matrix can be decomposed into the elastic and plastic components:

$$D^e_{ijkl} = D_{ijmn}; \qquad D^p_{ijkl} = \frac{D_{ijmn} \dfrac{\partial P}{\partial \sigma_{mn}} \dfrac{\partial F}{\partial \sigma_{pr}} D_{prkl}}{\dfrac{\partial F}{\partial \sigma_{mn}} D_{mnpr} \dfrac{\partial P}{\partial \sigma_{pr}}}. \tag{16.43}$$

16.4 Thermomechanics of a non-associated plastic flow

The hyperplastic framework is used below to demonstrate thermomechanical consistency of a non-associated plastic flow for frictional materials. For an isotropic linear elastic-plastic material, the Gibbs free energy potential:

$$g(p', q, \alpha_v, \alpha_s) = -\frac{p'^2}{2K} - \frac{q^2}{6G} - p'\alpha_v - q\alpha_s \tag{16.44}$$

leads to the *decomposition of strains*:

$$\varepsilon_v = -\frac{\partial g}{\partial p'} = \frac{p'}{K} + \alpha_v; \qquad \varepsilon_s = -\frac{\partial g}{\partial q} = \frac{q}{3G} + \alpha_s \tag{16.45}$$

and produces the following Ziegler's condition:

$$\chi_p = \frac{\partial d}{\partial \dot{\alpha}_v} = -\frac{\partial g}{\partial \alpha_v} = p'; \qquad \chi_q = \frac{\partial d}{\partial \dot{\alpha}_s} = -\frac{\partial g}{\partial \alpha_s} = q. \tag{16.46}$$

In the triaxial true stress space both the Mohr-Coulomb and Drucker-Prager failure criteria are given by:

$$F = q - Mp' - d_0 = 0 \tag{16.47}$$

As shown in Section 14.4, this failure surface corresponds to the following dissipation function:

$$d = \left[\left(M - M^* \right) p' + d_0 \right] \left| \dot{\alpha}_s \right| \ge 0 \tag{16.48}$$

complemented by a constant rate of dilation constraint:

$$c(\dot{\alpha}_v, \dot{\alpha}_s) = \dot{\alpha}_v + M^* \left| \dot{\alpha}_s \right| = 0 \tag{16.49}$$

The modified dissipation function then becomes:

$$d' = d + \Lambda c = \left[\left(M - M^* \right) p' + d_0 + M^* \Lambda \right] \left| \dot{\alpha}_s \right| + \Lambda \dot{\alpha}_v \tag{16.50}$$

with Λ as a Lagrangian multiplier, and the generalized stresses are given by

$$\chi_p = \frac{\partial d'}{\partial \dot{\alpha}_v} = \Lambda; \qquad \left| \chi_q \right| = \frac{\partial d'}{\partial \dot{\alpha}_s} = \left(M - M^* \right) p' + d_0 + M^* \Lambda. \tag{16.51}$$

Substitution of the first expression (16.51) into the second one produces the failure surface in the generalized stresses:

$$F = |\chi_q| - (M - M^*)p' - d_0 - M^*\chi_p = 0 \tag{16.52}$$

which, after substituting Ziegler's condition (16.46), produces expression (16.47) for the Mohr-Coulomb and Drucker-Prager failure criteria in the triaxial stress space.

As shown in Section 15.4, in hyperplastic formulation *normality* is satisfied in the generalized stress space and not in the true stress space:

$$\dot{\alpha}_v = \lambda \frac{\partial F(p', q, \chi_p, \chi_q)}{\partial \chi_p}; \qquad \dot{\alpha}_v = \lambda \frac{\partial F(p', q, \chi_p, \chi_q)}{\partial \chi_p} \tag{16.53}$$

automatically leading for the failure surface (16.52) to a non-associated flow rule

$$d\varepsilon_v^p = d\alpha_v = d\lambda \frac{\partial F}{\partial \chi_p} = -M^*; \qquad d\varepsilon_s^p = d\alpha_s = d\lambda \frac{\partial F}{\partial \chi_q} = 1. \tag{16.54}$$

It follows, *that for frictional materials, a non-associated flow rule does not only give a better fit to experimental behaviour, but also satisfies the Laws of Thermodynamics.* Because there is no restriction on the sign of M^*, a particular case of zero dilatancy with $M^* = 0$ at *the critical state* is also thermomechanically consistent, which will be discussed more in detail in Chapter 19.

In contrast, the associated flow rule, while being theoretically possible to accommodate within the hyperplastic framework by setting $M^* = M$, appears to be totally unrealistic, because in absence of cohesion $d_0 = 0$, it leads to the zero dissipation (16.48)!

Chapter 17

Plastic Yielding and Strain Hardening

TABLE OF CONTENTS

Chapter 17

Plastic Yielding and Strain Hardening

17.1 Introduction

In the previous chapters we learned how to model plastic flow and dilatancy at failure. In shearing, however, plastic flow starts long before the failure. This phenomenon is called the *plastic yielding*. In a triaxial compression test on slightly overconsolidated clay (Figure 17.1a), the yielding first occurs at the yield point $Y(p'_y, q_y)$, after which the stress-strain behaviour becomes irreversible (Figure 17.1b). After the unloading-reloading reversal, the yielding starts at a yield stress q_y^{new}, which is higher than the initial one (Figure 17.1b). This phenomenon is called *hardening* (if the new yield stress was lower than the initial one this would be called *softening*). Yielding causes both elastic and plastic strain increments, so that in the strain path (Figure 17.1c) it is responsible for the transition between the initial purely elastic and the post-failure purely plastic portions.

All possible yield stress states form a *yield surface* $Y(\sigma_{ij}, \chi_{ij}) = 0$, which is a function of stresses and the *hardening parameters* χ_{ij}, and can expand, contract or be shifted. The yield surface evolution is prescribed via its functional dependency on the hardening parameters, which is called the *hardening rule*.

The simplest yield surface in the effective triaxial stress space can be described by the straight line (Figure 17.1a):

$$Y(p', q, \varepsilon_s^p) = q - \mathrm{M}_y(\varepsilon_s^p) p' - d_y(\varepsilon_s^p) = 0 \qquad (17.1)$$

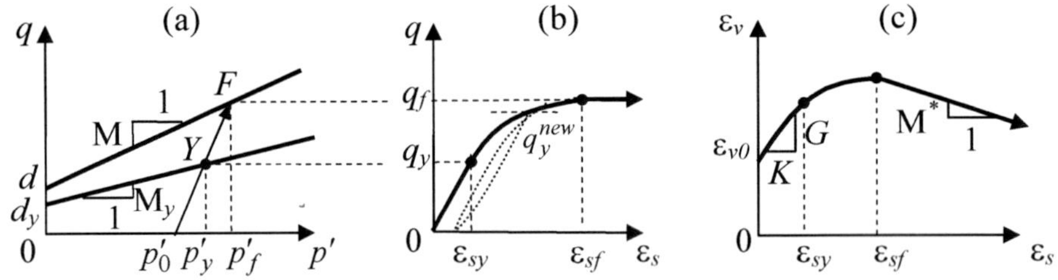

Figure 17.1: Yielding and hardening in a triaxial compression test: (a) the yield surface and the failure envelope; (b) deviatoric behaviour; (c) the strain path.

which, as the sample gets closer to failure, approaches the failure surface:

$$F(p',q) = q - \mathrm{M}p' - d = 0 \tag{17.2}$$

The hardening parameter in this case is represented by the plastic shear strain ε_s^p, which controls the yield surface evolution (i.e. expansion and/or shifting), via the hardening rule:

$$\mathrm{M}_y = \mathrm{M}_y(\varepsilon_s^p); \quad d_y = d_y(\varepsilon_s^p). \tag{17.3}$$

The functional form of this hardening rule is chosen to fit the deviatoric stress-strain curve (Figure 17.1b) in its yielding portion $\varepsilon_{sy} \leq \varepsilon_s \leq \varepsilon_{sf}$. The plastic flow during the yielding should be described by a flow rule, which should adequately model the strain path curve (Figure 17.1c) in its yielding portion $\varepsilon_{sy} \leq \varepsilon_s \leq \varepsilon_{sf}$.

This chapter introduces the basic concepts for pre-failure yielding and hardening, such as the yield surface, the hardening rule, the flow rule, etc., and derives an incremental stress-strain response for yielding both in triaxial and general stress spaces. It also explores the thermomechanical aspects of yielding and hardening.

17.2 Yielding in the triaxial stress space

17.2.1 Decomposition of strains

In yielding, like at failure, it is assumed that the total strain increments can be decomposed into elastic and plastic components (Figure 17.2a):

$$d\varepsilon_v = d\varepsilon_v^e + d\varepsilon_v^p; \qquad d\varepsilon_s = d\varepsilon_s^e + d\varepsilon_s^p, \tag{17.4}$$

where the elastic components are reversible unique functions of stress increments and stresses. In general, it is wrong to identify plastic strains ε^p with the irrecoverable component of the total strain ε^{ir} (Figure 17.2a),

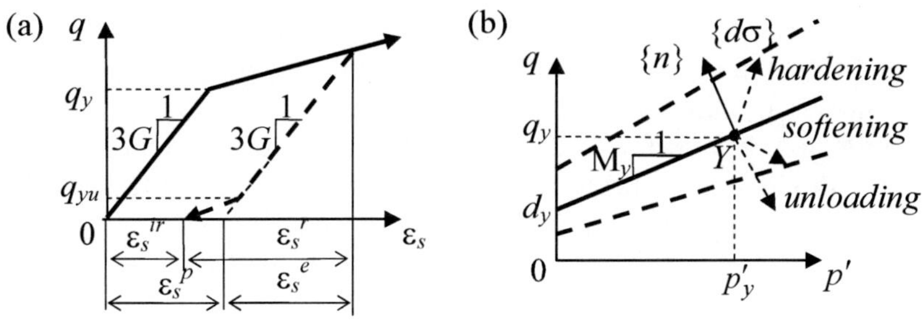

Figure 17.2: Yielding in a triaxial compression test: (a) decomposition of strains; (b) plastic loading.

because during an unloading a new yielding stress q_{yu} may be reached and "negative" plastic strains may develop, increasing the recovered strain ε^r.

17.2.2 Plastic loading and consistency condition

Because yielding is not the failure, there is no restriction on the direction of the incremental stress vector $\{d\sigma\}=\{dp',dq\}^T$: all stress states are statically admissible - inside, outside and on the current yield surface (Figure 17.2b). Unlike at failure, the yield surface may contract or expand, so that the stress state after the increment still stays on the (expanded/contracted) yield surface. This kind of the stress increment causes plastic strains and is called *plastic loading*. If however, similar to failure, the incremental stress vector brings the stress state inside the (expanded/contracted) yield surface interior, the soil element experiences an *unloading*, governed by the elastic law and plastic strains experience no change (Figure 17.2b).

Mathematically, the plastic loading condition for yielding can be expressed similarly to that for failure. Because both stress states before and after plastic loading should stay on the yield surface, one can write:

$$Y(p',q,\chi_i)=Y(p'+dp',q+dq,\chi_i+d\chi_i)=0 \tag{17. 5}$$

where $\chi_i, i=1,...,N$ is the vector of hardening parameters.

The *consistency condition* is then given by

$$dY=Y(p'+dp',q+dq,\chi_i+d\chi_i)-Y(p',q,\chi_i)=0 \tag{17.6}$$

which can be approximated as

$$dY=\frac{\partial Y}{\partial p'}dp'+\frac{\partial Y}{\partial q}dq+\frac{\partial Y}{\partial \chi_i}d\chi_i=\{n\}^T\cdot\{d\sigma\}+\frac{\partial Y}{\partial \chi_i}d\chi_i=0 \tag{17.7}$$

where

$$\{n\}=\left\{\frac{\partial Y}{\partial p'},\frac{\partial Y}{\partial q}\right\}^T \tag{17.8}$$

is the outward normal to the yield surface at the current stress state. It follows that, unlike at failure, the plastic loading can occur at any angle of the incremental stress vector to the outward normal to the yield surface (Figure 17.2b), provided the surface expands or contracts correspondingly. When $\{n\}^T\cdot\{d\sigma\}\geq 0$, the surface has to expand, so that:

$$\frac{\partial Y}{\partial \chi_i}d\chi_i\leq 0 \tag{17.9}$$

which is called *hardening*. When $\{n\}^T \cdot \{d\sigma\} < 0$, the surface has to contract

$$\frac{\partial Y}{\partial \chi_i} d\chi_i > 0 \tag{17.10}$$

which is called *softening*. Both hardening and softening are the cases of plastic loading, because the consistency condition $dF = 0$ is satisfied. It follows that the *unloading* can be defined by the condition:

$$dY = \frac{\partial Y}{\partial p'} dp' + \frac{\partial Y}{\partial q} dq + \frac{\partial Y}{\partial \chi_i} d\chi_i = \{n\}^T \cdot \{d\sigma\} + \frac{\partial Y}{\partial \chi_i} d\chi_i < 0 \tag{17.11}$$

For the yield surface (17.2), the plastic loading occurs when

$$dY = -\mathrm{M}_y\left(\varepsilon_s^p\right)dp' + dq - \left(\frac{\partial d_y\left(\varepsilon_s^p\right)}{\partial \varepsilon_s^p} + \frac{\partial \mathrm{M}_y\left(\varepsilon_s^p\right)}{\partial \varepsilon_s^p} p'\right) d\varepsilon_s^p = 0 \tag{17.12}$$

17.2.3 Hardening rule

Evolution of the yield surface during plastic loading is prescribed by the *hardening rule*. Depending on the functional dependency of the yield surface on the hardening parameters, we distinguish, normally, four types of hardening.

Isotropic hardening (Figure 17.3a)
In isotropic hardening, the yield surface is centred around the point $O(p'_O, q_O)$, keeps its shape, but increases or decreases its size depending on the hardening parameters:

$$Y(p', q, \chi_i) = y(p' - p'_O, q - q_O) - d(\chi_i) = 0 \tag{17. 13}$$

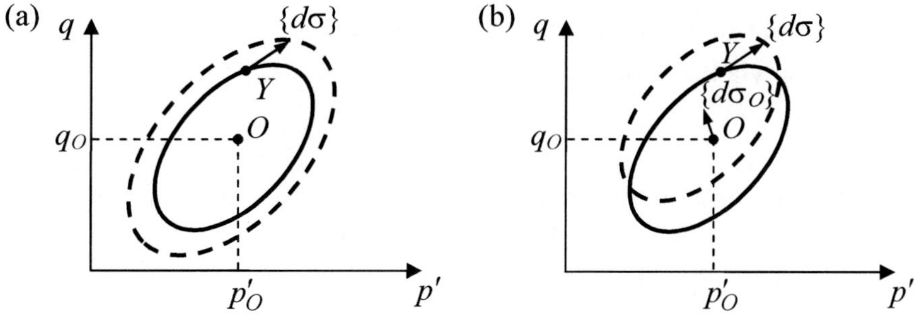

Figure 17.3: Hardening rule: (a) isotropic; (b) kinematic.

Kinematic hardening (Figure 17.3b)
In kinematic hardening, the yield surface keeps its shape and size, but its centre $O(p'_O, q_O)$ is shifted, depending on the hardening parameters:

$$Y(p', q, \chi_i) = y(p' - p'_O(\chi_i), q - q_O(\chi_i)) - d = 0 \qquad (17.\ 14)$$

where the functional dependency

$$p'_O = p'_O(\chi_i); \quad q_O = q_O(\chi_i) \qquad (17.\ 15)$$

is called the *translation rule*.

Mixed hardening
Mixed hardening is the combination of the kinematic and isotropic hardening:

$$Y(p', q, \chi_i) = y(p' - p'_O(\chi_i), q - q_O(\chi_i)) - d(\chi_i) = 0 \qquad (17.\ 16)$$

Perfect plasticity
In perfect plasticity, the yield surface is independent of hardening parameters:

$$Y(p', q, \chi_i) = y(p' - p'_O, q - q_O) - d = 0 \qquad (17.\ 17)$$

Perfect plasticity is normally used to describe *failure*.

When the hardening parameter is the plastic work, the phenomenon is called *work hardening*. More often, however, the hardening parameters are taken to be plastic strains, which can be then be described as *strain hardening*. The simplest yield surface in the effective triaxial stress space (17.1), described by the straight line in Figure 17.1a:

$$Y(p', q, \varepsilon_s^p) = q - \mathrm{M}_y(\varepsilon_s^p)p' - d_y(\varepsilon_s^p) = 0 \qquad (17.18)$$

is an example of mixed strain hardening with one hardening parameter $\chi_1 = \varepsilon_s^p$, the plastic shear strain, which controls the yield surface expansion and translation via the hardening rule:

$$\mathrm{M}_y = \mathrm{M}_y(\varepsilon_s^p); \quad d_y = d_y(\varepsilon_s^p). \qquad (17.19)$$

The functional form of this hardening rule is chosen to fit the deviatoric stress-strain curve (Figure 17.1b) in its yielding portion $\varepsilon_{sy} \le \varepsilon_s \le \varepsilon_{sf}$. A simple example is given by the linear hardening rule:

$$\mathrm{M}_y(\varepsilon_s^p) = \mathrm{M}_{y0} + \mathrm{M}'_y \varepsilon_s^p; \quad d_y(\varepsilon_s^p) = d_{y0} + d'_y \varepsilon_s^p, \qquad (17.20)$$

where d_{y0} and M_{y0} define the initial (pre-yielding) position of the yield surface; d'_y and M'_y define the change in this position for the unit plastic shear strain.

17.2.4 Associated flow rule

Similar to failure, the associated flow rule postulates that the incremental plastic strain vector should be parallel to the normal to the yield surface (Figure 17.4a):

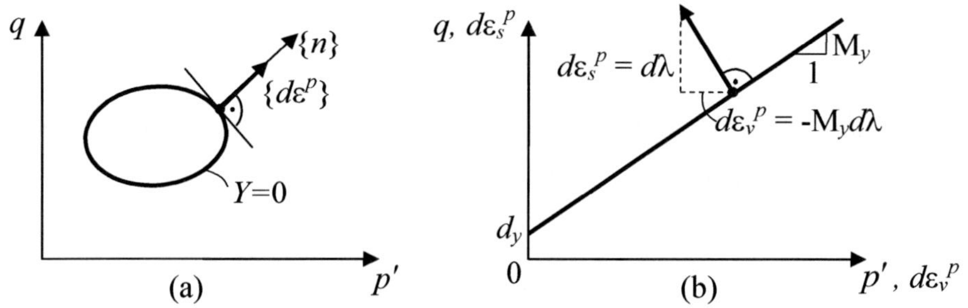

Figure 17.4: Associated flow rule: (a) normality; (b) the triaxial space example.

$$\left\{d\varepsilon^p\right\} = d\lambda\{n\} = d\lambda\left\{\frac{\partial Y}{\partial p'}, \frac{\partial Y}{\partial q}\right\}^T \tag{17.21}$$

where $d\lambda$ is an infinitesimal multiplier. For the yield surface (17.1), this gives (Figure 17.4b):

$$\begin{Bmatrix} d\varepsilon_v^p \\ d\varepsilon_s^p \end{Bmatrix} = d\lambda \begin{Bmatrix} -M_y \\ 1 \end{Bmatrix} \tag{17.22}$$

The associated flow rule, which also implies that the principal directions of the incremental plastic strain stress tensor $d\varepsilon_{ij}^p$ coincide with the principal directions of the stress tensor σ_{ij}, defines the direction (but not the length) of the incremental plastic strain vector.

17.2.5 Drucker's stability postulate

The *Drucker stability postulate* requires that the material behaviour at failure is stable, e.g., the second order plastic work is non-negative. In triaxial stress space this means:

$$\{d\sigma\}^T \cdot \left\{d\varepsilon^p\right\} = dp' d\varepsilon_v^p + dq d\varepsilon_s^p \geq 0 \tag{17.23}$$

At failure, the associated flow rule was shown to satisfy this condition always, provided the failure surface is convex. In yielding, however, the situation is more complex: substitution of the associated flow rule (17.21) into the consistency condition (17.7) gives

$$dY = \frac{\partial Y}{\partial p'}dp' + \frac{\partial Y}{\partial q}dq + \frac{\partial Y}{\partial \chi_i}d\chi_i = \frac{1}{d\lambda}\{d\sigma\}^T \cdot \{d\varepsilon^p\} + \frac{\partial Y}{\partial \chi_i}d\chi_i = 0 \tag{17.24}$$

Because in plastic loading $d\lambda \geq 0$, it follows that Drucker's postulate (17.23) can be satisfied only when

$$\frac{\partial Y}{\partial \chi_i}d\chi_i \leq 0 \tag{17.25}$$

i.e., when the yield surface is expanding, which is called *hardening*. In *softening*, when the yield surface contracts

$$\frac{\partial Y}{\partial \chi_i}d\chi_i > 0 \tag{17.26}$$

the behaviour is unstable even for the associated flow rule. In fact, as mentioned before, there is not a single thermodynamical reason why the material behaviour should be stable, and Drucker's postulate, therefore, does not have a power of a Thermodynamic Law.

17.2.6 Dilatancy and the Plastic Potential

Like at failure, accurate modelling of dilatancy in yielding may require that the incremental plastic strain vector is not normal to the yield surface. The best way to describe this mathematically is to introduce the *plastic potential* family of surfaces $P(p', q, x)$ with the following properties (Figure 17.5a):

(a) through each point of the yield surface passes one of the surfaces of the plastic potential family (this allows to define the parameter x);

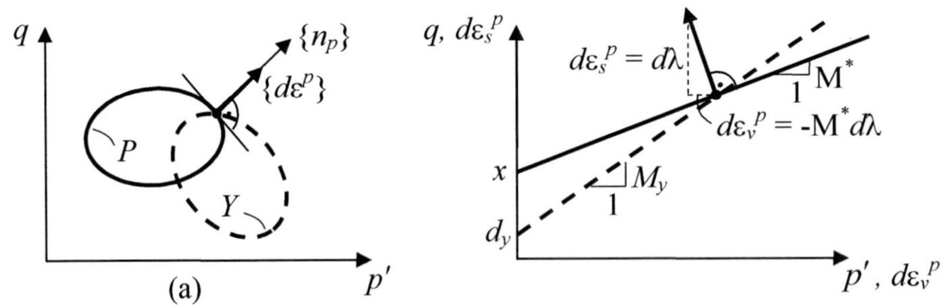

Figure 17.5: Non-associated flow rule: (a) the failure surface and the plastic potential; (b) the triaxial space example.

(b) the incremental plastic strain vector at this point is normal to the plastic potential and not to the failure surface.

Mathematically this can be described as a *non-associated flow rule*:

$$\{d\varepsilon^p\} = d\lambda\{n_p\} = d\lambda\left\{\frac{\partial P}{\partial p'}, \frac{\partial P}{\partial q}\right\}^T \tag{17.27}$$

where $d\lambda$ is an infinitesimal multiplier; $\{n_p\}$ is the normal to the plastic potential surface. Expression (17.27) defines the direction (but not the length) of the incremental plastic strain vector in the triaxial stress space. In general, it does not satisfy the Drucker stability postulate even for the hardening case.

For the yield surface (17.1), we can define the plastic potential also as a straight line (Figure 17.5b):

$$P(p', q, x) = q - \mathrm{M}^* p' - x = 0 \tag{17.28}$$

For each point (p'_y, q_y) on the yield surface (17.1), the parameter x is defined to ensure that the plastic potential passes through this point:

$$x = \left(\mathrm{M}_y - \mathrm{M}^*\right)p'_y + d_y \tag{17.29}$$

The corresponding non-associated flow rule is given by:

$$\begin{Bmatrix} d\varepsilon_v^p \\ d\varepsilon_s^p \end{Bmatrix} = d\lambda \begin{Bmatrix} -\mathrm{M}^* \\ 1 \end{Bmatrix} \tag{17.30}$$

17.2.7 Incremental elastic-plastic stress-strain response

The *flow rule* defines only the direction of the incremental plastic strain vector. In order to find its length, we can use the following procedure. Consider a strain hardening yield surface:

$$Y\left(p', q, \varepsilon_v^p, \varepsilon_s^p\right) = 0 \tag{17.31}$$

The consistency (plastic loading) condition for this surface is

$$dY = \frac{\partial Y}{\partial p'}dp' + \frac{\partial Y}{\partial q}dq + \frac{\partial Y}{\partial \varepsilon_v^p}d\varepsilon_v^p + \frac{\partial Y}{\partial \varepsilon_s^p}d\varepsilon_s^p = 0 \tag{17.32}$$

Substituting plastic strain increments from the flow rule (17.27), we obtain the following expression for the infinitesimal multiplier $d\lambda$:

$$d\lambda = \frac{1}{H}\left(\frac{\partial Y}{\partial p'}dp' + \frac{\partial Y}{\partial q}dq\right) \tag{17.33}$$

where

$$H = -\frac{\partial Y}{\partial \varepsilon_v^p}\frac{\partial P}{\partial p'} - \frac{\partial Y}{\partial \varepsilon_s^p}\frac{\partial P}{\partial q} \tag{17.34}$$

is the *hardening modulus*.

Substitution of the multiplier $d\lambda$ into the flow rule (17.27) gives expressions for plastic strain increments:

$$\begin{Bmatrix} d\varepsilon_v^p \\ d\varepsilon_s^p \end{Bmatrix} = \frac{1}{H}\begin{bmatrix} \frac{\partial Y}{\partial p'}\frac{\partial P}{\partial p'} & \frac{\partial Y}{\partial q}\frac{\partial P}{\partial p'} \\ \frac{\partial Y}{\partial p'}\frac{\partial P}{\partial q} & \frac{\partial Y}{\partial q}\frac{\partial P}{\partial q} \end{bmatrix}\begin{Bmatrix} dp' \\ dq \end{Bmatrix} \tag{17.35}$$

Note that, unlike at failure, zero stress increments in yielding do lead to zero plastic strains increments. In general, for the smooth yield surface and plastic potential, the plastic strain increments in yielding are uniquely defined by the stress state and the stress increments. Substitution of these plastic strain increments together with isotropic elastic strain increments:

$$\begin{Bmatrix} d\varepsilon_v^e \\ d\varepsilon_s^e \end{Bmatrix} = \begin{bmatrix} 1/K & 0 \\ 0 & 1/(3G) \end{bmatrix}\begin{Bmatrix} dp' \\ dq \end{Bmatrix} \tag{17.36}$$

into the strain decomposition expressions (17.4) produces the total incremental stress-strain response for yielding with non-associated flow rule:

$$\begin{Bmatrix} d\varepsilon_v \\ d\varepsilon_s \end{Bmatrix} = \frac{1}{H}\begin{bmatrix} \frac{H}{K} + \frac{\partial Y}{\partial p'}\frac{\partial P}{\partial p'} & \frac{\partial Y}{\partial q}\frac{\partial P}{\partial p'} \\ \frac{\partial Y}{\partial p'}\frac{\partial P}{\partial q} & \frac{H}{3G} + \frac{\partial Y}{\partial q}\frac{\partial P}{\partial q} \end{bmatrix}\begin{Bmatrix} dp' \\ dq \end{Bmatrix} \tag{17.37}$$

Note that the compliance matrix is not symmetric, unless we assume the associated flow rule $P = Y$. The asymmetry of the compliance (and, hence, stiffness) matrix may cause numerical difficulties in FE/FD calculations.

For the case of the yield surface:

$$Y\left(p', q, \varepsilon_s^p\right) = q - \mathrm{M}_y\left(\varepsilon_s^p\right)p' - d_y\left(\varepsilon_s^p\right) = 0 \tag{17.38}$$

the plastic potential:

$$P\left(p', q\right) = q - \mathrm{M}^* p' - x = 0 \tag{17.39}$$

and the hardening rule:

$$\mathrm{M}_y\left(\varepsilon_s^p\right) = \mathrm{M}_{y0} + \mathrm{M}'_y\varepsilon_s^p; \qquad d_y\left(\varepsilon_s^p\right) = d_{y0} + d'_y\varepsilon_s^p, \tag{17.40}$$

the incremental stress strain response becomes:

$$\begin{Bmatrix} d\varepsilon_v \\ d\varepsilon_s \end{Bmatrix} = \frac{1}{H} \begin{bmatrix} \dfrac{H}{K} + \mathrm{M}_y \mathrm{M}^* & -\mathrm{M}^* \\ -\mathrm{M}_y & \dfrac{H}{3G} + 1 \end{bmatrix} \begin{Bmatrix} dp' \\ dq \end{Bmatrix} \tag{17.41}$$

where the hardening modulus:

$$H = -\frac{\partial Y}{\partial \varepsilon_s^p} \frac{\partial P}{\partial q} = d'_y + \mathrm{M}'_y p' \tag{17.42}$$

which explains its physical meaning.

Let us consider two particular cases of the hardening rule (17.40). The first case, with $\mathrm{M}'_y = 0$ and $\mathrm{M}_y = \mathrm{M} = const$, produces the *cohesive hardening*. The corresponding incremental response is linear, and this is why the model is described as elastic-plastic with *linear hardening*.

The behaviour of this model in a triaxial compression test is shown in Figure 17.6, where

$$G^{ep} = \frac{1}{3} \frac{dq}{d\varepsilon_s} = \frac{Gd'_y}{d'_y + (3 - \mathrm{M})G} \tag{17.43}$$

is the elastic-plastic shear modulus in yielding;

$$\mathrm{M}^{ep} = -\frac{d\varepsilon_v}{d\varepsilon_s} = -\frac{G}{K} \frac{d'_y - \mathrm{M}^*(3 - \mathrm{M})K}{d'_y + (3 - \mathrm{M})G} \tag{17.44}$$

is the elastic-plastic dilation in yielding.

The second case, with $d'_y = 0$ and $d_y = d = const$, gives the *frictional hardening*. The corresponding incremental response in triaxial compression is non-linear and the deviatoric plastic strain increment is given by

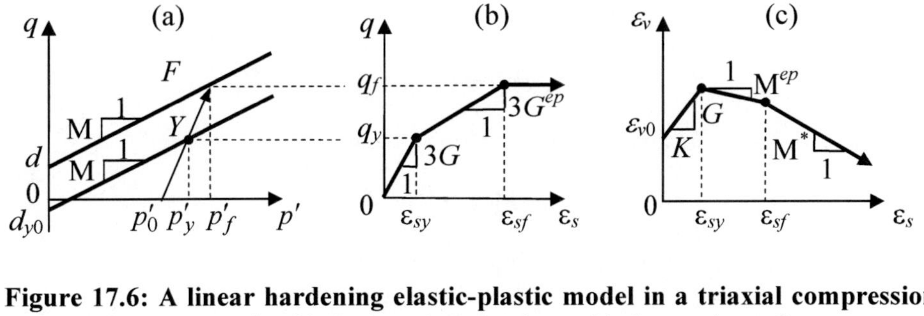

Figure 17.6: A linear hardening elastic-plastic model in a triaxial compression test: (a) the stress path; (b) deviatoric behaviour; (c) the strain path.

$$d\varepsilon_s^p = \left(\frac{3\text{-M}_{y0}}{\text{M}'_y} - \varepsilon_s^p \right) \frac{dp'}{p'} \tag{17.45}$$

which, after integration with initial condition $\varepsilon_s^p \Big|_{p'=p'_y} = 0$, produces:

$$\varepsilon_s^p = \frac{3\text{-M}_{y0}}{\text{M}'_y} \left(1 - \frac{p'_y}{p'} \right) \tag{17.46}$$

Using the equation of the triaxial stress path $q = 3(p' - p'_0)$ and the elastic law $\varepsilon_s^e = q/3G$, equation (17.46) produces the non-linear yielding deviatoric stress-strain curve :

$$\varepsilon_s = \frac{q}{3G} + \frac{3\text{-M}_{y0}}{\text{M}'_y} \left(1 - \frac{q_y + 3p'_0}{q + 3p'_0} \right) \tag{17.47}$$

In general, the irreversible constitutive behaviour of soils in yielding is best described via the incremental stress strain response, where the strains are obtained by integrating strain increments along the stress path until the failure is reached.

17.3 Incremental response in the general stress space

17.3.1 Assumptions

Derivation of the strain hardening incremental response in the general space follows the same steps as in the previous section. The main assumptions are:

- decomposition of the strain increment:

$$d\varepsilon_{ij} = d\varepsilon_{ij}^e + d\varepsilon_{ij}^p \tag{17.48}$$

- elastic relationship:

$$d\varepsilon_{ij}^e = C_{ijkl} d\sigma_{kl} \tag{17.49}$$

- the yield surface and the hardening rule:

$$Y\left(\sigma_{ij}, \varepsilon_{ij}^p\right) = 0 \tag{17.50}$$

- consistency (plastic loading) condition:

$$dY = \frac{\partial Y}{\partial \sigma_{ij}} d\sigma_{ij} + \frac{\partial Y}{\partial \varepsilon_{ij}^p} d\varepsilon_{ij}^p = 0 \tag{17.51}$$

- the plastic potential:

$$P\left(\sigma_{ij}, \varepsilon_{ij}^{p}\right) = 0 \tag{17.52}$$

- and the flow rule:

$$d\varepsilon_{ij}^{p} = d\lambda \frac{\partial P}{\partial \sigma_{ij}} \tag{17.53}$$

17.3.2 Derivation

Substituting plastic strain increments from the flow rule (17.49) into the consistency condition (17.47), we obtain the following expression for the infinitesimal multiplier $d\lambda$:

$$d\lambda = \frac{1}{H} \frac{\partial Y}{\partial \sigma_{ij}} d\sigma_{ij} \tag{17.54}$$

where

$$H = -\frac{\partial Y}{\partial \varepsilon_{ij}^{p}} \frac{\partial P}{\partial \sigma_{ij}} \tag{17.55}$$

is the *hardening modulus*.

Substitution of the multiplier $d\lambda$ into the flow rule (17.49) gives expressions for plastic strain increments:

$$d\varepsilon_{ij}^{p} = \frac{1}{H} \frac{\partial Y}{\partial \sigma_{kl}} d\sigma_{kl} \frac{\partial P}{\partial \sigma_{ij}} \tag{17.56}$$

Substitution of these plastic strain increments together with isotropic elastic strain increments (17.45) into the strain decomposition expressions (17.44) produces the total incremental stress-strain response for strain hardening with non-associated flow rule:

$$d\varepsilon_{ij} = C_{ijkl}^{ep} d\sigma_{kl} = \left[C_{ijkl}^{e} + C_{ijkl}^{p}\right] d\sigma_{kl} \tag{17.57}$$

where the elastic-plastic compliance matrix C_{ijkl}^{ep} can be decomposed into the elastic and plastic components:

$$C_{ijkl}^{e} = D_{ijmn}; \qquad C_{ijkl}^{p} = \frac{1}{H} \frac{\partial Y}{\partial \sigma_{kl}} \frac{\partial P}{\partial \sigma_{ij}}. \tag{17.58}$$

Note that the compliance matrix is not symmetric, unless we assume the associated flow rule $P = Y$.

17.4 Thermomechanics of yielding and hardening

The hyperplastic framework is used below to validate thermomechanical consistency of plastic yielding and strain hardening.

17.4.1 Perfect plasticity

Following the formulation in Section 15.4, for a linear elastic - perfectly plastic material with no yielding and hardening, the Gibbs free energy potential has a simple linear dependency on kinematic variables:

$$g(p',q,\alpha_v,\alpha_s) = -\frac{p'^2}{2K} - \frac{q^2}{6G} - p'\alpha_v - q\alpha_s \tag{17.59}$$

while the dissipation function does not depend on kinematic variables at all, resulting in a *failure surface* independent of plastic strains:

$$\lambda F(p',q,\chi_p,\chi_q) = \chi_p\dot{\alpha}_v + \chi_q\dot{\alpha}_s - d(p',q,\dot{\alpha}_v,\dot{\alpha}_s) = 0 \tag{17.60}$$

where λ is a Lagrangian multiplier, and

$$\chi_p = \frac{\partial d}{\partial \dot{\alpha}_v} \qquad \chi_q = \frac{\partial d}{\partial \dot{\alpha}_s} \tag{17.61}$$

are generalized dissipative stresses.

Expression (17.59) leads to the *decomposition of strains*:

$$\varepsilon_v = -\frac{\partial g}{\partial p'} = \frac{p'}{K} + \alpha_v = \frac{p'}{K} + \varepsilon_v^p; \qquad \varepsilon_s = -\frac{\partial g}{\partial q} = \frac{q}{3G} + \alpha_s = \frac{q}{3G} + \varepsilon_s^p \tag{17.62}$$

and produces the following Ziegler's condition:

$$\chi_p = \frac{\partial d}{\partial \dot{\alpha}_v} = -\frac{\partial g}{\partial \alpha_v} = p'; \qquad \chi_q = \frac{\partial d}{\partial \dot{\alpha}_s} = -\frac{\partial g}{\partial \alpha_s} = q, \tag{17. 63}$$

which being substituted into the expression (17.60) results in the *perfectly plastic failure surface* in the true stresses:

$$F(p',q) = 0 \tag{17.64}$$

17.4.2 Isotropic hardening

In order to ensure thermomechanical consistency for an isotropic hardening plastic material, both the dissipation function and the yield surface should depend on kinematic variables:

$$\lambda Y(p',q,\chi_p,\chi_q,\alpha_v,\alpha_s) = \chi_p\dot{\alpha}_v + \chi_q\dot{\alpha}_s - d(p',q,\dot{\alpha}_v,\dot{\alpha}_s,\alpha_v,\alpha_s) = 0 \tag{17.65}$$

As above, expression for the Gibbs free energy (17.59) leads to the decomposition of strains (17.62) and the Ziegler's condition (17.63), which being substituted into the expression (17.65) result in the *isotropic strain hardening yield surface* in the true stresses:

$$Y\left(p', q, \varepsilon_v^p, \varepsilon_s^p\right) = 0 \tag{17.66}$$

17.4.3 Kinematic hardening

For a purely kinematic hardening plastic material, the dissipation function and the yield surface should not depend on kinematic variables.

$$\lambda Y\left(\chi_p, \chi_q\right) = \chi_p \dot{\alpha}_v + \chi_q \dot{\alpha}_s - d\left(\dot{\alpha}_v, \dot{\alpha}_s\right) = 0 \tag{17.67}$$

The Gibbs free energy potential in this case, should have an additional term:

$$g\left(p', q, \alpha_v, \alpha_s\right) = -\frac{p'^2}{2K} - \frac{q^2}{6G} - p'\alpha_v - q\alpha_s + g_2\left(\alpha_v, \alpha_s\right) \tag{17.68}$$

which does not affect decomposition of strains (17.62), but changes the Ziegler's condition (17.63):

$$\begin{aligned} \chi_p &= \frac{\partial d}{\partial \dot{\alpha}_v} = -\frac{\partial g}{\partial \alpha_v} = p' - p'_O\left(\alpha_v, \alpha_s\right) \\ \chi_q &= \frac{\partial d}{\partial \dot{\alpha}_s} = -\frac{\partial g}{\partial \alpha_s} = q - q_O\left(\alpha_v, \alpha_s\right) \end{aligned} \tag{17. 69}$$

where

$$p'_O = \frac{\partial g_2\left(\alpha_v, \alpha_s\right)}{\partial \alpha_v}; \qquad q_O = \frac{\partial g_2\left(\alpha_v, \alpha_s\right)}{\partial \alpha_s} \tag{17.70}$$

is the *translation rule*.

Equations (17.62) and (17.69) being substituted into the expression (17.67) result in the *kinematic strain hardening yield surface* in the true stresses:

$$Y\left(p' - p'_O\left(\varepsilon_v^p, \varepsilon_s^p\right), q - q_O\left(\varepsilon_v^p, \varepsilon_s^p\right)\right) = 0 \tag{17.71}$$

17.4.4 Mixed hardening

Finally, the thermomechanical consistency of a more general case of a mixed hardening plastic material will be ensured if its behaviour can be defined by the generalized dissipation function or the yield surface (17.65) and the generalized Gibbs free energy potential (17.68). In this case, equations (17.62) and (17.69) being substituted into the expression (17.65) will result in the *mixed strain hardening yield surface* in the true stresses:

$$Y\left(p' - p'_O\left(\varepsilon_v^p, \varepsilon_s^p\right), q - q_O\left(\varepsilon_v^p, \varepsilon_s^p\right), \varepsilon_v^p, \varepsilon_s^p\right) = 0 \tag{17.72}$$

17.4.5 Example

The simple model considered in the section 17.2.7, with the yield surface:

$$Y\left(p', q, \varepsilon_s^p\right) = q - \mathrm{M}_y\left(\varepsilon_s^p\right)p' - d_y\left(\varepsilon_s^p\right) = 0 \qquad (17.73)$$

the plastic potential:

$$P(p', q) = q - \mathrm{M}^* p' - x = 0 \qquad (17.74)$$

and the hardening rule:

$$\mathrm{M}_y\left(\varepsilon_s^p\right) = \mathrm{M}_{y0} + \mathrm{M}'_y \varepsilon_s^p; \qquad d_y\left(\varepsilon_s^p\right) = d_{y0} + d'_y \varepsilon_s^p \qquad (17.75)$$

can be expressed within the hyperplastic formulation with the Gibbs free energy potential:

$$g(p', q, \alpha_v, \alpha_s) = -\frac{p'^2}{2K} - \frac{q^2}{6G} - p'\alpha_v - q\alpha_s \qquad (17.76)$$

and the dissipation function:

$$d = \left[\left(\mathrm{M}_{y0} + \mathrm{M}'_y \alpha_s - M^*\right)p' + d_{y0} + d'_y \alpha_s\right]\left|\dot{\alpha}_s\right| \geq 0 \qquad (17.77)$$

complemented by a constant rate of dilation constraint:

$$c(\dot{\alpha}_v, \dot{\alpha}_s) = \dot{\alpha}_v + M^*\left|\dot{\alpha}_s\right| = 0 \qquad (17.78)$$

Therefore, this model satisfies the Laws of Thermodynamics.

Chapter 18

Pre-consolidation

TABLE OF CONTENTS

Chapter 18

Pre-consolidation

18.1 Introduction

In the previous chapters we learned how to model failure, yielding and hardening in triaxial compression. The failure and yielding will also take place for any straight effective stress path emanating from the initial pre-shear stress state $(p'_0,0)$, provided inclination of this stress path is steeper than M_y (Figure 18.1a). If the slope of this stress path is smaller that M_y, the failure will never occur, but experiments show that the yielding will still take place, with the main plastic strain component being *volumetric*.

Consider, for example, the isotropic consolidation and swelling stress path in Figure (18.1b). In this test, the sample is first consolidated to the pre-consolidation stress p'_{c0} and then unloaded to p'_0, similarly to the first stage of the triaxial compression test on a slightly overconsolidated clay. In this test, however, instead of triaxial shearing, the sample is reconsolidated isotropically to a larger pre-consolidation stress p'_{c1}, unloaded back to p'_0,

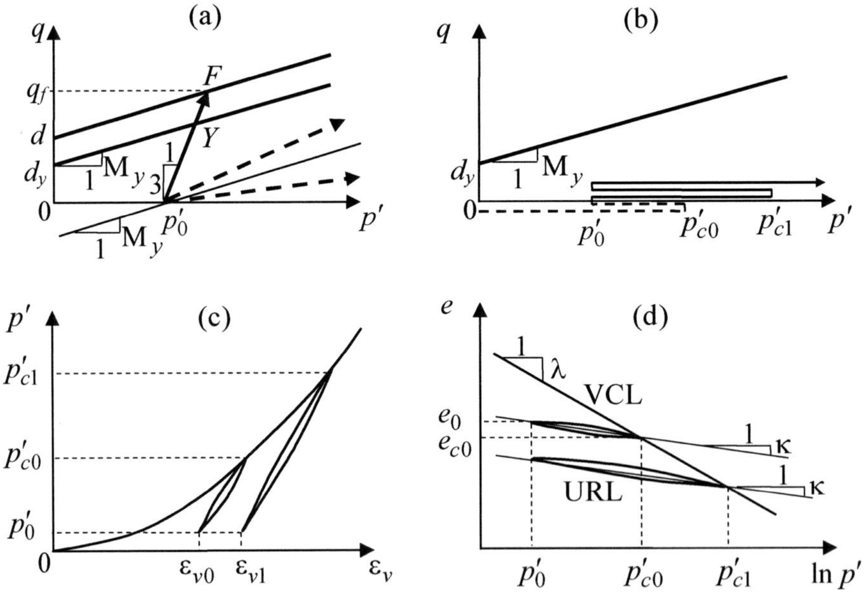

Figure 18.1: Yielding in isotropic consolidation: (a) the triaxial compression stress path; (b) the isotropic consolidation/swelling stress path; (c) volumetric stress-strain curves; (d) semi-logarithmic plot.

and loaded again beyond the current pre-consolidation stress p'_{c1}. The corresponding volumetric stress-strain behaviour clearly indicates such features as non-linearity, irreversibility, yielding and hardening (Figure 18.1c). Indeed, non-linear elastic unloading-reloading curves differ from the non-linear elastic-plastic initial loading curve. The *plastic yielding* always begins when the current *pre-consolidation stress* (i.e. the largest mean effective stress before the unloading) is exceeded in reloading. This yielding takes place along the initial loading curve, until the next unloading, so that the new pre-consolidation (yield) stress is always larger than the previous one, indicating *hardening*.

The non-linearity of the volumetric stress-strain behaviour is successfully dealt with by plotting the experimental results from isotropic compression and swelling tests in a semi-logarithmic scale of the void ratio e vs. $\ln p'$. In this plot (Figure 18.1d), both initial loading (also called virgin compression line, or VCL) and unloading-reloading (URL) curves are well approximated by the straight lines with the slopes λ and κ, respectively:

VCL: $$e = e_{c0} - \lambda \ln p'/p'_{c0} \tag{18.1}$$

URL: $$e = e_0 - \kappa \ln p'/p'_0 \tag{18.2}$$

This chapter shows how the above described pre-consolidation phenomenon can be modelled using the concepts of plastic yielding and hardening. It will be also shown how the volumetric yielding can be combined with the yielding in shearing, leading to so called cap models. The thermomechanical aspects of pre-consolidation are also explored.

18.2 Elastic-plastic model with volumetric strain hardening

Let us first construct the simplest model of pre-consolidation using the tools of the theory of plasticity, which we have learned so far.

18.2.1 Decomposition of strains

We start with the usual decomposition of the strain increments into elastic and plastic components:

$$d\varepsilon_v = d\varepsilon_v^e + d\varepsilon_v^p; \qquad d\varepsilon_s = d\varepsilon_s^e + d\varepsilon_s^p. \tag{18.3}$$

18.2.2 Elastic volumetric behaviour

Elastic components of strain increment are defined via the following isotropic relationship:

$$\begin{Bmatrix} d\varepsilon_v^e \\ d\varepsilon_s^e \end{Bmatrix} = \begin{bmatrix} 1/K(p') & 0 \\ 0 & 1/(3G) \end{bmatrix} \begin{Bmatrix} dp' \\ dq \end{Bmatrix} \tag{18.4}$$

The volumetric part of this elastic relationship is non-linear, and in order to fit the experimentally observed elastic URL (18.2) in Figure 18.1d, the tangent bulk modulus has to be defined as follows:

$$K(p') = \frac{(1+e_0)p'}{\kappa} \tag{18.5}$$

Indeed, substituting this bulk modulus into the elastic relationship (18.4):

$$d\varepsilon_v^e = \frac{\kappa}{1+e_0}\frac{dp'}{p'} \tag{18.6}$$

and integrating it, and assuming that at $p = p'_0$: $\varepsilon_v^e = 0$, we obtain

$$\varepsilon_v^e = \frac{\kappa}{1+e_0}\ln\frac{p'}{p'_0} \tag{18.7}$$

The specific volume can be defined as the ratio between the total volume of the soil element (i.e., the sum of the volumes of the solid particles v_s and the voids between them v_v) and the volume of the solid particles v_s :

$$v = \frac{v_s + v_v}{v_s} = 1 + e \tag{18.8}$$

Then, assuming that at $e = e_0$: $\varepsilon_v = 0$, and that positive strain corresponds to compaction, we obtain:

$$\varepsilon_v = -\frac{v - v_0}{v_0} = \frac{e_0 - e}{1+e_0} \tag{18.9}$$

Considering that along the URL elastic strain increments are equal to the total ones, substitution of equation (18.9) into (18.7) gives the URL equation:

$$e = e_0 - \kappa \ln p'/p'_0 \tag{18.10}$$

Note, that expression (18.5) leading to the URL equation, is very similar to the hypoelastic volumetric relationship defining pressure dependency of stiffness in Chapter 12. There is, however, one very important difference: in Chapter 12 we used

$$k = \frac{1+e_0}{\kappa} = const \tag{18.11}$$

while here it will change depending on the initial void ratio e_0. This dependency will be later shown to affect its thermodynamic consistency.

18.2.3 Yield surface

We choose the simplest yield surface represented by the vertical line in the triaxial stress space (Figure 18.2a):

$$Y\left(p',q,\varepsilon_v^p\right)=p'-p_c'\left(\varepsilon_v^p\right)=0 \qquad (18.12)$$

where $p_c' = f\left(\varepsilon_v^p\right)$ is the pre-consolidation pressure.

18.2.4 Consistency condition

The consistency condition (and plastic loading criterion) is then defined by

$$dY = dp' - \frac{\partial p_c'}{\partial \varepsilon_v^p} d\varepsilon_v^p = 0 \qquad (18.13)$$

as is seen it does not depend on the deviator stress increment dq.

18.2.5 Flow rule

By assuming the associated flow rule (Figure 18.2a):

$$\begin{Bmatrix} d\varepsilon_v^p \\ d\varepsilon_s^p \end{Bmatrix} = d\lambda \begin{Bmatrix} \partial Y/\partial p' \\ \partial Y/\partial q \end{Bmatrix} = d\lambda \begin{Bmatrix} 1 \\ 0 \end{Bmatrix} \qquad (18.14)$$

we postulate that only volumetric plastic strains can develop.

18.2.6 Hardening rule

Hardening rule defines functional dependency of the pre-consolidation stress on the hardening parameter – volumetric plastic strain ε_v^p:

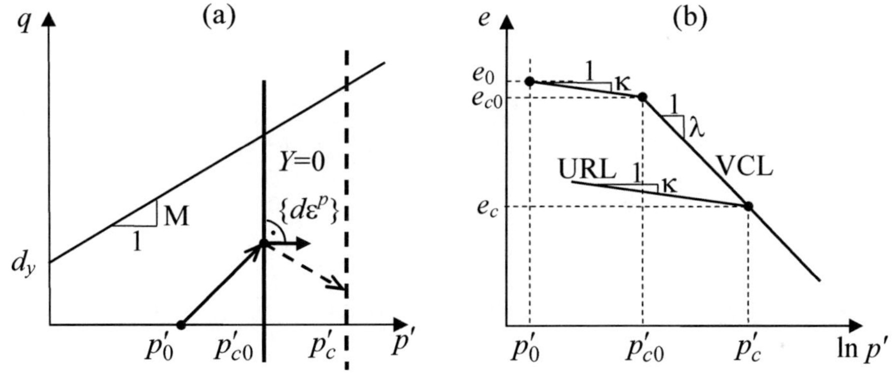

Figure 18.2: Elastic-plastic model with volumetric strain hardening: (a) the yield surface and the flow rule; (b) elastic-plastic volumetric behaviour.

$$p_c'\left(\varepsilon_v^p\right) = p_{c0}' \exp\left(\frac{1+e_0}{\lambda-\kappa}\varepsilon_v^p\right) \tag{18.15}$$

so that $\varepsilon_v^p = 0$ at $p_c' = p_{c0}'$. As shown below, this hardening rule has been calibrated to fit the experimental elastic-plastic VCL expression (18.1) in Figure 18.1d.

18.2.7 Incremental response

The incremental stress-strain response is obtained by substituting the flow (18.14) and hardening (18.15) rules into the consistency condition (18.13):

$$dY = dp' - \frac{\partial p_c'}{\partial \varepsilon_v^p} d\varepsilon_v^p = dp' - \frac{1+e_0}{\lambda-\kappa} p_c' d\lambda = 0 \tag{18.16}$$

Resolving it with respect to $d\lambda$ and substituting into the flow rule, we obtain:

$$d\varepsilon_v^p = \frac{\lambda-\kappa}{1+e_0}\frac{dp'}{p_c'} = \frac{\lambda-\kappa}{1+e_0}\frac{dp'}{p'} \tag{18.17}$$

where the last equality is derived by using the equation of the yield surface $Y = p' - p_c' = 0$. The total stress increment is then obtained by substituting elastic (18.6) and plastic (18.17) strain increments into the strain decomposition (18.3):

$$d\varepsilon_v = d\varepsilon_v^e + d\varepsilon_v^p = \frac{\kappa}{1+e_0}\frac{dp'}{p'} + \frac{\lambda-\kappa}{1+e_0}\frac{dp'}{p'} = \frac{\lambda}{1+e_0}\frac{dp'}{p'} \tag{18.18}$$

Upon integration, this produces

$$\varepsilon_v - \varepsilon_{vc0} = \frac{\lambda}{1+e_0}\ln\frac{p'}{p_{c0}'} \tag{18.19}$$

where

$$\varepsilon_v - \varepsilon_{vc0} = \frac{e_0 - e}{1+e_0} - \frac{e_0 - e_{c0}}{1+e_0} = \frac{e_{c0} - e}{1+e_0} \tag{18.20}$$

leading to the VCL equation

$$e = e_{c0} - \lambda \ln p'/p_{c0}' \tag{18.21}$$

as shown in Figure 18.2b, together with the elastic URL response (18.10).

18.3 Double hardening or "cap" models

Now that we learnt how to model shear yielding and pre-consolidation separately, let us combine it into a so called *double hardening* model. Such models are often called *cap models*, because in the principal stress space

(Chapter 14) the volumetric yield surface can be visualized as a cap put on the top of the shear yield surface cone (Figure 18.3a).

A simple double hardening model in a triaxial stress space can be formulated as follows (Figure 18.3b). The yield surface is built of two components, one with isotropic cohesive shear hardening, another one with isotropic volumetric hardening:

$$Y_s = q - \mathrm{M}p' - d_y\left(\varepsilon_s^p\right) = 0 \tag{18.22}$$

$$Y_v = p' - p_c'\left(\varepsilon_v^p\right) = 0 \tag{18.23}$$

Depending on which surface the yielding takes place, three zones can be identified (Figure 18.3b): Zone I – *shear yielding*; Zone II – *volumetric yielding*; Zone III – *mixed yielding*. The corresponding consistency (plastic loading) conditions for each yielding zone are:

Zone I: $\quad Y_s = 0; \quad dY_s = -\mathrm{M}dp' + dq - \dfrac{\partial d_y\left(\varepsilon_s^p\right)}{\partial \varepsilon_s^p} d\varepsilon_s^p = 0 \qquad (18.24)$

Zone II: $\quad Y_v = 0; \quad dY_v = dp' - \dfrac{\partial p_c'}{\partial \varepsilon_v^p} d\varepsilon_v^p = 0 \qquad (18.25)$

Zone III: $\quad Y_s = 0$ and $Y_v = 0$;

$$dY_s = -\mathrm{M}dp' + dq - \frac{\partial d_y\left(\varepsilon_s^p\right)}{\partial \varepsilon_s^p} d\varepsilon_s^p = 0 \quad \text{and} \quad dY_v = dp' - \frac{\partial p_c'}{\partial \varepsilon_v^p} d\varepsilon_v^p = 0 \tag{18.26}$$

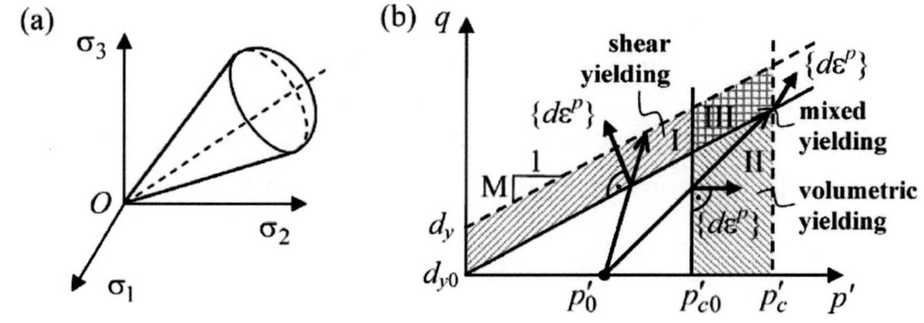

Figure 18.3: Double hardening model: (a) schematic view in principal stresses space; (b) a simple triaxial example.

The associated flow rule for each zone gives (Figure 18.3b):

Zone I:
$$\begin{Bmatrix} d\varepsilon_v^p \\ d\varepsilon_s^p \end{Bmatrix} = d\lambda_s \begin{Bmatrix} \partial Y_s/\partial p' \\ \partial Y_s/\partial q \end{Bmatrix} = d\lambda_s \begin{Bmatrix} -\mathrm{M} \\ 1 \end{Bmatrix} \tag{18.27}$$

Zone II:
$$\begin{Bmatrix} d\varepsilon_v^p \\ d\varepsilon_s^p \end{Bmatrix} = d\lambda_v \begin{Bmatrix} \partial Y_v/\partial p' \\ \partial Y_v/\partial q \end{Bmatrix} = d\lambda_v \begin{Bmatrix} 1 \\ 0 \end{Bmatrix} \tag{18.28}$$

Zone III:
$$\begin{Bmatrix} d\varepsilon_v^p \\ d\varepsilon_s^p \end{Bmatrix} = \begin{Bmatrix} -\mathrm{M}d\lambda_s + d\lambda_v \\ d\lambda_s \end{Bmatrix} \tag{18.29}$$

And, finally, the hardening rule is given by

$$d_y\left(\varepsilon_s^p\right) = d_{y0} + d'_y \varepsilon_s^p \tag{18.30}$$

$$p'_c\left(\varepsilon_v^p\right) = p'_{c0} \exp\left(\frac{1+e_0}{\lambda-\kappa}\varepsilon_v^p\right) \tag{18.31}$$

The incremental response for the Zones I and II has been already derived in Sections 17.2.7 and 18.2.7, respectively. A special feature of the double hardening model, however, is that during the shear yielding in the Zone I, the volumetric yield surface $Y_v = p - p'_c = 0$ shrinks! Indeed, according to the flow rule (18.27), the volumetric plastic strain decreases, affecting the pre-consolidation stress p'_c via the hardening rule (18.31).

Another special feature of the double hardening model is the incremental plastic response in the Zone III, which is obtained by substituting the flow and hardening rules (18.29-31) into the consistency condition (18.26):

$$\begin{cases} -\mathrm{M}dp' + dq - d'_y d\lambda_s = 0 \\ dp' - \dfrac{1+e_0}{\lambda-\kappa} p'_c\left(d\lambda_v - \mathrm{M}d\lambda_s\right) = 0 \end{cases} \tag{18.32}$$

Solving these two equations with respect to $d\lambda_v$ and $d\lambda_s$, and using the equation of the yield surface $Y = p' - p'_c = 0$, we obtain:

$$\begin{cases} d\lambda_v - \mathrm{M}d\lambda_s = \dfrac{\lambda-\kappa}{1+e_0}\dfrac{dp'}{p'} \\ d\lambda_s = \dfrac{1}{d'_y}\left(-\mathrm{M}dp' + dq\right) \end{cases} \tag{18.33}$$

By substituting the above expressions into the flow rule (18.29), we obtain the plastic strain increment in the Zone III:

$$\begin{Bmatrix} d\varepsilon_v^p \\ d\varepsilon_s^p \end{Bmatrix} = \frac{1}{d'_y} \begin{bmatrix} \frac{\lambda-\kappa}{1+e_0} \frac{d'_y}{p'} & 0 \\ -\mathrm{M} & 1 \end{bmatrix} \begin{Bmatrix} dp' \\ dq \end{Bmatrix} \tag{18.34}$$

which together with the elastic relationship

$$\begin{Bmatrix} d\varepsilon_v^e \\ d\varepsilon_s^e \end{Bmatrix} = \begin{bmatrix} \frac{\kappa}{1+e_0} \frac{1}{p'} & 0 \\ 0 & \frac{1}{3G} \end{bmatrix} \begin{Bmatrix} dp' \\ dq \end{Bmatrix} \tag{18.35}$$

produces the total incremental stress strain response in the Zone III:

$$\begin{Bmatrix} d\varepsilon_v \\ d\varepsilon_s \end{Bmatrix} = \begin{bmatrix} \frac{\lambda}{1+e_0} \frac{1}{p'} & 0 \\ -\frac{\mathrm{M}}{d'_y} & \frac{1}{d'_y} + \frac{1}{3G} \end{bmatrix} \begin{Bmatrix} dp' \\ dq \end{Bmatrix} \tag{18.36}$$

Note that in spite of the fact that the associated flow rule has been chosen for the both yield surfaces, in the Zone III the compliance matrix is not symmetric. This is a typical problem of a “corner” between the two adjacent yield surfaces – the flow at this corner is always non-associated. Indeed, depending on the direction of the incremental stress vector $\{dp',dq\}$, the incremental plastic strain vector $\{dp',dq\}$ can have any direction between the two normal vectors to the two surfaces at this corner. This may cause certain numerical difficulties in the FE applications.

18.4 Example of a cap model

In order to better describe yielding and avoid creating “corners”, the cap in the double hardening models is often curved, and sometimes includes an additional surface to provide a smooth transition between the cap and the shear cone. Example of such a surface, adopted as a built-in model in ABAQUS, is briefly described below.

The yield surface in triaxial space consists of three parts (Figure 18.4). The shear yield surface is given by the Drucker-Prager failure surface:

$$F_s = q - \mathrm{M}p' - d = 0 \tag{18.37}$$

where d is a constant (i.e., no hardening takes place).

The cap yield surface is an ellipse:

$$Y_v = \sqrt{(p' - p'_a)^2 + \left[\frac{Rq}{1+\alpha-\alpha/\cos\beta}\right]^2} - R(d + p'_a\mathrm{M}) = 0 \tag{18.38}$$

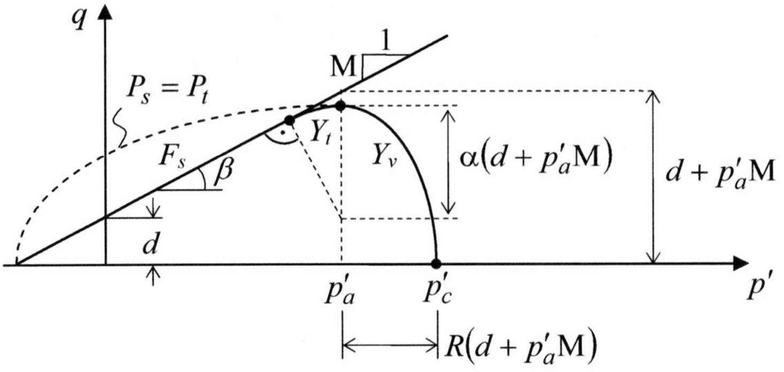

Figure 18.4: Example of a "smooth" cap model.

where $\tan\beta = \mathrm{M}$; α and R are material constants; and

$$p'_a = \frac{p'_c - Rd}{1 + R\mathrm{M}} \tag{18.39}$$

where p'_c is the pre-consolidation pressure.

The transition between the above two surfaces is achieved by the transition yield surface, which is also elliptical:

$$Y_t = \sqrt{(p' - p'_a)^2 + \left[q - \left(1 - \frac{\alpha}{\cos\beta}\right)(d + p'_a\mathrm{M})\right]^2} - \alpha(d + p'_a\mathrm{M}) = 0 \tag{18.40}$$

The flow rule for the cap is associated, with the plastic potential having the same shape as the yield surface:

$$P_v = \sqrt{(p' - p'_a)^2 + \left[\frac{Rq}{1 + \alpha - \alpha/\cos\beta}\right]^2} - x_v \tag{18.41}$$

While for the shear and transition surfaces a non-associated flow rule is adopted with the same elliptical plastic potential surface:

$$P_s = P_t = \sqrt{[(p'_a - p')\mathrm{M}]^2 + \left[\frac{q}{1 + \alpha - \alpha/\cos\beta}\right]^2} - x_s \tag{18.42}$$

Finally, the hardening rule controls the size of the cap and the transition yield surfaces by relating the pre-consolidation pressure to the volumetric plastic strain in a usual way:

$$p'_c = p'_{c0} \exp\left(\frac{1 + e_0}{\lambda - \kappa}\varepsilon_v^p\right) \tag{18.43}$$

The incremental response of this model can be derived from the above equations using the corresponding plastic loading conditions. An advantage of this model is that it does not have corners and the plastic flow direction is changing smoothly. The disadvantages are the complexity of the formulation and the non-associated flow rule, leading to non-symmetric compliance matrices.

18.5 Thermomechanical aspects of the pre-consolidation

In order to ensure thermomechanical consistency for a pre-consolidation model, both the dissipation function and the yield surface should depend on the volumetric kinematic variable:

$$\lambda Y\left(p',q,\chi_p,\chi_q,\alpha_v\right)=\chi_p\dot{\alpha}_v+\chi_q\dot{\alpha}_s-d\left(p',q,\dot{\alpha}_v,\dot{\alpha}_s,\alpha_v\right)=0 \tag{18.44}$$

For example, the simple model considered in the section 18.2, with the non-linear volumetric elastic behaviour

$$\varepsilon_v^e=\frac{\kappa}{1+e_0}\ln\frac{p'}{p_0'}; \qquad \varepsilon_s^e=\frac{q}{3G}, \tag{18.45}$$

the yield surface:

$$Y\left(p',q,\varepsilon_v^p\right)=p'-p_c'\left(\varepsilon_v^p\right)=0 \tag{18.46}$$

the associated flow rule

$$\begin{Bmatrix} d\varepsilon_v^p \\ d\varepsilon_s^p \end{Bmatrix}=d\lambda\begin{Bmatrix} \partial Y/\partial p' \\ \partial Y/\partial q \end{Bmatrix}=d\lambda\begin{Bmatrix} 1 \\ 0 \end{Bmatrix} \tag{18.47}$$

and the hardening rule:

$$p_c'\left(\varepsilon_v^p\right)=p_{c0}'\exp\left(\frac{1+e_0}{\lambda-\kappa}\varepsilon_v^p\right) \tag{18.48}$$

automatically satisfies the Laws of Thermodynamics, because it can be expressed within the hyperplastic formulation with the Gibbs free energy potential:

$$g\left(p',q,\alpha_v,\alpha_s\right)=-\frac{\kappa p'}{1+e_0}\left(\ln\left(\frac{p'}{p_0'}\right)-1\right)-\frac{q^2}{6G}-p'\alpha_v-q\alpha_s \tag{18.49}$$

and the dissipation function:

$$d\left(\alpha_v,\dot{\alpha}_v\right)=p_{c0}'e^{\frac{1+e_0}{\lambda-\kappa}\alpha_v}\dot{\alpha}_v\geq 0 \tag{18.50}$$

complemented by the zero deviatoric plastic strain rate constraint:

$$c(\dot{\alpha}_v, \dot{\alpha}_s) = \dot{\alpha}_s = 0 \tag{18.51}$$

In order to prove that the two formulations are equivalent, we modify the dissipation function as:

$$d' = d + \Lambda c = p'_{c0} e^{\frac{1+e_0}{\lambda-\kappa}\alpha_v} \dot{\alpha}_v + \Lambda \dot{\alpha}_s \tag{18.52}$$

with Λ as a Lagrangian multiplier, and obtain the generalized stresses:

$$\chi_p = \frac{\partial d'}{\partial \dot{\alpha}_v} = p'_{c0} e^{\frac{1+e_0}{\lambda-\kappa}\alpha_v}; \qquad \chi_q = \frac{\partial d'}{\partial \dot{\alpha}_s} = \Lambda . \tag{18.53}$$

The first expression (18.52) produces the yield surface in the generalized stresses:

$$Y(\chi_p, \alpha_v) = \chi_p - p'_{c0} e^{\frac{1+e_0}{\lambda-\kappa}\alpha_v} \tag{18.54}$$

with the flow rule

$$\dot{\alpha}_v = \lambda \frac{\partial Y(\chi_p, \alpha_v)}{\partial \chi_p} = \lambda; \qquad \dot{\alpha}_s = \lambda \frac{\partial Y(\chi_p, \alpha_v)}{\partial \chi_q} = 0 . \tag{18.55}$$

Next, because the Gibbs free energy potential (18.48) leads to the following decomposition of strains:

$$\begin{aligned} \varepsilon_v &= -\frac{\partial g}{\partial p'} = \frac{\kappa}{1+e_0} \ln \frac{p'}{p'_0} + \alpha_v = \varepsilon_v^e + \varepsilon_v^p \\ \varepsilon_s &= -\frac{\partial g}{\partial q} = \frac{q}{3G} + \alpha_s = \varepsilon_s^e + \varepsilon_s^p \end{aligned} \tag{18.56}$$

and produces the Ziegler's condition:

$$\chi_p = \frac{\partial d}{\partial \dot{\alpha}_v} = -\frac{\partial g}{\partial \alpha_v} = p'; \qquad \chi_q = \frac{\partial d}{\partial \dot{\alpha}_s} = -\frac{\partial g}{\partial \alpha_s} = q \tag{18.57}$$

substitution of equations (18.56)-(18.57) into the expressions (18.54)-(18.55) results in the yield surface, flow and hardening rules of the original formulation (18.45)-(18.48), thus confirming its thermomechanical consistency.

Chapter 19

Critical State

TABLE OF CONTENTS

Chapter 19

Critical State

19.1 Introduction

In the previous chapter we made an attempt to model both volumetric and shear yielding separately by using double hardening, or cap models. This approach leads to complicated multi-component yield surfaces and non-associated flow rules, which may cause numerical difficulties. More importantly, while the cap models are capable of reasonably accurate modelling of the strain hardening behaviour of normally to slightly overconsolidated clays (Figure 19.1a,b), they largely fail to predict the strain softening behaviour of the heavily overconsolidated clays (Figure 19.1d,e). Also, although the cap models are capable of modelling contraction of slightly overconsolidated clays (Figure 19.1c) and dilation of heavily overconsolidated clays (Figure 19.1f) in yielding, they cannot simulate for both cases the constant volume shearing at failure using associated flow rule. This constant volume shearing at failure is called the *critical state*. The so called Critical State models - original and modified Cam-Clay models - successfully overcome all the above limitations.

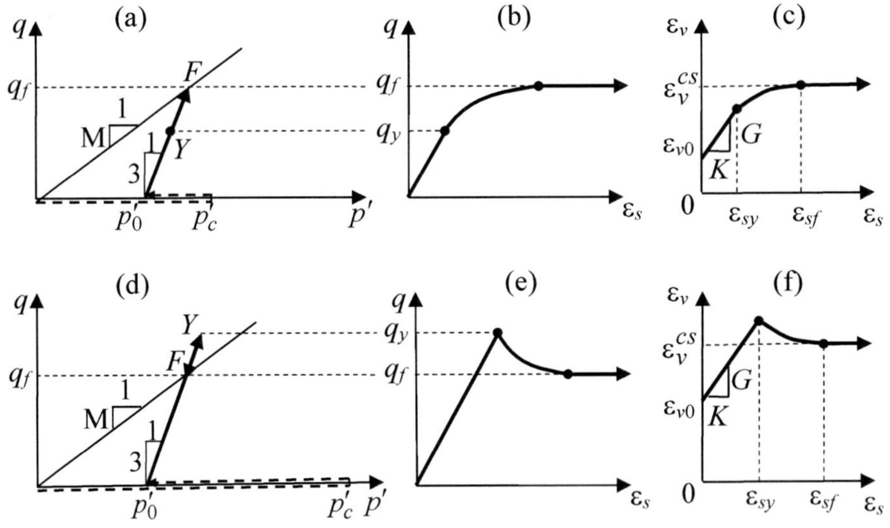

Figure 19.1: Triaxial compression tests on slightly and heavily overconsolidated clays with (a) and (d) the stress paths; (b) and (e) deviatoric stress-strain curves; (c) and (f) the strain paths, respectively.

This chapter shows how the above described pre-consolidation, strain hardening, strain softening, and critical state phenomena can be modelled using the Critical State models. The models are introduced within the formal framework of the theory of plasticity and not in the classical way they were developed. The thermomechanical aspects of pre-consolidation are also explored.

19.2 The original Cam-Clay model formulation

In this section the original Cam-Clay model (Schofield and Wroth, 1968) is introduced within the framework of the theory of plasticity and its incremental elastic-plastic response is derived.

19.2.1 Decomposition of strains

Decomposition of the strain increments into elastic and plastic components gives:

$$d\varepsilon_v = d\varepsilon_v^e + d\varepsilon_v^p\,; \qquad d\varepsilon_s = d\varepsilon_s^e + d\varepsilon_s^p\,. \tag{19.1}$$

19.2.2 Elastic volumetric behaviour

Elastic components of strain increment are defined via the following isotropic relationship:

$$\begin{Bmatrix} d\varepsilon_v^e \\ d\varepsilon_s^e \end{Bmatrix} = \begin{bmatrix} 1/K(p') & 0 \\ 0 & 0 \end{bmatrix} \begin{Bmatrix} dp' \\ dq \end{Bmatrix} \tag{19.2}$$

The original Cam-Clay model assumes zero elastic shear strains, while the volumetric part of this elastic relationship is non-linear, with the tangent bulk modulus:

$$K(p') = \frac{(1+e_0)p'}{\kappa} \tag{19.3}$$

which is defined to fit the experimentally observed elastic unloading-reloading line (URL) in isotropic compression:

$$\varepsilon_v^e = \frac{\kappa}{1+e_0} \ln \frac{p'}{p_0'} \tag{19.4}$$

The latter URL definition is easier recognized when expressed via the void ratio (Figure 19.2a):

$$e = e_0 - \kappa \ln p'/p_0' \tag{19.5}$$

$$\varepsilon_v = -\frac{v - v_0}{v_0} = \frac{e_0 - e}{1+e_0} \tag{19.6}$$

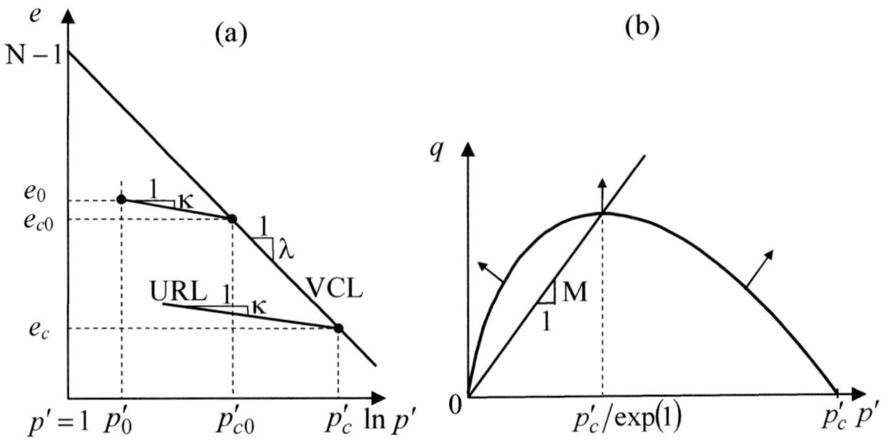

Figure 19.2: The original Cam-Clay model: (a) elastic-plastic volumetric behaviour; (b) yield surface and the associated flow rule.

19.2.3 Yield surface

In the triaxial stress space the model has a bullet shaped yield surface (Figure 19.2b):

$$Y\left(p', q, \varepsilon_v^p\right) = q - \mathrm{M}p' \ln p'_c + \mathrm{M}p' \ln p' = 0 \tag{19.7}$$

where $p'_c = f\left(\varepsilon_v^p\right)$ is the pre-consolidation pressure. The surface has its maximum at ($dq/dp' = 0$):

$$p' = p'_c/\exp(1); \qquad q = Mp' = Mp'_c/\exp(1). \tag{19.8}$$

19.2.4 Consistency condition

The consistency condition (and plastic loading criterion) is then defined by

$$dY = dq - dp'\mathrm{M}\ln p'_c - \frac{p'}{p'_c}\mathrm{M}dp'_c + dp'\mathrm{M}\ln p' + \mathrm{M}dp' = 0 \tag{19.9}$$

which, using the equation of the yield surface, can also be written in a more compact form

$$dY = dq + dp'\left(\mathrm{M} - \frac{q}{p'}\right) - \frac{p'}{p'_c}\mathrm{M}dp'_c = 0 \tag{19.10}$$

19.2.5 Flow rule

The associated flow rule (Figure 19.2b) produces:

$$\begin{Bmatrix} d\varepsilon_v^p \\ d\varepsilon_s^p \end{Bmatrix} = d\lambda \begin{Bmatrix} \partial Y/\partial p' \\ \partial Y/\partial q \end{Bmatrix} = d\lambda \begin{Bmatrix} \mathrm{M}\left(1 - \ln \dfrac{p'_c}{p'}\right) \\ 1 \end{Bmatrix} = d\lambda \begin{Bmatrix} \mathrm{M} - \dfrac{q}{p'} \\ 1 \end{Bmatrix} \quad (19.11)$$

where the last equality is derived by using the equation of the yield surface. It follows that the sign of the volumetric plastic strain depends whether the stress point lies below or above the $q = \mathrm{M}p'$ line, leading to contraction and dilation, respectively.

19.2.6 Hardening rule

A hardening rule defines the functional dependency of the pre-consolidation stress on the hardening parameter – volumetric plastic strain ε_v^p:

$$p'_c\left(\varepsilon_v^p\right) = p'_{c0} \exp\left(\frac{1+e_0}{\lambda - \kappa} \varepsilon_v^p\right) \quad (19.12)$$

so that $\varepsilon_v^p = 0$ at $p'_c = p'_{c0}$. As will be shown below, this hardening rule has been calibrated to fit the experimental elastic-plastic virgin compression line (VCL) in isotropic compression (Figure 19.2):

$$e = e_{c0} - \lambda \ln p'/p'_{c0} = N - 1 - \lambda \ln p' \quad (19.13)$$

19.2.7 Incremental response

The incremental stress-strain response is obtained by substituting the flow (19.11) and hardening (19.12) rules into the expression for dp'_c:

$$dp'_c = \frac{\partial p'_c}{\partial \varepsilon_v^p} d\varepsilon_v^p = \frac{1+e_0}{\lambda - \kappa} p'_c d\lambda \left(\mathrm{M} - \frac{q}{p'}\right) \quad (19.14)$$

Substitution of this expression into the consistency condition (19.9) gives:

$$dY = dq + dp'\left(\mathrm{M} - \frac{q}{p'}\right) - \frac{1+e_0}{\lambda - \kappa} p' d\lambda \mathrm{M}\left(\mathrm{M} - \frac{q}{p'}\right) = 0 \quad (19.15)$$

Resolving it with respect to $d\lambda$ and substituting into the flow rule, we obtain:

$$d\varepsilon_v^p = \frac{\lambda - \kappa}{1+e_0} \frac{dq + dp'(\mathrm{M} - q/p')}{\mathrm{M}p'} \quad (19.16)$$

$$d\varepsilon_s^p = \frac{\lambda - \kappa}{1+e_0} \frac{dq + dp'(\mathrm{M} - q/p')}{\mathrm{M}p'(\mathrm{M} - q/p')} \tag{19.17}$$

The total stress increment is then obtained by substituting elastic (19.2) and plastic (19.16-17) strain increments into (19.1):

$$d\varepsilon_v = d\varepsilon_v^e + d\varepsilon_v^p = \frac{\lambda}{1+e_0}\frac{dp'}{p'}\left(1 + \left(\frac{\lambda-\kappa}{\lambda \mathrm{M}}\right)\left(\frac{dq}{dp'} - \frac{q}{p'}\right)\right) \tag{19.18}$$

$$d\varepsilon_s = d\varepsilon_s^e + d\varepsilon_s^p = \frac{\lambda-\kappa}{1+e_0}\frac{dp'}{p'}\frac{dq/dp' + (\mathrm{M} - q/p')}{\mathrm{M}(\mathrm{M} - q/p')} \tag{19.19}$$

19.3 Behaviour of the original Cam-Clay model

19.3.1 Triaxial compression

The behaviour of the original Cam-Clay model captures special features of the yielding in triaxial compression of both slightly and heavily overconsolidated clays presented in Figure 19.1.

Indeed, for smaller $OCR = \frac{p_c'}{p_0'} < \frac{3\exp(1)}{3-\mathrm{M}}$, the model exhibits a behaviour called "wet" of critical (Figure 19.3a-c). In this case the yielding starts at a point Y below the $q = \mathrm{M}p'$ line (Figure 19.3a). Because of the associated flow rule, the volumetric component of the incremental plastic strain vector $d\varepsilon_v^p$ below the $q = \mathrm{M}p'$ line is positive. Therefore, the volumetric plastic strain increases (Figure 19.3b), leading via the hardening rule (19.12) to the increase in p_c', so that the yield surface expands producing strain hardening (Figure 19.3c). As the stress path approaches the $q = \mathrm{M}p'$ line, the positive volumetric component of the incremental plastic strain vector $d\varepsilon_v^p$ becomes smaller and smaller, until it vanishes completely, as stress path reaches the $q = \mathrm{M}p'$ line at the point F (Figure 19.3a). At this point, the volumetric plastic strain remains constant (Figure 19.3b), leading via the hardening rule to the constant p_c', so that the yield surface cannot expand any more, and the sample fails at constant volume, representing the *critical state* (Figure 19.3c). The $q = \mathrm{M}p'$ line is called the critical state line (CSL). It is also the failure surface, but it has never been defined explicitly as such – as it is reached, the sample fails automatically. This is what makes the original Cam-Clay model so elegant and convenient: the failure at the critical state is modelled using associated flow rule, in spite of the fact that the incremental plastic strain vector is not normal to the failure surface!

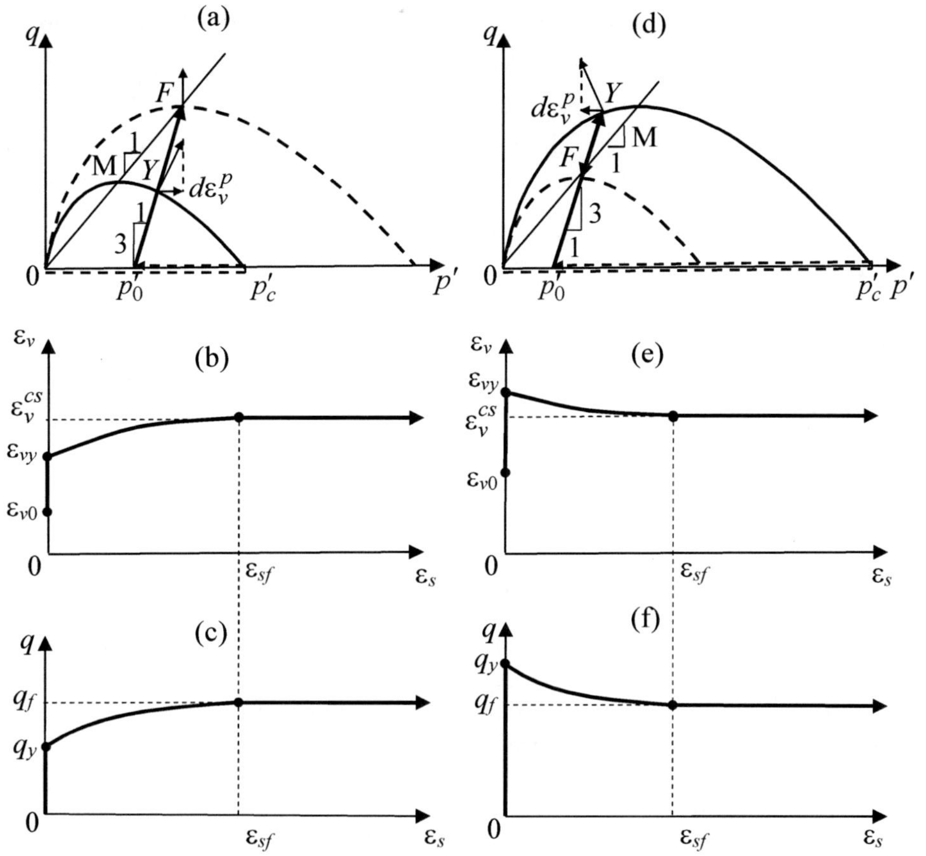

Figure 19.3: "Wet" and "dry" behaviour of the original Cam-Clay model in drained triaxial compression tests with (a) and (d) the stress paths; (b) and (e) the strain paths; (c) and (f) deviatoric stress-strain curves, respectively.

For larger $OCR = \dfrac{p_c'}{p_0'} > \dfrac{3\exp(1)}{3-\mathrm{M}}$, the model exhibits a behaviour called "dry" of critical (Figure 19.3d-f). In this case the yielding starts at a point Y above the CSL $q = \mathrm{M}p'$ (Figure 19.3d). Because of the associated flow rule, the volumetric component of the incremental plastic strain vector $d\varepsilon_v^p$ above the CSL $q = \mathrm{M}p'$ is negative. Therefore, the volumetric plastic strain decreases (Figure 19.3e), leading via the hardening rule (19.12) to the decrease in p_c', so that the yield surface contracts producing strain softening (Figure 19.3f). As the stress path approaches the CSL $q = \mathrm{M}p'$, the absolute value of the negative volumetric component of the incremental plastic strain vector $d\varepsilon_v^p$ becomes smaller and smaller, until it vanishes completely, as stress path reaches the CSL $q = Mp'$ at the point F (Figure 19.3d). At this point, the volumetric plastic strain remains constant (Figure 19.3e), leading via the hardening rule to the constant p'_c, so that the yield surface cannot expand any more, and the sample also fails at constant volume, representing the *critical state* (Figure 19.3f). The deviatoric stress at failure q_f represents in the "dry" case the *residual strength*, in contrast to the *peak strength* q_y, at which the softening begins.

19.3.2 Isotropic compression

For isotropic compression $dq = q = 0$, integration of equation (19.18), produces

$$\varepsilon_v - \varepsilon_{vc0} = \frac{\lambda}{1+e_0} \ln \frac{p'}{p'_{c0}} \tag{19.20}$$

where
$$\varepsilon_v - \varepsilon_{vc0} = \frac{e_0 - e}{1+e_0} - \frac{e_0 - e_{c0}}{1+e_0} = \frac{e_{c0} - e}{1+e_0} \tag{19.21}$$

leads to the VCL equation (19.13).

This is not, unfortunately, the only strain component, which develops in isotropic compression. Integration of equation (19.19), produces

$$\varepsilon_s - \varepsilon_{sc0} = \frac{1}{M} \frac{\lambda - \kappa}{1+e_0} \ln \frac{p'}{p'_{c0}} \tag{19.22}$$

Development of the shear strains in isotropic compression is a shortcoming of the original Cam-clay model bullet shape with a cusp at $q = 0$.

19.3.3 K₀-consolidation

For K₀-consolidation $d\varepsilon_3 = 0$ and $\sigma'_3/\sigma'_1 = K_0 = const$, so that

$$\frac{d\varepsilon_v}{d\varepsilon_s} = \frac{3}{2}; \qquad \frac{dq}{dp'} = \frac{q}{p'} = 3\frac{1-K_0}{1+2K_0}. \tag{19.23}$$

Using these relationships in equations (19.18) and (19.19) gives

$$\frac{d\varepsilon_v}{d\varepsilon_s} = \frac{M - q/p'}{1-k/\lambda} = \frac{M - 3(1-K_0)/(1+2K_0)}{1-k/\lambda} = \frac{3}{2} \tag{19.24}$$

so that K_0 is indeed constant and equal to

$$K_0 = \frac{1}{2}\left(\frac{3}{2M/3 + k/\lambda} - 1\right) \tag{19.25}$$

For typical values of $\varphi' = 30^\circ$ and $\frac{\kappa}{\lambda} = 0.2$, $M = \frac{6\sin\varphi'}{3-\sin\varphi'} = 1.2$ and from the above equations it follows that $K_0 = 1.0$ This estimate is significantly higher than the value $K_0 = 1 - \sin\varphi' = 0.5$ typical for normally consolidated clays. This is another shortcoming of the bullet shape of the original Cam-clay model.

19.3.4 Discussion

As is seen, the original Cam-Clay model predicts such important phenomena as pre-consolidation, strain hardening, strain softening, failure and critical state using a simple yield surface and the associated flow rule. It represents, therefore, a considerable step forward compared to the cap models. Unfortunately, its bullet like shape is responsible for overpredicting: (i) shear strains in isotropic compression and (ii) the at rest earth pressure coefficient in K_0-consolidation. In addition, the model neglects elastic shear strains. These problems can be solved by using the modified Cam-Clay model.

19.4 The modified Cam-Clay model

In this Section the modified Cam-Clay model (Roscoe and Burland, 1968) is introduced within the framework of the theory of plasticity and its incremental elastic-plastic response is derived.

19.4.1 Decomposition of strains

Decomposition of the strain increments into elastic and plastic components gives:

$$d\varepsilon_v = d\varepsilon_v^e + d\varepsilon_v^p; \qquad d\varepsilon_s = d\varepsilon_s^e + d\varepsilon_s^p. \tag{19.26}$$

19.4.2 Elastic volumetric behaviour

Elastic components of strain increment are defined via the following isotropic relationship:

$$\begin{Bmatrix} d\varepsilon_v^e \\ d\varepsilon_s^e \end{Bmatrix} = \begin{bmatrix} 1/K(p') & 0 \\ 0 & 1/3G \end{bmatrix} \begin{Bmatrix} dp' \\ dq \end{Bmatrix} \tag{19.27}$$

The modified Cam-Clay model assumes a constant shear modulus, while the volumetric part of this elastic relationship is non-linear, with the tangent bulk modulus:

$$K(p') = \frac{(1+e_0)p'}{\kappa} \tag{19.28}$$

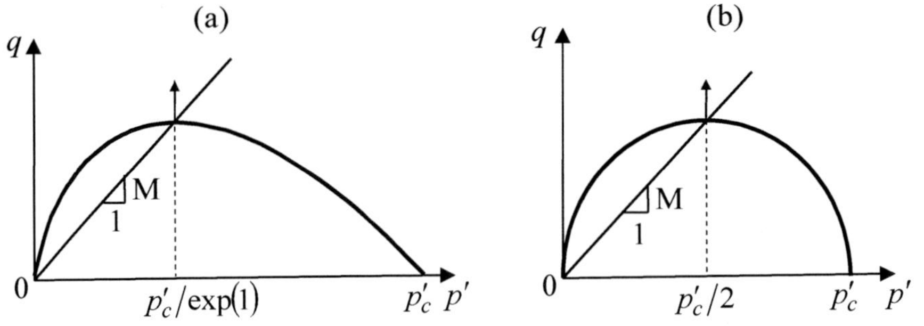

Figure 19.4: The yield surfaces of the (a) original and (b) modified Cam-Clay Models.

19.4.3 Yield surface

In contrast to the bullet shaped yield surface of the original Cam-Clay model (Figure 19.4a), the modified Cam-Clay model has an elliptical shape (Figure 19.4b):

$$Y\left(p', q, \varepsilon_v^p\right) = q^2 + \mathrm{M}^2\left(p' - \frac{p_c'}{2}\right)^2 - \left(\frac{p_c'}{2}\right)^2 \mathrm{M}^2 = 0 \qquad (19.29)$$

or

$$Y\left(p', q, \varepsilon_v^p\right) = q^2 + \mathrm{M}^2 p'^2 - \mathrm{M}^2 p' p'_c = 0 \qquad (19.30)$$

where $p'_c = f\left(\varepsilon_v^p\right)$ is the pre-consolidation pressure. The surface has its maximum at ($dq/dp' = 0$):

$$p' = p'_c/2\,; \qquad q = \mathrm{M}p' = \mathrm{M}p'_c/2\,. \qquad (19.31)$$

19.4.4 Consistency condition

The consistency condition (and plastic loading criterion) is then defined by

$$dY = 2q\,dq + 2\mathrm{M}^2\left(p' - \frac{p'_c}{2}\right)dp' - \mathrm{M}^2 p'\,dp'_c = 0 \qquad (19.32)$$

which using the equation of the yield surface

$$\mathrm{M}^2 p'_c = p'\left(q^2/p'^2 + \mathrm{M}^2\right) \qquad (19.33)$$

can also be written as

$$dY = 2q\,dq + \left(\mathrm{M}^2 - q^2/p'^2\right)p'\,dp' - \mathrm{M}^2 p'\,dp'_c = 0 \qquad (19.34)$$

19.4.5 Flow rule

The associated flow rule produces:

$$\begin{Bmatrix} d\varepsilon_v^p \\ d\varepsilon_s^p \end{Bmatrix} = d\lambda \begin{Bmatrix} \partial Y/\partial p' \\ \partial Y/\partial q \end{Bmatrix} = d\lambda \begin{Bmatrix} 2\mathrm{M}^2(p' - p'_c/2) \\ 2q \end{Bmatrix} = d\lambda \begin{Bmatrix} p'\left(\mathrm{M}^2 - q^2/p'^2\right) \\ 2q \end{Bmatrix} \tag{19.35}$$

where the last equality is derived by using the equation of the yield surface. Like for the original Cam-Clay model, the sign of the volumetric plastic strain depends upon whether the stress point lies below or above the $q = \mathrm{M}p'$ line, leading to contraction and dilation, respectively.

19.4.6 Hardening rule

A hardening rule defines the functional dependency of the pre-consolidation stress on the hardening parameter – volumetric plastic strain ε_v^p:

$$p'_c\left(\varepsilon_v^p\right) = p'_{c0} \exp\left(\frac{1+e_0}{\lambda - \kappa}\varepsilon_v^p\right) \tag{19.36}$$

so that $\varepsilon_v^p = 0$ at $p'_c = p'_{c0}$.

19.4.7 Incremental response

The incremental stress-strain response is obtained by substituting the flow (19.35) and hardening (19.36) rules and the equation for the yield surface (19.33) into the expression for dp'_c:

$$dp'_c = \frac{\partial p'_c}{\partial \varepsilon_v^p} d\varepsilon_v^p = d\lambda \frac{1+e_0}{\lambda - \kappa} p'_c p'\left(\mathrm{M}^2 - q^2/p'^2\right) = \frac{d\lambda}{M^2}\frac{1+e_0}{\lambda - \kappa} p'^2\left(\mathrm{M}^4 - q^4/p'^4\right) \tag{19.37}$$

Substitution of this expression into the consistency condition (19.34) we obtain:

$$dY = 2qdq + \left(\mathrm{M}^2 - q^2/p'^2\right)p'dp' - p'^3 d\lambda \frac{1+e_0}{\lambda - \kappa}\left(\mathrm{M}^4 - q^4/p'^4\right) = 0 \tag{19.38}$$

Resolving it with respect to $d\lambda$ and substituting into the flow rule, we obtain:

$$d\varepsilon_v^p = \frac{\lambda - \kappa}{1 + e_0} \frac{2\eta dq + \left(M^2 - \eta^2\right)dp'}{\left(M^2 + \eta^2\right)p'} \tag{19.39}$$

$$d\varepsilon_s^p = \frac{\lambda - \kappa}{1 + e_0} \frac{4\eta^2 dq + 2\eta\left(M^2 - \eta^2\right)dp'}{\left(M^4 - \eta^4\right)p'} \tag{19.40}$$

where $\eta = q/p'$.

The total stress increment is then obtained by substituting elastic (19.27) and plastic (19.38-39) strain increments into (19.26):

$$d\varepsilon_v = \frac{\lambda}{1 + e_0} \frac{dp'}{p'} \left(\frac{\kappa}{\lambda} + \left(1 - \frac{\kappa}{\lambda}\right) \frac{2\eta dq/dp' + M^2 - \eta^2}{M^2 + \eta^2} \right) \tag{19.41}$$

$$d\varepsilon_s = \frac{dq}{3G} + \frac{\lambda - \kappa}{1 + e_0} \frac{4\eta^2 dq + 2\eta\left(M^2 - \eta^2\right)dp'}{\left(M^4 - \eta^4\right)p'} \tag{19.42}$$

19.4.8 Triaxial compression

The behaviour of the modified Cam-Clay model also captures special features of the yielding in triaxial compression of both slightly and heavily overconsolidated clays presented in Figure 19.1.

Indeed, for smaller $OCR = \frac{p_c'}{p_0'} < \frac{6}{3 - M}$, the model exhibits the behaviour "wet" of critical (Figure 19.5a-c). For larger $OCR = \frac{p_c'}{p_0'} > \frac{6}{3 - M}$, the model exhibits the behaviour "dry" of critical (Figure 19.5d-f). The major difference with the original Cam-Clay model here is the existence of elastic shear strains.

19.4.9 Isotropic compression

For isotropic compression $dq = q = \eta = 0$, and $p' = p_c'$ integration of equation (19.41), produces

$$\varepsilon_v - \varepsilon_{vc0} = \frac{\lambda}{1 + e_0} \ln \frac{p'}{p_{c0}'} \tag{19.43}$$

leading to the VCL equation (19.13).

From equation (19.42) it follows that the shear strain component in isotropic compression is equal to zero.

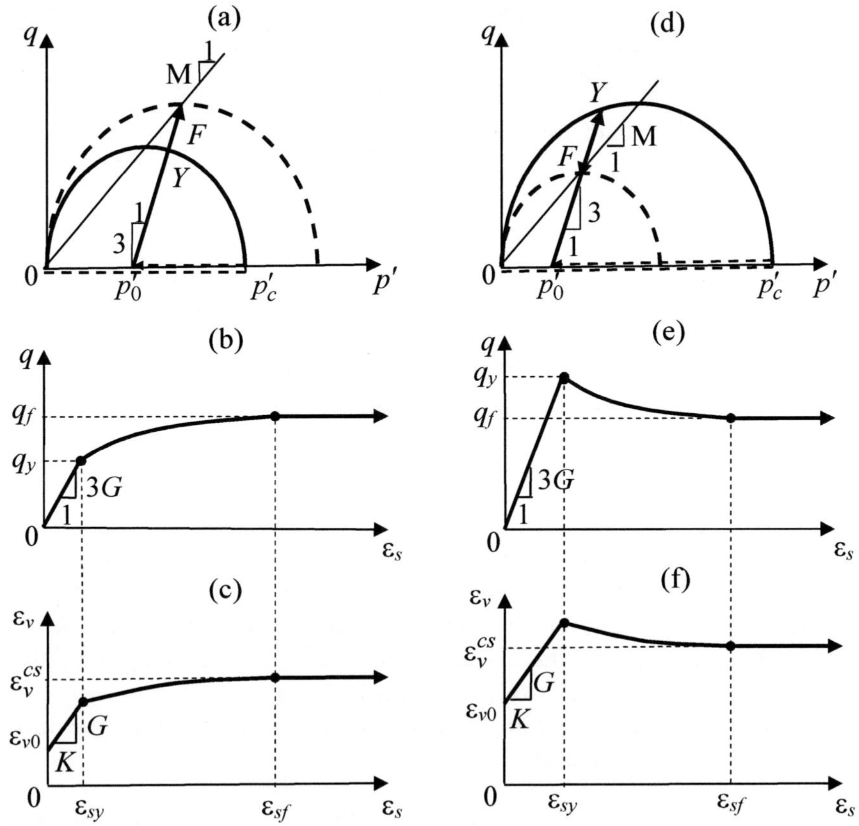

Figure 19.5: "Wet" and "dry" behaviour of the modified Cam-Clay model in drained triaxial compression tests with (a) and (d) the stress paths; (b) and (e) the strain paths; (c) and (f) deviatoric stress-strain curves, respectively.

19.4.10 K$_0$-consolidation

For K$_0$-consolidation $d\varepsilon_3 = 0$ and $\sigma'_3/\sigma'_1 = K_0 = const$, so that

$$\frac{d\varepsilon_v}{d\varepsilon_s} = \frac{3}{2} \qquad \frac{dq}{dp'} = \frac{q}{p'} = \eta = 3\frac{1-K_0}{1+2K_0} \tag{19.44}$$

Ignoring elastic shear strains $(G \to \infty)$ and using the above relationships in equations (19.41) and (19.42) gives

$$\frac{d\varepsilon_v}{d\varepsilon_s} = \frac{M^2 - \eta^2}{2\eta(1-k/\lambda)} = \frac{3}{2} \tag{19.45}$$

which can be resolved with respect to η:

$$\eta = \sqrt{M^2 + \frac{9}{4}(1 - k/\lambda)^2} - \frac{3}{2}(1 - k/\lambda) \tag{19.46}$$

so that K_0 is indeed constant since it is equal to

$$K_0 = \frac{3 - \eta}{3 + 2\eta} \tag{19.47}$$

For typical values of $\varphi' = 30^\circ$ and $\frac{\kappa}{\lambda} = 0.2$, $\mathrm{M} = \frac{6\sin\varphi'}{3 - \sin\varphi'} = 1.2$ and from the above equations it follows that $\eta \approx 0.5$ and $K_0 \approx 0.63$ which is much closer to the value $K_0 = 1 - \sin\varphi' = 0.5$ typical for normally consolidated clays.

19.4.11 Discussion

As is seen, like the original Cam-Clay model, the modified model also predicts such important phenomena as pre-consolidation, strain hardening, strain softening, failure and critical state using a simple yield surface and the associated flow rule. However, its elliptical shape has advantages allowing for predicting: (i) zero shear strains in isotropic compression and (ii) the at rest earth pressure coefficient in K_0-consolidation. In addition, the model does not neglect elastic shear strains.

19.5 Thermomechanical consistency of the Critical State models

19.5.1 The original Cam-Clay model

The original Cam-Clay model presented in Section 19.2 automatically satisfies the Laws of Thermodynamics, because it can be expressed within the hyperplastic formulation with the Gibbs free energy potential:

$$g(p', q, \alpha_v, \alpha_s) = -\frac{\kappa p'}{1 + e_0}\left(\ln\left(\frac{p'}{p'_0}\right) - 1\right) - p'\alpha_v - q\alpha_s \tag{19.48}$$

and the dissipation function:

$$d(\alpha_v, \dot{\alpha}_s, \dot{\alpha}_v) = M\dot{\alpha}_s p'_{c0} e^{\frac{1+e_0}{\lambda-\kappa}\alpha_v} e^{\frac{\dot{\alpha}_v}{M\dot{\alpha}_s} - 1} \geq 0 \tag{19.49}$$

In order to prove that the two formulations are equivalent, expressions for the generalized stresses:

$$\chi_p = \frac{\partial d}{\partial \dot{\alpha}_v} = p'_{c0} e^{\frac{1+e_0}{\lambda-\kappa}\alpha_v} e^{\frac{\dot{\alpha}_v}{M\dot{\alpha}_s}-1}$$

$$\chi_q = \frac{\partial d}{\partial \dot{\alpha}_s} = Mp'_{c0} e^{\frac{1+e_0}{\lambda-\kappa}\alpha_v} e^{\frac{\dot{\alpha}_v}{M\dot{\alpha}_s}-1}\left(1-\frac{\dot{\alpha}_v}{M\dot{\alpha}_s}\right) \tag{19.50}$$

have to be combined to produce the yield surface in the generalized stresses:

$$Y(\chi_p, \chi_q, \alpha_v) = \chi_q - M\chi_p \ln\left(p_{c0}' e^{\frac{1+e_0}{\lambda-\kappa}\alpha_v}\right) + M\chi_p \ln \chi_p = 0 \tag{19.51}$$

with the corresponding flow rule given by

$$\dot{\alpha}_v = \Lambda \frac{\partial Y(\chi_p, \chi_q, \alpha_v)}{\partial \chi_p} = \Lambda\left(1 - \ln\left(\frac{p_{c0}'}{p'} e^{\frac{1+e_0}{\lambda-\kappa}\alpha_v}\right)\right)$$

$$\dot{\alpha}_s = \Lambda \frac{\partial Y(\chi_p, \chi_q, \alpha_v)}{\partial \chi_q} = \Lambda \tag{19.52}$$

The Gibbs free energy potential (19.48) leads to the following decomposition of strains:

$$\varepsilon_v = -\frac{\partial g}{\partial p'} = \frac{\kappa}{1+e_0} \ln \frac{p'}{p_0'} + \alpha_v = \varepsilon_v^e + \varepsilon_v^p; \qquad \varepsilon_s = -\frac{\partial g}{\partial q} = \alpha_s = \varepsilon_s^p \tag{19.53}$$

and produces the Ziegler's condition:

$$\chi_p = \frac{\partial d}{\partial \dot{\alpha}_v} = -\frac{\partial g}{\partial \alpha_v} = p'; \qquad \chi_q = \frac{\partial d}{\partial \dot{\alpha}_s} = -\frac{\partial g}{\partial \alpha_s} = q, \tag{19.54}$$

which after substitution into the expressions (19.51)-(19.52) result in the yield surface (19.7), flow rule (19.11) and hardening rule (19.12) of the original Cam-Clay model, confirming its thermomechanical consistency.

19.5.2 The modified Cam-Clay model

The modified Cam-Clay model presented in Section 19.4 automatically satisfies the Laws of Thermodynamics, because it can be expressed within the hyperplastic formulation with the Gibbs free energy potential:

$$g(p', q, \alpha_v, \alpha_s) = -\frac{\kappa p'}{1+e_0}\left(\ln\left(\frac{p'}{p_0'}\right) - 1\right) - \frac{q^2}{6G} - p'\alpha_v - q\alpha_s \tag{19.55}$$

and the dissipation function:

$$d(\alpha_v,\dot{\alpha}_s,\dot{\alpha}_v)=\frac{1}{2}p'_{c0}e^{\frac{1+e_0}{\lambda-\kappa}\alpha_v}\left(\dot{\alpha}_v+\sqrt{\dot{\alpha}_v^2+M^2\dot{\alpha}_s^2}\right)\geq 0 \qquad (19.56)$$

In order to prove that the two formulations are equivalent, the generalized stresses:

$$\begin{aligned}\chi_p&=\frac{\partial d}{\partial\dot{\alpha}_v}=\frac{1}{2}p'_{c0}e^{\frac{1+e_0}{\lambda-\kappa}\alpha_v}\left(1+\frac{\dot{\alpha}_v}{\sqrt{\dot{\alpha}_v^2+M^2\dot{\alpha}_s^2}}\right)\\ \chi_q&=\frac{\partial d}{\partial\dot{\alpha}_s}=\frac{1}{2}p'_{c0}e^{\frac{1+e_0}{\lambda-\kappa}\alpha_v}\frac{M^2\dot{\alpha}_s}{\sqrt{\dot{\alpha}_v^2+M^2\dot{\alpha}_s^2}}\end{aligned} \qquad (19.57)$$

have to be combined to produce the yield surface in the generalized stresses:

$$Y(\chi_p,\chi_q,\alpha_v)=\frac{\chi_q^2}{\mathrm{M}^2}+\left(\chi_p-\frac{p_{c0}'}{2}e^{\frac{1+e_0}{\lambda-\kappa}\alpha_v}\right)^2-\left(\frac{p_{c0}'}{2}e^{\frac{1+e_0}{\lambda-\kappa}\alpha_v}\right)^2=0 \qquad (19.58)$$

with the corresponding flow rule given by

$$\begin{aligned}\dot{\alpha}_v&=\Lambda\frac{\partial Y(\chi_p,\chi_q,\alpha_v)}{\partial\chi_p}=2\Lambda\chi_p-\Lambda p_{c0}'e^{\frac{1+e_0}{\lambda-\kappa}\alpha_v}\\ \dot{\alpha}_s&=\Lambda\frac{\partial Y(\chi_p,\chi_q,\alpha_v)}{\partial\chi_q}=\frac{2\Lambda\chi_q}{M^2}\end{aligned} \qquad (19.59)$$

The Gibbs free energy potential (19.55) leads to the decomposition of strains:

$$\begin{aligned}\varepsilon_v&=-\frac{\partial g}{\partial p'}=\frac{\kappa}{1+e_0}\ln\frac{p'}{p_0'}+\alpha_v=\varepsilon_v^e+\varepsilon_v^p\\ \varepsilon_s&=-\frac{\partial g}{\partial q}=\frac{q}{3G}+\alpha_s=\varepsilon_s^e+\varepsilon_s^p\end{aligned} \qquad (19.60)$$

and produces the Ziegler's condition (19.54), which after substitution into the expressions (19.58)-(19.59) result in the yield surface (19.29), flow rule (19.35) and hardening rule (19.35) of the modified Cam-Clay model, confirming its thermomechanical consistency.

Chapter 20

Irreversibility in Cyclic Behaviour

TABLE OF CONTENTS

Chapter 20

Irreversibility in Cyclic Behaviour

20.1 Introduction

In a typical drained triaxial compression test on slightly overconsolidated clay, after isotropic consolidation up to the pre-consolidation pressure p'_c with subsequent unloading down to the initial mean effective stress p'_0, the sample is sheared along the straight effective stress path (Figure 20.1a).

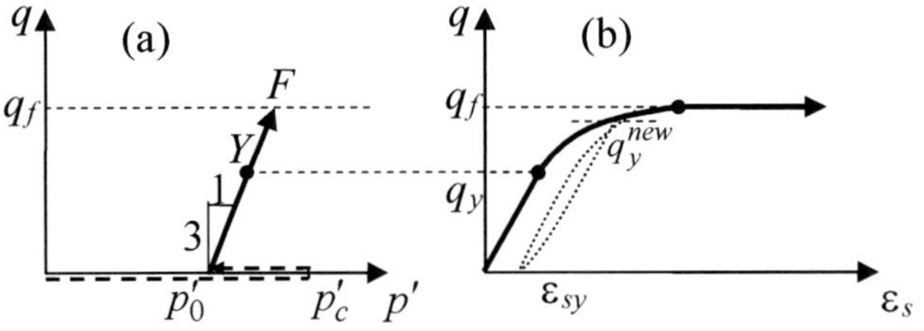

Figure 20.1: Results of a triaxial compression test on a slightly overconsolidated clay: (a) the stress path; (b) deviatoric behaviour.

In the initial part of this stress path, the deviatoric behaviour of the sample (Figure 20.1b) is normally assumed as being almost *reversible*. According to Chapter 13, however, recent experimental research has demonstrated that even at the very small strains the stress-strain behaviour is not entirely reversible, i.e. it exhibits some very small permanent strains in a closed loading cycle (Figure 20.2).

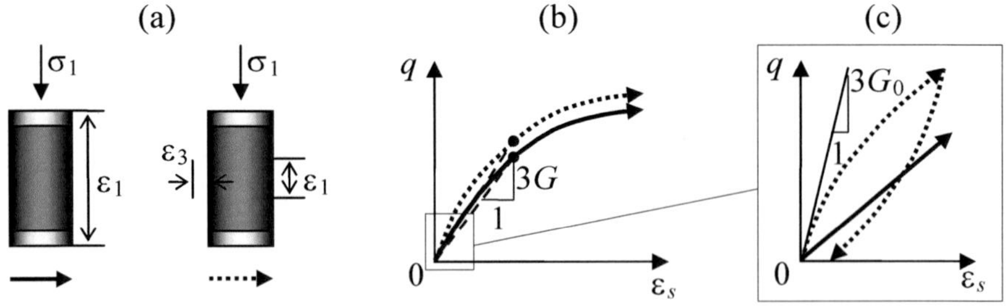

Figure 20.2: External vs. local deformation measurements in a triaxial test: (a) schematic setup; (b) deviatoric stress-strain curves; (c) zoom into small strains.

Further evidence of this irreversibility is the cyclic soil behaviour below the yield stress state $Y(p'_y, q_y)$, which defines a boundary of the large scale plastic yielding (Figure 20.1b), so that the descending dotted line, representing unloading, does not follow the stress-strain curve for initial loading. When the sample is reloaded (the ascending dotted line in Figure 20.1b), the yield stress q_y^{new} is higher than the one reached in the initial loading, i.e. the material is *strain-hardening*. A more advanced analysis of the stress cycle in Figure 20.1b indicates that even before the yield stress q_y^{new} is reached, the reloading stress-strain curve does not follow the unloading curve exactly, i.e., the soil behaviour is *irreversible* even before the large scale yielding takes place.

20.2 Consolidation surface plasticity

The preceding chapters describe the models formulated using basic plasticity theories, culminating in the so called *consolidation surface plasticity.* Cap models and Critical State soil models (original Cam-Clay and modified Cam-Clay) are among those based on this kind of plasticity. In the consolidation surface plasticity, the isotropic hardening yield surface (Figure 20.3a) is defined by equation:

$$Y\left(p', q, \varepsilon_v^p\right) = 0 \tag{20.1}$$

Volumetric and deviatoric strain increments can be separated into elastic and plastic components:

$$d\varepsilon_v = d\varepsilon_v^e + d\varepsilon_v^p; \quad d\varepsilon_s = d\varepsilon_s^e + d\varepsilon_s^p. \tag{20.2}$$

It is assumed that if the stress state is inside the yield surface, or if it is on the yield surface but the incremental stress vector is directed inside the surface, the behaviour is elastic and governed by Hooke's law:

$$d\varepsilon_v^e = \frac{dp'}{K}; \quad d\varepsilon_s^p = \frac{dq}{3G}, \tag{20.3}$$

where G is the small strain shear modulus, K is the bulk modulus. If the stress state is on the yield surface and the incremental stress vector is directed outwards, the behaviour is elastic-plastic, and the plastic components of the incremental strain vector are defined by equations:

$$d\varepsilon_v^p = \lambda \frac{\partial Y}{\partial p'}; \quad d\varepsilon_s^p = \lambda \frac{\partial Y}{\partial q}, \tag{20.4}$$

where λ is an infinitisemal parameter. Equations (20.4) are based on the assumption that the plastic potential coincides with the yield surface (associated flow rule). Parameter λ can then be obtained from the consistency condition:

$$dY(p', q, \varepsilon_v^p) = 0 \tag{20.5}$$

resulting from the assumption that during plastic flow, the stress state stays on the yield surface. It follows,

$$\lambda = \frac{1}{H}\left(\frac{\partial Y}{\partial p'}dp' + \frac{\partial Y}{\partial q}dq\right) \tag{20.6}$$

where

$$H = -\frac{\partial Y}{\partial \varepsilon_v^p}\frac{\partial Y}{\partial p'} \tag{20.7}$$

is the hardening plastic modulus. Subsequent substitutions of equation (20.7) into (20.6) and (20.4) yield unique expressions for plastic strain increments.

As is seen from the above formulation, when the stress state tends to move outside the surface, it leads to an increment in volumetric plastic strain. The stress state stays on the yield surface and the yield surface expands isotropically, because it depends on the internal variable ε_v^p (i.e., degree of material consolidation, so that this surface is often regarded as a *consolidation surface*).

Significant advantage of the consolidation surface plasticity is that it is capable of an accurate and consistent treatment of the large scale yielding before failure. However, as mentioned above, the response within the yield surface is perfectly elastic; this is hardly consistent with the real behaviour of soils exhibiting the irreversibility of the small strain behaviour and the hysteresis during cyclic stress reversals inside the consolidation yield surface. In order to introduce plastic deformation within the yield surface, two alternative approaches can be used: *nested surface plasticity* and *bounding surface plasticity*.

20.3 Nested loading or yield surface plasticity

A kinematic hardening multiple surface plasticity formulation for soils can be most easily illustrated by the two surfaces model (Mroz et al., 1979). Assume that the yield surface, if it exists, is smaller than the consolidation surface defined in the previous section, and may translate and expand or contract in the stress space (Figure 20.3b); thus

$$y(p' - \rho_p, q - \rho_q, \varepsilon_v^p) = 0 \tag{20.8}$$

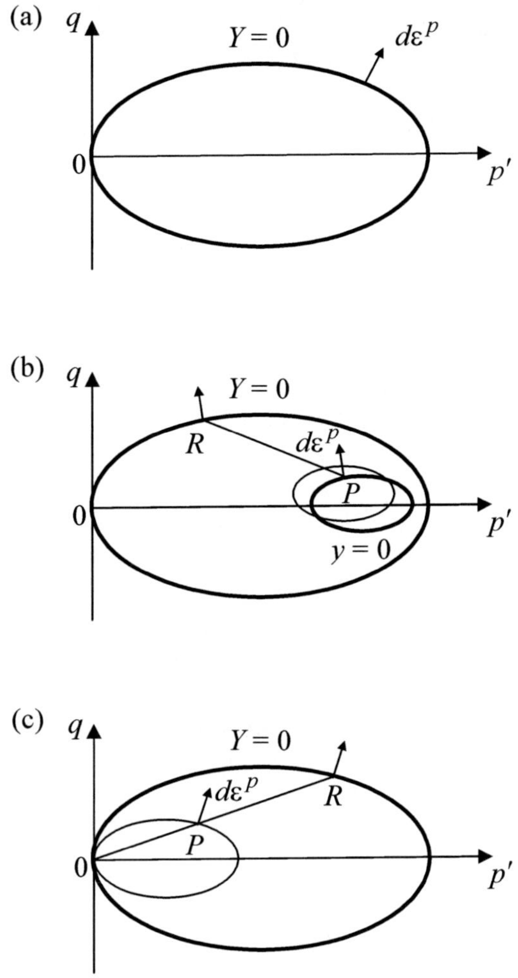

Figure 20.3: Schematic illustration of different plasticity formulations: a) isotropic hardening; b) nested surfaces; c) bounding surface.

and the associated flow rule applies. Parameters ρ_p, ρ_q define the position of the yield surface and can be interpreted as the stress coordinates of its "centre". In order to construct the complete set of equations, the translation rule of the yield surface (defining changes in these parameters) and its dependence on the plastic volumetric strain should be specified.

It is assumed that if the stress state is inside the yield surface, or if it is on the yield surface but the incremental stress vector is directed inwards, the behaviour is elastic, and governed by equations (20.3). If the stress state is on the yield surface and the incremental stress vector is directed outwards, the behaviour is elastic-plastic, and the plastic components of the incremental strain vector are defined from the equations:

$$d\varepsilon_v^p = \frac{1}{h}\left(\frac{\partial y}{\partial p'}dp' + \frac{\partial y}{\partial q}dq\right)\frac{\partial y}{\partial p}$$
$$d\varepsilon_s^p = \frac{1}{h}\left(\frac{\partial y}{\partial p'}dp' + \frac{\partial y}{\partial q}dq\right)\frac{\partial y}{\partial q} \qquad (20.9)$$

which satisfy the associated flow rule for surface y (20.8), with h standing for the hardening plastic modulus.

Assume that besides the yield surface y (20.8), enclosing the elastic domain, there exists a larger consolidation or boundary surface Y, defined by the degree of material consolidation (Figure 20.3b). The domain enclosed by the consolidation surface in this formulation is not elastic and for stress trajectories within this surface plastic flow occurs once the yield condition (20.8) is satisfied. Assuming that the consolidation surface expands or contracts isotropically, we can identify it with the isotropic hardening surface (20.1) discussed in the previous section. It is further assumed that the hardening modulus H on the consolidation surface varies according to (20.7), whereas the hardening modulus h on the yield surface depends on the relative configuration of these two surfaces, and becomes equal to H when these two surfaces get in contact with each other. In the case of kinematic hardening, the consistency condition:

$$dy(p' - \rho_p, q - \rho_q, \varepsilon_v^p) = 0 \qquad (20.10)$$

is not sufficient to define the hardening modulus h for equations (20.9), because equation (20.10) contains two other unknown variables $d\rho_p$ and $d\rho_q$. This problem is usually dealt with by postulating non-intersection of the yield surfaces, hardening rule and translation rule.

Non-intersection condition

In order to preserve a smooth transition between the deformation processes inside and outside of the consolidation surface, it is usually postulated that the surfaces $y = 0$ and $Y = 0$ do not intersect but engage each other along a common normal. As discussed later, this condition is not necessary, but sometimes may be convenient. Mathematically, this can be expressed by associating with any stress state $P(p', q)$ on the yield surface y a conjugate point $R(p'_R, q_R)$ on the consolidation surface Y, characterized by the same direction of exterior normal (Figure 20.3b).

Hardening rule

The rule for variation of hardening modulus h in the course of plastic deformation has to be specified first. One possibility is to introduce a *set of nested loading surfaces* within the domain contained between $y=0$ and $Y=0$. These surfaces can translate and expand or contract due to plastic volumetric strain variation. With each surface, a certain value of hardening modulus is associated, defining plastic flow when this particular surface becomes active. This variation for each nested loading surface is defined by an interpolation rule between an initial hardening modulus h_0 on the yield surface and the hardening modulus H on the consolidation surface. For instance, it can be assumed for the k-th surface:

$$h_k = H + \left(\frac{n-k}{n}\right)^{\gamma}(h_0 - H); \qquad k=0,...,n \tag{20.11}$$

The configuration of nested loading surfaces and the interpolation rule (20.11) define the field of hardening moduli, and the material response for any loading history may be studied by following the evolution of this field. This model possesses a multi-level memory structure, since for cyclically varying stress only a certain number of surfaces undergo translation; the other surfaces may change only isotropically. *In some models the nested loading surfaces are also assumed to be true yield surfaces, each contributing individually to the plastic strains.*

For practical purposes this model can be simplified (on account of the memory structure). Instead of the interpolation rule (20.11) specifying the hardening modulus on consecutive nested loading surfaces, the field of hardening moduli can be described by prescribing the variation of h with the distance $\delta = PR$ between the stress point P on $y=0$ and the conjugate point R on $Y=0$. If the maximal distance between the surfaces at the end of initial consolidation is δ_0 and initial hardening modulus is h_0, then for instance

$$h(\delta,\delta_0) = H + \left(\frac{\delta}{\delta_0}\right)^{\gamma}(h_0 - H) \tag{20.12}$$

so that $h(0,\delta_0)=H$ and $h(\delta_0,\delta_0)=h_0$. This formulation has a restricted memory and it is very close to bounding surface plasticity formulations described in the following section.

A compromise between these two extreme kinematic hardening formulations is a model using two yield surfaces and a hardening rule with an infinite number of nested loading surfaces. This model has a better memory than the two surfaces model with the hardening rule (20.12), still being simpler that the multiple loading surface model. It does not require that location of all the loading surfaces is memorized, the memory being erased as soon as a stress-reversal surface becomes active again.

Translation rule

In order to complete the kinematic hardening formulation, a translation rule, defining changes $d\rho_p$ and $d\rho_q$ in the stress coordinates of the "centre" of the yield surface, has to be specified. Most of the existing models satisfy the condition that the surfaces $y=0$ and $Y=0$ do not intersect but engage each other along a common normal, the relative motion of the point *P* with respect to conjugate point *R* has to occur along the vector *PR*:

$$\begin{aligned} dp' - dp'_R &= \mu(p' - p'_R) \\ dq - dq_R &= \mu(q - q_R) \end{aligned} \tag{20.13}$$

Assume that the surfaces $y=0$ and $Y=0$ are similar in shape, which does not change during isotropic and kinematic hardening. Then using the proportionality condition, the coordinates of the conjugate point $R(p'_R, q_R)$ can be found, and the stress increments in (20.13) can be expressed through the changes $d\rho_p$ and $d\rho_q$ in the stress coordinates of the "centre" of the yield surface. Then parameter μ can be found by introducing $d\rho_p$ and $d\rho_q$ into the consistency condition (20.10) where the hardening rule (20.12) is utilized. When the stress point reaches the consolidation surface, the yield surface translates in such a way that the two surfaces stay tangent to each other at the stress point.

20.4 Bounding surface plasticity

In bounding surface plasticity formulation (Dafalias and Herrmann, 1982), no yield surface is assumed to exist. As in the previous section, there exists a consolidation, or *bounding surface*, defined by the degree of material consolidation (Figure 20.3c). The domain enclosed by the bounding surface is not elastic, though for some stress trajectories within this surface, elastic response can be obtained. Assuming that the bounding surface expands or contracts isotropically, we can identify it with the isotropic hardening surface (20.1). The actual stress point $P(p', q)$ always lies within or on the bounding surface. For each $P(p', q)$, a unique "image" point $R(p'_R, q_R)$ on the bounding surface, $Y=0$, is defined, according to a specific mapping rule which is a part of the constitutive relations. One possibility is the simple "radial" rule: assuming that the origin *O* lies always within a convex bounding surface, $R(p'_R, q_R)$ is defined as the intersection of $F=0$ and the straight line connecting the origin with $P(p', q)$, (Figure 20.3c). Analytically this can be expressed by

$$p'_R = \beta(p', q, \varepsilon_v^p)p'; \quad q_R = \beta(p', q, \varepsilon_v^p)q, \tag{20.14}$$

with β obtained from $F\left(\beta p',\beta q,\varepsilon_v^p\right)=0$.

As a result of this mapping rule, it follows that at each point $P(p',q)$, a surface homeothetic to the bounding surface with respect to the origin is indirectly defined. It is assumed that if the incremental stress vector at the point $P(p',q)$ is directed inside this surface, the behaviour is elastic and governed by equations (20.3). Mathematically, this unloading condition can be expressed through the gradient of this quasi-surface at the point $P(p',q)$ which is simply equal to the gradient of the bounding surface $Y=0$ at the "image" point $R(p'_R,q_R)$. If the incremental stress vector at each point $P(p',q)$ is directed outwards, the behaviour is elastic-plastic, and the plastic components of the incremental strain vector are defined by the equations:

$$\begin{aligned} d\varepsilon_v^p &= \frac{1}{h}\left(\left.\frac{\partial Y}{\partial p'}\right|_R dp' + \left.\frac{\partial Y}{\partial q}\right|_R dq\right)\left.\frac{\partial Y}{\partial p'}\right|_R \\ d\varepsilon_s^p &= \frac{1}{h}\left(\left.\frac{\partial Y}{\partial p'}\right|_R dp' + \left.\frac{\partial Y}{\partial q}\right|_R dq\right)\left.\frac{\partial Y}{\partial q}\right|_R \end{aligned} \tag{20.15}$$

where h is the hardening modulus defined from the hardening rule:

$$h(\delta,\delta_0)=H+\left(\frac{\delta}{\delta_0-\delta}\right)\bar{h}\left(p',q,\varepsilon_v^p\right) \tag{20.16}$$

where H is the hardening modulus on the bounding surface defined by (20.7); $\bar{h}$ is a positive "shape" hardening function of the state requiring the identification and experimental determination of certain material parameters; $\delta = PR$ is the distance between the stress point P and its image point R on $Y=0$; δ_0 is a properly chosen reference stress or distance in the corresponding space. An extreme case of equation (20.16) is $h(0,\delta_0)=H$, providing a smooth transition in plastic deformation process when the stress state reaches the bounding surface.

The quasi-surface passing through the point $P(p',q)$ defines a quasi-elastic domain, but it is not a yield surface since the stress point may move first elastically inwards and then cause plastic loading before it reaches this surface again. As a matter of fact it never enters this formulation explicitly.

20.5 Discussion

As is seen from the above examples of different plasticity formulations, the common feature to all of them is the existence of the isotropic hardening boundary surface $Y=0$ defined by the degree of material consolidation. In isotropic hardening models (e.g. modified Cam-Clay), this surface is assumed to be a yield surface, containing an elastic domain. To incorporate plastic flow within this surface, nested surface models and bounding surface models were developed.

In nested surface models this surface is treated as a consolidation surface enclosing a smaller yield surface and a set of nesting yield or loading surfaces, with postulated hardening and translation rules depending on relative configuration of these surfaces. In bounding surface models the $Y=0$ surface is incorporated as a bounding surface, and a quasi-surface passing through the current stress state is defined using a specific mapping rule. This mapping rule also defines the distance in the stress space between the stress state and the bounding surface; the postulated hardening rule depends on this distance.

For kinematic hardening models with nesting surfaces, the stress history of cyclic loading may be followed, but enormous amounts of computer memory and calculation time are required and predicted stress-strain curves are not smooth. Bounding surface models, and similar kinematic hardening, two-surface models, are much simpler, but do not have a memory of multiple stress-reversals. Another limitation of the bounding surface models is that they may accumulate finite plastic strains during infinitesimal cyclic loading, the effect known as *ratcheting*. A compromise between these two approaches is a model using a hardening rule with an infinite number of loading surfaces.

20.6 Thermomechanics of irreversible cyclic behaviour

20.6.1 Multisurface hyperplasticity

In cases when irreversible cyclic behaviour is modelled using nested surface plasticity, its thermomechanical consistency will be guaranteed if the nested loading surfaces can be considered as true kinematic hardening *yield surfaces* each contributing a component to the total plastic strain, and if the corresponding model can be formulated within a framework of *multisurface hyperplasticity* (Houlsby and Puzrin, 2006).

<u>Potential functions</u>

In *multisurface hyperplasticity*, the specific Gibbs free energy becomes a function of the stresses and a finite number of internal variables $\alpha_v^{(n)}, \alpha_s^{(n)}, (n=1,\ldots,N)$, where N is the total number of the yield surfaces, and is written in the following form:

$$g\left(p',q,\alpha_v^{(1)},\ldots,\alpha_v^{(N)},\alpha_s^{(1)},\ldots,\alpha_s^{(N)}\right)=g_1(p',q)-\\ -p'\sum_{n=1}^{N}\alpha_v^{(n)}-q\sum_{n=1}^{N}\alpha_s^{(n)}+\sum_{n=1}^{N}g_2^{(n)}\left(\alpha_v^{(n)},\alpha_s^{(n)}\right) \tag{20.17}$$

The dissipation function also becomes a function of the finite number of internal variables and their rates $\dot{\alpha}_v^{(n)}, \dot{\alpha}_s^{(n)}, (n=1,\ldots,N)$:

$$d^g\left(p',q,\alpha_v^{(1)},\ldots,\alpha_v^{(N)},\alpha_s^{(1)},\ldots,\alpha_s^{(N)},\dot{\alpha}_v^{(1)},\ldots,\dot{\alpha}_v^{(N)},\dot{\alpha}_s^{(1)},\ldots,\dot{\alpha}_s^{(N)}\right)= \\ =\sum_{n=1}^{N} d^{g(n)}\left(p',q,\alpha_v^{(1)},\ldots,\alpha_v^{(N)},\alpha_s^{(1)},\ldots,\alpha_s^{(N)},\dot{\alpha}_v^{(n)},\dot{\alpha}_s^{(n)}\right)\geq 0 \tag{20.18}$$

where we assume that the dissipation can be decomposed into additive terms involving each individual plastic strain component. For a rate independent material, the dissipation is a homogeneous first order function of the plastic strain rates. For the case of associated plasticity, the dissipation function is independent of the stress.

Link to conventional plasticity

In the conventional formulation of multiple yield surface kinematic hardening plasticity, calculation of incremental stress-strain response requires the equations for all the yield surfaces to be defined explicitly. Then, for each yield surface the following rules are specified:

- the flow rule (or more usually the plastic potential function);
- the strain hardening rule;
- the translation rule.

All these equations, and the resulting incremental stress-strain response, can in fact be derived from the potentials (20.17) and (20.18) using Legendre transformations and their properties for active and passive variables (see Houlsby and Puzrin, 2006). Important difference to classical multiple yield surface plasticity is that within the multisurface hyperplasticity the *non-intersection condition* is not necessary.

20.6.2 Continuous hyperplasticity

The advantage of the multiple surface models is clearly that they are able to fit more accurately the non-linear cyclic behaviour of geotechnical materials across a wide range of strain amplitudes. The disadvantage is of course that a large number of material parameters (associated with each yield surface) are necessary. When the modelling of nonlinearity is taken to its logical conclusion by introducing, in effect, an infinite number of yield surfaces this reduces the number of material parameters required to specify the models, although at the expense that certain functions also have to be chosen. Thermomechanical consistency of such a model will be guaranteed if it can be formulated within a framework of *continuous hyperplasticity* (Houlsby and Puzrin, 2006).

Potential functionals

The multisurface hyperplastic formulation can be further extended to describe the case of a continuous field of kinematic hardening yield surfaces, of the type originally suggested by Mroz and Norris (1982). The continuous hyperplastic models are capable of reproducing decoupled associated kinematic hardening plasticity with a continuous field of yield surfaces. For this case the specific Gibbs free energy is a *functional* (rather than *function*) of the stress and an internal variable function $\hat{\alpha}_v(\eta), \hat{\alpha}_s(\eta)$:

$$\begin{aligned} g[p', q, \hat{\alpha}_v(\eta), \hat{\alpha}_s(\eta)] = g_1(p', q) - \\ - p' \int_{\mathrm{Y}} \hat{\alpha}_v(\eta)\Gamma(\eta) d\eta - q \int_{\mathrm{Y}} \hat{\alpha}_s(\eta)\Gamma(\eta) d\eta + \\ + \int_{\mathrm{Y}} \hat{g}_2(\hat{\alpha}_v(\eta), \hat{\alpha}_s(\eta), \eta)\Gamma(\eta) d\eta \end{aligned} \tag{20.19}$$

where Y is the domain of η. The function $\Gamma(\eta)$ is a weighting function, such that $\Gamma(\eta)d\eta$ is the fraction of the total number of the yield surfaces having a dimensionless size parameter between η and $\eta + d\eta$.

The dissipation functional, which is a functional of internal variable functions and their rates $\dot{\hat{\alpha}}_v(\eta)$, $\dot{\hat{\alpha}}_s(\eta)$, is also required:

$$\begin{aligned} d^g\left[p', q, \hat{\alpha}_v(\eta), \hat{\alpha}_s(\eta), \dot{\hat{\alpha}}_v(\eta), \dot{\hat{\alpha}}_s(\eta)\right] = \\ = \int_{\mathrm{Y}} \hat{d}^g\left(p', q, \hat{\alpha}_v(\eta), \hat{\alpha}_s(\eta), \dot{\hat{\alpha}}_v(\eta), \dot{\hat{\alpha}}_s(\eta), \eta\right)\Gamma(\eta) d\eta \geq 0 \end{aligned} \tag{20.20}$$

Dissipation functionals with no dependence on stress lead to models in which the flow rule (in the conventional sense in plasticity theory) is associated.

Link to conventional plasticity

The field of yield functions, with the corresponding flow, hardening and translation rules, and the resulting incremental stress-strain response, can be derived from the potentials (20.19) and (20.20) using Frechet derivatives and Legendre transformations and their properties for active and passive variables (see Houlsby and Puzrin, 2006).

Chapter 21

Rate and Time Dependency

TABLE OF CONTENTS

Chapter 21

Rate and Time Dependency

21.1 Introduction

All the constitutive models in the previous chapters were rate and time independent. In real soils, however, the rate dependency of both strength and stiffness, as well as time dependency of stresses and strains, play an important role.

Triaxial compression test is normally performed at a constant deviatoric strain rate (Figure 21.1a). Faster rate normally leads to a stiffer deviatoric stress-strain response with a larger strength (Figure 21.1b). This phenomenon is called *rate dependency.*

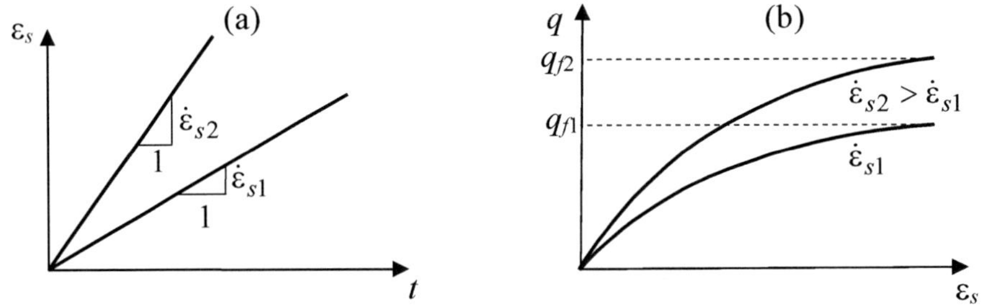

Figure 21.1: Triaxial compression tests at a constant strain rate: (a) the strain history; (b) deviatoric stress-strain curves.

Another interesting phenomenon is observed when a triaxial test is terminated at a certain deviatoric stress $q_c < q_f$, which is then maintained constant (Figure 21.2a). The deviatoric strain in this case will be increasing in time, at a decaying, constant or accelerating rate, depending on the q_c/q_f ratio (Figure 21.2b). This phenomenon, representing *time dependency of strains*, is called *creep*.

A similar phenomenon is observed when a triaxial test is terminated at a certain deviatoric strain ε_{sc}, which is then maintained constant (Figure 21.3a). The deviatoric stress in this case will be decreasing in time, partially or totally, at a decaying rate (Figure 21.3b). This phenomenon, representing *time dependency of stresses*, is called *stress relaxation*.

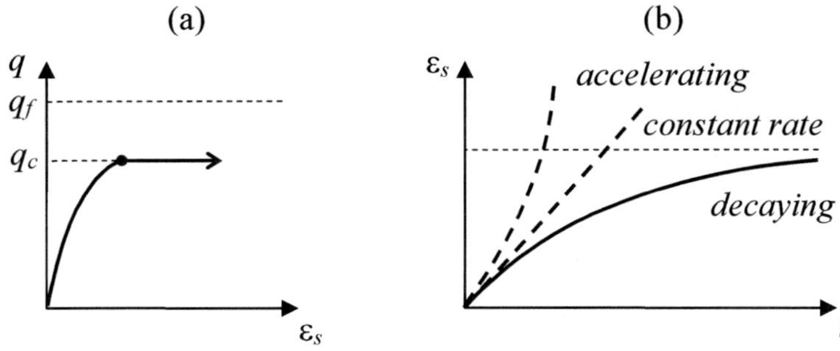

Figure 21.2: Creep in a triaxial compression test: (a) the deviatoric stress-strain curve; (b) the strain history.

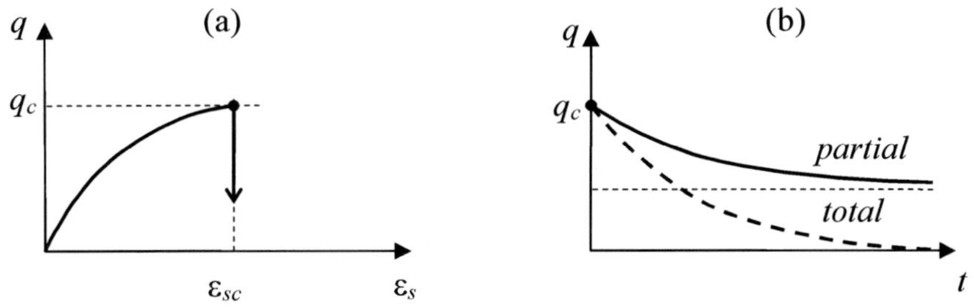

Figure 21.3: Stress relaxation in a triaxial compression test: (a) the deviatoric stress-strain curve; (b) the stress history.

This Chapter shows how the above described dependency of both elastic and plastic components of deviatoric soil behaviour on time and rate can be modelled using various viscous-elastic and viscous-plastic rheological models. The thermomechanical aspects of rate dependency are also explored.

21.2 Fundamental rheological models

The basic features of the elastic, plastic and viscous deviatoric behaviour can be described via the following fundamental rheological models, respectively.

21.2.1 The Hookean model

The *perfectly elastic* deviatoric behaviour can be visualised with the help of the Hookean model – a spring with the elastic constant $3G$ (Figure 21.4a), so that the deviatoric strain is defined by:

$$\varepsilon_s = \frac{q}{3G} \tag{21.1}$$

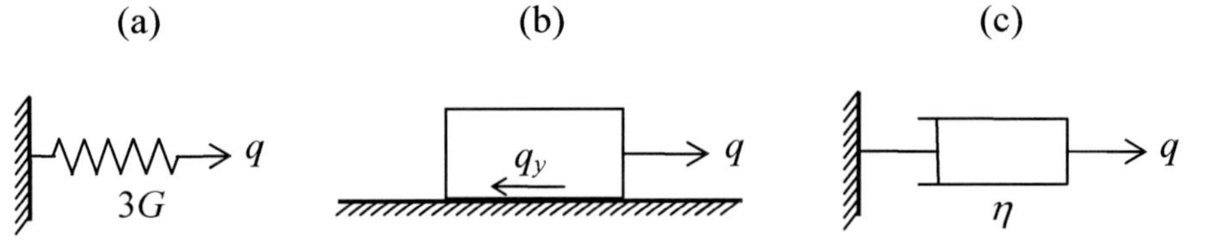

Figure 21.4: Fundamental rheological models: (a) the Hookean; (b) the yield stress; (c) the Newtonian.

21.2.2 The yield stress model

The *perfectly plastic* deviatoric behaviour can be described using the yield stress model – a sliding block with the sliding stress q_y (Figure 21.4b), so that the deviatoric strain is defined by:

$$\varepsilon_s = 0 \quad \text{when} \quad q < q_y \tag{21.2}$$

$$\dot{\varepsilon}_s > 0 \quad \text{when} \quad q = q_y \tag{21.3}$$

where $\dot{\varepsilon}_s = d\varepsilon_s / dt$.

21.2.3 The Newtonian model

Finally, the *perfectly viscous* deviatoric behaviour can be described using the Newtonian model of the perfectly viscous fluid – a dashpot with the viscosity constant η (Figure 21.4c), so that the deviatoric strain rate is defined by:

$$\dot{\varepsilon}_s = \frac{q}{\eta} \tag{21.4}$$

For $\dot{\varepsilon}_s = \dot{\varepsilon}_{sc} = const$ the model describes the *rate dependency of strength*:

$$q = \eta \dot{\varepsilon}_{sc} \tag{21.5}$$

For $\varepsilon_s = 0$ at $t = 0$ and $q = q_c = const$ it describes a *constant rate creep*:

$$\varepsilon_s = \frac{q_c}{\eta} t \tag{21.6}$$

21.3 Composite rheological models

Assembling different fundamental rheological models in different parallel and series combinations produces the following well known composite rheological models.

21.3.1 The St. Venant model

The St. Venant model consists of the Hookean and yield stress models connected in series (Figure 21.5a). Its deviatoric stress strain response is linear elastic until the yield stress q_y is reached, after which it becomes perfectly plastic (Figure 21.5b).

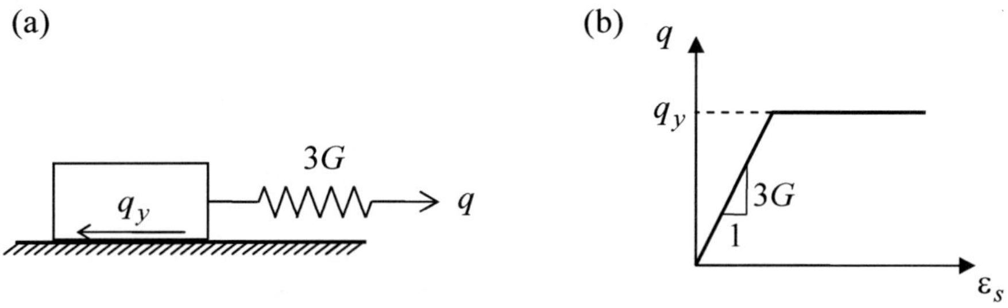

Figure 21.5: The St. Venant model: (a) the mechanical model; (b) the deviatoric stress-strain response.

21.3.2 The linear hardening model

The linear hardening model consists of the Hookean and Yield Stress models connected in parallel, and another Hookean model connected in series (Figure 21.6a). Its deviatoric stress strain response is linear elastic until the yield stress q_y is reached, after which it becomes plastic with linear hardening (Figure 21.6b):

$$\varepsilon_s = \frac{q}{3G} + \frac{q - q_y}{H} \text{ and } \varepsilon_{sy} = \frac{q_y}{3G}, \Rightarrow \frac{q - q_y}{\varepsilon_s - \varepsilon_{sy}} = \frac{3GH}{3G + H} \tag{21.7}$$

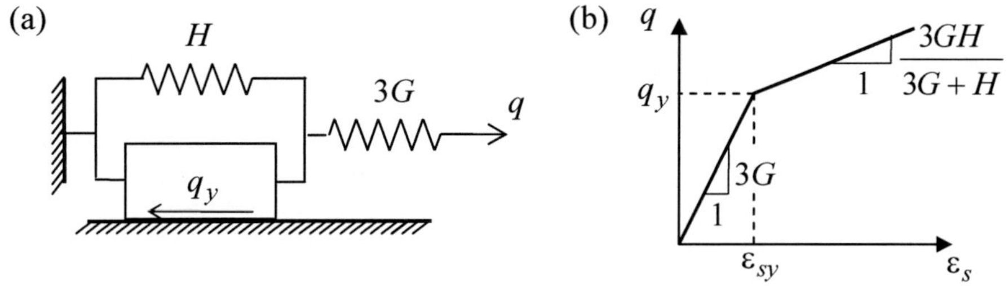

Figure 21.6: The linear hardening model: (a) the mechanical model; (b) the deviatoric stress-strain response.

21.3.3 The Kelvin (Voigt) model

The Kelvin (Voigt) model consists of the Hookean and Newtonian models connected in parallel (Figure 21.7a). Its shear strain response to the deviatoric stress step loading exhibits the *decaying creep*, with all the shear strain being *recoverable* (Figure 21.7b).

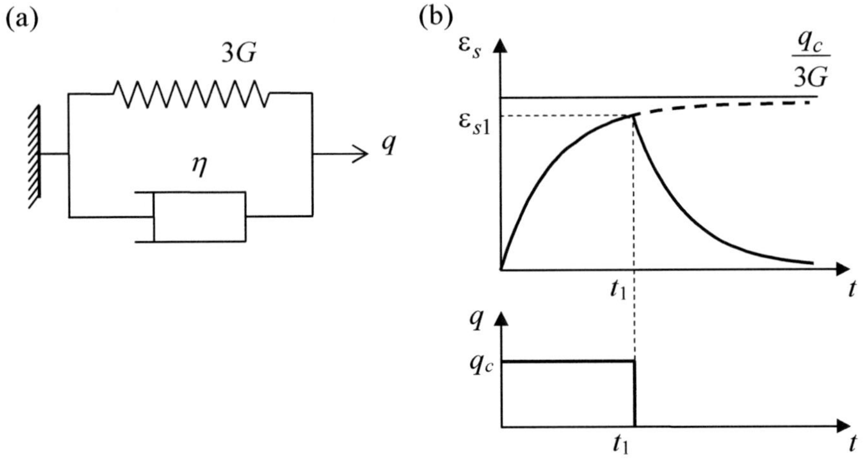

Figure 21.7: The Kelvin (Voigt) model: (a) the mechanical model; (b) the shear strain response to the deviatoric stress step loading.

This response can be derived by considering the following equations for stresses and strains resulting from the combination in parallel of the Hookean (H) and Newtonian (N) elements:

$$q = q_N + q_H ; \qquad \varepsilon_{sN} = \varepsilon_{sH} = \varepsilon_s . \tag{21.8}$$

Substitution of the constitutive equations (21.1) and (21.4) gives:

$$q = 3G\varepsilon_s + \eta\dot{\varepsilon}_s \quad \text{or} \quad \frac{q}{\eta} = \frac{3G}{\eta}\varepsilon_s + \dot{\varepsilon}_s \tag{21.9}$$

The latter equation is a particular case of the equation

$$N(t) = M(t)y(t) + \frac{dy}{dt} \tag{21.10}$$

with the solution

$$y(t) = e^{-\int M(t)dt}\left[\int_0^t N(\tau)e^{\int M(\tau)d\tau}d\tau + C\right] \tag{21.11}$$

where the integration constant C can be found from the initial conditions (e.g., for $y(0)=0$ we obtain $C=0$).

For the Kelvin model:

$$y(t)=\varepsilon_s(t); \qquad N(t)=\frac{q(t)}{\eta}; \qquad M(t)=\frac{3G}{\eta}=const. \tag{21.12}$$

And assuming zero immediate shear strain $\varepsilon_s(0)=0$, we obtain:

$$\varepsilon_s=e^{-\frac{3G}{\eta}t}\left[\int_0^t \frac{q(\tau)}{\eta}e^{\frac{3G}{\eta}\tau}d\tau\right] \tag{21.13}$$

For the constant deviatoric stress $q=q_c$:

$$\varepsilon_s=\frac{q_c}{3G}\left(1-e^{-\frac{3G}{\eta}t}\right) \tag{21.14}$$

so that $\lim_{t\to\infty}\varepsilon_s=\frac{q_c}{3G}$ (see Figure 21.7b), i.e. the creep is *decaying*.

At some $t=t_1$ (Figure 21.7b):

$$\varepsilon_{s1}=e^{-\frac{3G}{\eta}t_1}\int_0^{t_1}\frac{q_c}{\eta}e^{\frac{3G}{\eta}\tau}d\tau=\frac{q_c}{3G}\left(1-e^{-\frac{3G}{\eta}t_1}\right) \tag{21.15}$$

If at this moment $t=t_1$ the stress is instantaneously reduced to $q=0$, for $t>t_1$ we can decompose the time integral (21.13) into two parts:

$$\varepsilon_s=e^{-\frac{3G}{\eta}t}\left[\int_0^{t_1}\frac{q(t)}{\eta}e^{\frac{3G}{\eta}\tau}d\tau+\int_{t_1}^{t}\frac{q(t)}{\eta}e^{\frac{3G}{\eta}\tau}d\tau\right] \tag{21.16}$$

And taking into account the loading history (Figure 21.7b) we obtain:

$$\varepsilon_s=e^{-\frac{3G}{\eta}t}\left[\int_0^{t_1}\frac{q_c}{\eta}e^{\frac{3G}{\eta}\tau}d\tau+\int_{t_1}^{t}\frac{0}{\eta}e^{\frac{3G}{\eta}\tau}d\tau\right]=\varepsilon_{s1}e^{-\frac{3G}{\eta}(t-t_1)} \tag{21.17}$$

so that $\lim_{t\to\infty}\varepsilon_s=0$ (see Figure 21.7b), i.e. the strain is *fully recoverable*.

21.3.4 The Maxwell model

The Maxwell model consists of the Hookean and Newtonian models connected in series (Figure 21.8a). Its shear strain response to the deviatoric stress step loading exhibits the *constant rate creep*, with the shear strain being partially *irrecoverable* (Figure 21.8b). At the constant shear strain the Maxwell model produces the *total stress relaxation* (Figure 21.9a), while at constant shear strain rate it simulates *the rate dependency of the shear strength* (Figure 21.9b).

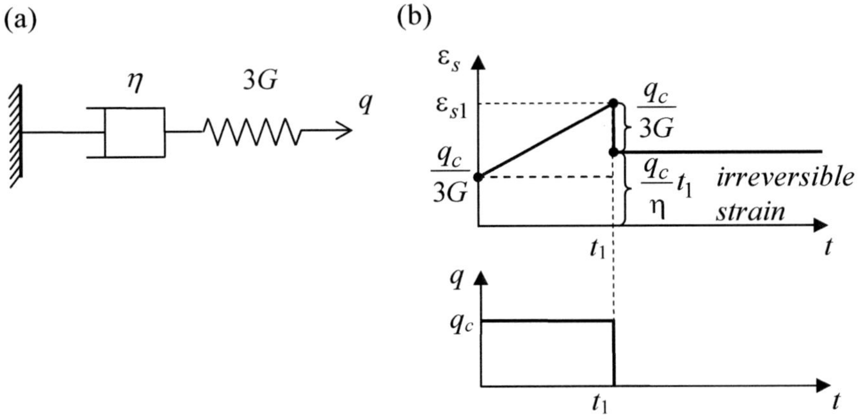

Figure 21.8: The Maxwell model: (a) the mechanical model; (b) the shear strain response to the deviatoric stress step loading.

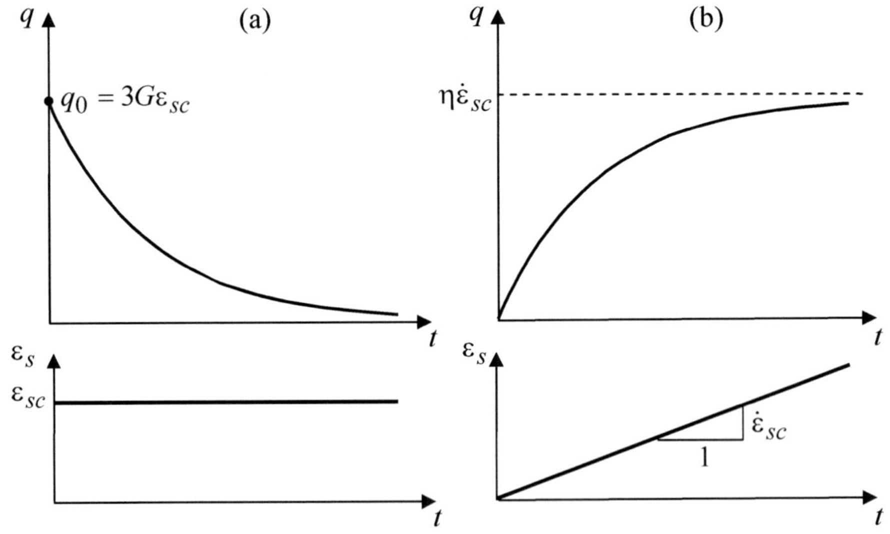

Figure 21.9: The deviatoric stress response of the Maxwell model to the: (a) constant shear strain loading; (b) constant shear strain rate loading.

These responses can be derived by considering the following equations for stresses and strains resulting from the combination in series of the Hookean (H) and Newtonian (N) elements:

$$q = q_N = q_H\,; \qquad \dot{\varepsilon}_{sN} + \dot{\varepsilon}_{sH} = \dot{\varepsilon}_s\,. \tag{21.18}$$

Substitution of the constitutive equations (21.1) and (21.4) gives:

$$\dot{\varepsilon}_s = \frac{\dot{q}}{3G} + \frac{q}{\eta} \quad \text{or} \quad 3G\dot{\varepsilon}_s = \frac{3G}{\eta}q + \dot{q} \tag{21.19}$$

From the first equation (21.19) it follows that for the constant stress $q = q_c$:

$$\dot{q} = 0 \text{ and } \dot{\varepsilon}_s = \frac{q_c}{\eta}, \quad \text{so that } \varepsilon_s = \frac{q_c}{\eta}t + \frac{q_c}{3G} \tag{21.20}$$

i.e. the creep takes place at the *constant rate* after the immediate strain $\frac{q_c}{3G}$ (see Figure 21.8b). At some $t = t_1$ (Figure 21.8b):

$$\varepsilon_{s1} = \frac{q_c}{\eta}t_1 + \frac{q_c}{3G} \tag{21.21}$$

If at this moment $t = t_1$ the stress is instantaneously reduced to $q = 0$, there is a release of the immediate strain $\frac{q_c}{3G}$, and for $t > t_1$:

$$\dot{\varepsilon}_s = \frac{0}{\eta} = 0, \text{ so that } \varepsilon_s = \varepsilon_{s1} - \frac{q_c}{3G} = \frac{q_c}{\eta}t_1 = const \tag{21.22}$$

and the strain in Figure 21.8b is partially *irrecoverable.*

The second equation (21.19) is also a particular case of the equation (21.10), with the following definitions for the Maxwell model:

$$y(t) = q(t); \qquad N(t) = 3G\dot{\varepsilon}_s; \qquad M(t) = \frac{3G}{\eta} = const\,. \tag{21.23}$$

Assuming the immediate stress $q(0) = q_0$, we obtain:

$$q(t) = e^{-\frac{3G}{\eta}t}\left[\int_0^t 3G\dot{\varepsilon}_s e^{\frac{3G}{\eta}\tau}\,d\tau + q_0\right] \tag{21.24}$$

For the constant shear strain $\varepsilon_s = \varepsilon_{sc}$ we have $\dot{\varepsilon}_s = 0$, $q_0 = 3G\varepsilon_{sc}$ and

$$q(t) = 3G\varepsilon_{sc}e^{-\frac{3G}{\eta}t} \tag{21.25}$$

so that $\lim_{t\to\infty} q = 0$ (see Figure 21.9a), i.e. the stress relaxation is *total.*

For the constant shear strain rate $\dot{\varepsilon}_s = \dot{\varepsilon}_{sc}$ we have $q_0 = 0$ and

$$q(t) = e^{-\frac{3G}{\eta}t} \int_0^t 3G\dot{\varepsilon}_{sc}\, e^{\frac{3G}{\eta}\tau}\, d\tau = \eta\dot{\varepsilon}_{sc}\left(1 - e^{-\frac{3G}{\eta}t}\right) \tag{21.26}$$

so that $\lim_{t\to\infty} q = \eta\dot{\varepsilon}_{sc}$ (see Figure 21.9b), i.e. the shear strength is *rate dependent*.

21.3.5 The Bingham model

The Bingham model differs from the Maxwell model in that it has an additional sliding element in between the Hookean and Newtonian models connected in series (Figure 21.10). Like in the Maxwell model, its shear strain response to the deviatoric stress step loading exhibits the *constant rate creep*, with the shear strain being partially *irrecoverable* (Figure 21.11a). Only for the Bingham model the magnitude of this irreversible shear strain can be controlled by adjusting the yield stress q_y. At the constant shear strain the Maxwell model produces the *partial stress relaxation* (Figure 21.11b), which can also be controlled by adjusting the yield stress q_y.

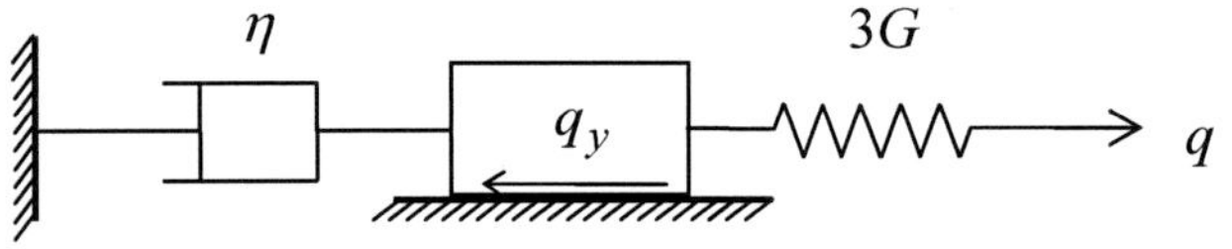

Figure 21.10: The Bingham model.

Considering the following equations for stresses and strains resulting from the combination in series of the Hookean (H) and Newtonian (N) elements (for $q > q_y$):

$$q = q_H = q_N + q_y; \qquad \dot{\varepsilon}_{sN} + \dot{\varepsilon}_{sH} = \dot{\varepsilon}_s. \tag{21.27}$$

Substitution of the constitutive equations (21.1) and (21.4) gives:

$$\dot{\varepsilon}_s = \frac{\dot{q}}{3G} + \frac{q - q_y}{\eta} \quad \text{or} \quad 3G\dot{\varepsilon}_s = \frac{3G}{\eta}(q - q_y) + (\dot{q} - \dot{q}_y) \tag{21.28}$$

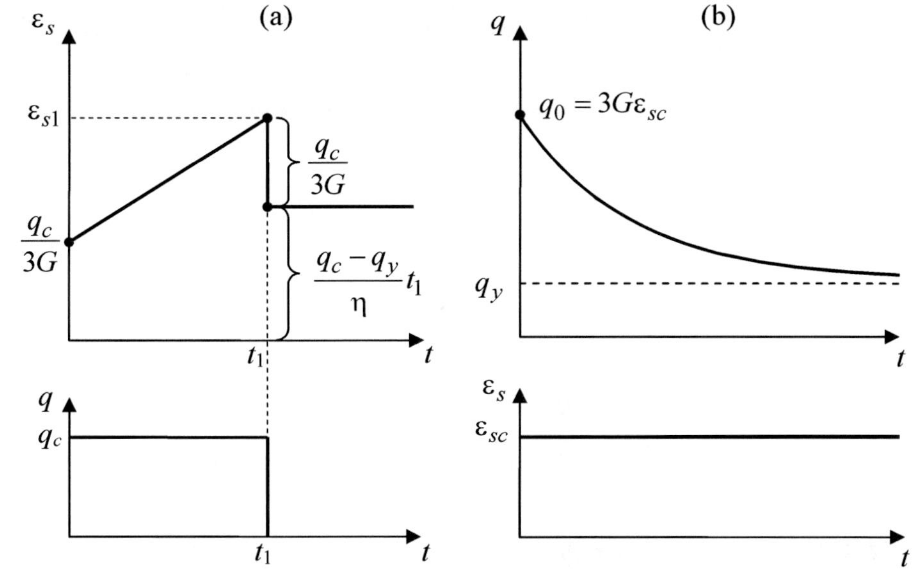

Figure 21.11: The Bingham model: (a) the shear strain response to the deviatoric stress step loading; (b) the deviatoric stress response to the constant shear strain loading.

From the first equation (21.28) it follows that for the constant stress $q = q_c$:

$$\dot{q} = 0 \text{ and } \dot{\varepsilon}_s = \frac{q_c - q_y}{\eta}, \quad \text{so that } \varepsilon_s = \frac{q_c - q_y}{\eta} t + \frac{q_c}{3G} \tag{21.29}$$

i.e. the creep takes place at the *constant rate* after the immediate strain $\frac{q_c}{3G}$ (see Figure 21.11a). At some $t = t_1$ (Figure 21.11a):

$$\varepsilon_{s1} = \frac{q_c - q_y}{\eta} t_1 + \frac{q_c}{3G} \tag{21.30}$$

If at this moment $t = t_1$ the stress is instantaneously reduced to $q = 0$, there is a release of the immediate strain $\frac{q_c}{3G}$, and for $t > t_1$:

$$\dot{\varepsilon}_s = \frac{0}{\eta} = 0, \text{ so that } \varepsilon_s = \varepsilon_{s1} - \frac{q_c}{3G} = \frac{q_c - q_y}{\eta} t_1 = const \tag{21.31}$$

and the strain in Figure 21.11a is partially *irrecoverable*.

The second equation (21.28) is a particular case of the equation (21.10), with the following definitions for the Bingham model:

$$y(t) = q(t) - q_y; \qquad N(t) = 3G\dot{\varepsilon}_s; \qquad M(t) = \frac{3G}{\eta} = const. \tag{21.32}$$

Assuming the immediate stress $q(0) = q_0 > q_y$, we obtain:

$$q(t) = q_y + e^{-\frac{3G}{\eta}t}\left[\int_0^t 3G\dot{\varepsilon}_s e^{\frac{3G}{\eta}\tau} d\tau + q_0\right] \tag{21.33}$$

For the constant shear strain $\varepsilon_s = \varepsilon_{sc}$ we have $\dot{\varepsilon}_s = 0$, $q_0 = 3G\varepsilon_{sc}$ and

$$q(t) = q_y + 3G\varepsilon_{sc} e^{-\frac{3G}{\eta}t} \tag{21.34}$$

so that $\lim_{t\to\infty} q = q_y$ (see Figure 21.11b), i.e. the stress relaxation is *partial*.

21.4 The two-layer model

An example of a realistic viscous-elastic-plastic rheological model is the two-layer model implemented in ABAQUS. The two-layer model consists of the Maxwell and St. Venant models connected in parallel (Figure 21.12a). At a constant shear strain rate it simulates *the rate dependency of the shear strength* (Figure 21.12b).

The deviatoric stress-strain response of the model changes when the deviatoric stress in the St. Venant model reaches the yield stress q_y, which takes place at the following total shear strain:

$$\varepsilon_s = \frac{q_y}{3G_2} \tag{21.35}$$

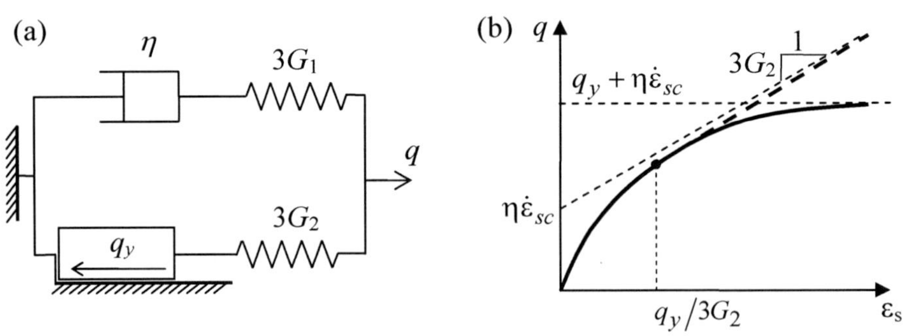

Figure 21.12: The two-layer model: (a) the mechanical model; (b) the deviatoric stress-strain response to the constant shear strain rate loading.

For the constant shear strain rate $\dot{\varepsilon}_{sc}$, from equation (21.26) of the Maxwell model it follows that (Figure 21.12b):

$$\text{For } \varepsilon_s < q_y/3G_2: \qquad q = \eta\dot{\varepsilon}_{sc}\left(1 - e^{-\frac{3G_1}{\eta}\frac{\varepsilon_s}{\dot{\varepsilon}_{sc}}}\right) + 3G_2\varepsilon_s \tag{21.36}$$

$$\text{For } \varepsilon_s \geq q_y/3G_2: \qquad q = \eta\dot{\varepsilon}_{sc}\left(1 - e^{-\frac{3G_1}{\eta}\frac{\varepsilon_s}{\dot{\varepsilon}_{sc}}}\right) + q_y \tag{21.37}$$

This model approaches asymptotically the shear strength of $q_y + \eta\dot{\varepsilon}_{sc}$, which has a plastic and viscous (*rate dependent*) part, respectively.

Adding another Hookean element in parallel to the sliding block leads to the two-layer model with linear hardening (Figure 21.13a). Its deviatoric stress-strain response to the constant shear strain rate $\dot{\varepsilon}_{sc}$ loading is described by (Figure 21.13b):

$$\varepsilon_s < q_y/3G_2: \; q = \eta\dot{\varepsilon}_{sc}\left(1 - e^{-\frac{3G_1}{\eta}\frac{\varepsilon_s}{\dot{\varepsilon}_{sc}}}\right) + 3G_2\varepsilon_s \tag{21.38}$$

$$\varepsilon_s \geq q_y/3G_2: \; q = \eta\dot{\varepsilon}_{sc}\left(1 - e^{-\frac{3G_1}{\eta}\frac{\varepsilon_s}{\dot{\varepsilon}_{sc}}}\right) + q_a'\varepsilon_s + q_a \tag{21.39}$$

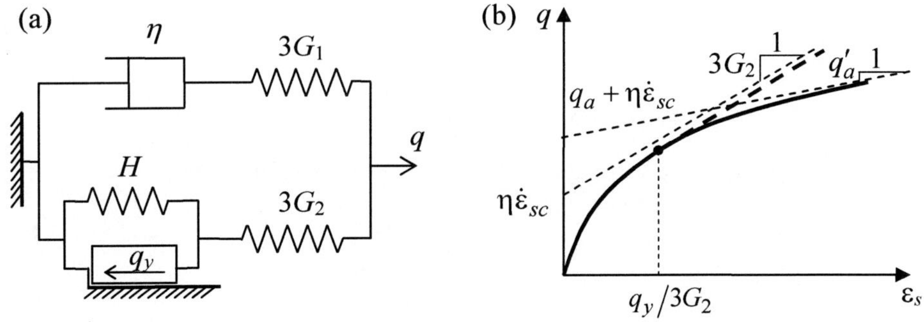

Figure 21.13: The two-layer model with linear hardening: (a) the mechanical model; (b) the deviatoric stress-strain response to the constant shear strain rate loading.

where $$q_a = \frac{3G_2 q_y}{3G_2 + H}; \qquad q'_a = \frac{3G_2 H}{3G_2 + H}. \tag{21.40}$$

This model approaches asymptotically the yield stress of $q_a + q'_a \varepsilon_s + \eta \dot{\varepsilon}_{sc}$, which also has a plastic and viscous (*rate dependent*) part, respectively.

21.5 The thermomechanical aspects of rate dependency

21.5.1 Rate dependent hyperplasticity

The hyperplastic framework can be also used to validate thermomechanical consistency of a *rate dependent* elastic and elastic-plastic behaviour. For a particular case of modelling isothermal, rate dependent, stress controlled one-dimensional deviatoric stress-strain behaviour, the *sufficient* (but not necessary) conditions for the model to satisfy the First and the Second Laws of Thermodynamics are summarized below:

1. Existence of a Gibbs free energy potential $g(q, \alpha_s)$ for strains:

$$\varepsilon_s = -\frac{\partial g}{\partial q} \tag{21.41}$$

   where α_s is a kinematic internal variable, normally associated with irreversible deviatoric strain.

2. Existence of a non-negative dissipation function $d(q, \alpha_s, \dot{\alpha}_s) \geq 0$, which in contrast to a rate independent behaviour is not first-order homogeneous in rates of internal variables, but can be expressed as:

$$d = \frac{\partial z}{\partial \dot{\alpha}_s} \dot{\alpha}_s \geq 0 \tag{21.42}$$

   where $z(q, \alpha_s, \dot{\alpha}_s)$ is a *flow potential.*

3. Functions g and z satisfy Ziegler's orthogonality condition:

$$-\frac{\partial g}{\partial \alpha_s} = \frac{\partial z}{\partial \dot{\alpha}_s} \tag{21.43}$$

21.5.2 Composite rheological models

Simple rheological models considered in the Section 21.3 satisfy the Laws of Thermodynamics, because they can be expressed within the rate dependent hyperplastic formulation with the Gibbs free energy potential and the dissipation function or flow potential.

St. Venant and linear hardening models

The linear hardening model can be formulated using the Gibbs free energy potential:

$$g(q,\alpha_s)=-\frac{q^2}{6G}-q\alpha_s+\frac{H}{2}\alpha_s^2 \tag{21.44}$$

and the dissipation function (rate independent behaviour):

$$d(\dot{\alpha}_s)=z(\dot{\alpha}_s)=q_y|\dot{\alpha}_s|\geq 0 \tag{21.45}$$

with Ziegler's orthogonality condition (21.43) (for $\dot{\alpha}_s \geq 0$):

$$q-H\alpha_s=q_y \tag{21.46}$$

which being substituted into the expression for strains

$$\varepsilon_s=-\frac{\partial g}{\partial q}=\frac{q}{3G}+\alpha_s \tag{21.47}$$

produces the constitutive equation (21.7) for the linear hardening model:

$$\varepsilon_s=\frac{q}{3G}+\frac{q-q_y}{H} \tag{21.48}$$

The particular case of $H=0$ corresponds to the St. Venant model.

The Kelvin (Voigt) model

The Kelvin (Voigt) model can be formulated using the Gibbs free energy potential:

$$g(q,\alpha_s)=-q\alpha_s+\frac{3G}{2}\alpha_s^2 \tag{21.49}$$

and the dissipation function (rate dependent behaviour):

$$d(\dot{\alpha}_s)=\eta\dot{\alpha}_s^2\geq 0 \tag{21.50}$$

or the flow potential:

$$z(\dot{\alpha}_s)=\frac{\eta}{2}\dot{\alpha}_s^2 \tag{21.51}$$

with Ziegler's orthogonality condition (21.43):

$$q-3G\alpha_s=\eta\dot{\alpha}_s \tag{21.52}$$

Substituting the kinematic variable α_s from the expression for strains

$$\varepsilon_s = -\frac{\partial g}{\partial q} = \alpha_s \tag{21.53}$$

into equation (21.52) produces the constitutive equation (21.9) for the Kelvin (Voigt) model:

$$q = 3G\varepsilon_s + \eta\dot{\varepsilon}_s \tag{21.54}$$

The Maxwell model

The Maxwell model can be formulated using the Gibbs free energy potential:

$$g(q,\alpha_s) = -\frac{q^2}{6G} - q\alpha_s \tag{21.55}$$

and the dissipation function (rate dependent behaviour):

$$d(\dot{\alpha}_s) = \eta\dot{\alpha}_s^2 \geq 0 \tag{21.56}$$

or the flow potential:

$$z(\dot{\alpha}_s) = \frac{\eta}{2}\dot{\alpha}_s^2 \tag{21.57}$$

with Ziegler's orthogonality condition (21.43):

$$q = \eta\dot{\alpha}_s \tag{21.58}$$

Substituting the kinematic variable α_s from the expression for strains:

$$\varepsilon_s = -\frac{\partial g}{\partial q} = \frac{q}{3G} + \alpha_s \tag{21.59}$$

into equation (21.58) produces the constitutive equation (21.19) for the Maxwell model:

$$\dot{\varepsilon}_s = \frac{\dot{q}}{3G} + \frac{q}{\eta} \tag{21.60}$$

The Bingham model

The Bingham model can be formulated using the Gibbs free energy potential:

$$g(q,\alpha_s) = -\frac{q^2}{6G} - q\alpha_s \tag{21.61}$$

and the dissipation function (rate dependent behaviour):

$$d(\dot{\alpha}_s) = q_y|\dot{\alpha}_s| + \eta\dot{\alpha}_s^2 \geq 0 \tag{21.62}$$

or the flow potential:

$$z(\dot{\alpha}_s) = q_y|\dot{\alpha}_s| + \frac{\eta}{2}\dot{\alpha}_s^2 \tag{21.63}$$

with Ziegler's orthogonality condition (21.43) (for $\dot{\alpha}_s \geq 0$):

$$q = q_y + \eta\dot{\alpha}_s \tag{21.64}$$

Substituting the kinematic variable α_s from the expression for strains:

$$\varepsilon_s = -\frac{\partial g}{\partial q} = \frac{q}{3G} + \alpha_s \tag{21.65}$$

into equation (21.64) produces the constitutive equation (21.28) for the Bingham model:

$$\dot{\varepsilon}_s = \frac{\dot{q}}{3G} + \frac{q - q_y}{\eta} \tag{21.66}$$

21.5.3 The two-layer models

For modelling isothermal, rate dependent, stress controlled one-dimensional deviatoric stress-strain behaviour, the *sufficient* (but not necessary) conditions for *a two layer model* to satisfy the First and the Second Laws of Thermodynamics are summarized below:

1. Existence of a Helmholtz free energy potential $f(\varepsilon_s, \alpha_{s1}, \alpha_{s2})$ for stresses:

$$q = \frac{\partial f}{\partial \varepsilon_s} \tag{21.67}$$

   where α_s are two independent kinematic internal variables, normally associated with irreversible deviatoric strains in each layer of the model.

2. Existence of a non-negative dissipation function $d(\dot{\alpha}_{s1}, \dot{\alpha}_{s2}) \geq 0$, which in contrast to a rate independent behaviour is not first-order homogeneous in rates of internal variables, but can be expressed as:

$$d = \frac{\partial z}{\partial \dot{\alpha}_{s1}}\dot{\alpha}_{s1} + \frac{\partial z}{\partial \dot{\alpha}_{s2}}\dot{\alpha}_{s2} \geq 0 \tag{21.68}$$

where $z(q,\dot{\alpha}_{s1},\dot{\alpha}_{s2})$ is a *flow potential.*

3. Functions *f* and *z* satisfy Ziegler's orthogonality condition:

$$-\frac{\partial f}{\partial \alpha_{s1}}=\frac{\partial z}{\partial \dot{\alpha}_{s1}} \qquad -\frac{\partial f}{\partial \alpha_{s2}}=\frac{\partial z}{\partial \dot{\alpha}_{s2}} \tag{21.69}$$

The two-layer models of the Section 21.4 satisfy the Laws of Thermodynamics, because they can be formulated using the Helmholtz free energy potential:

$$f(\varepsilon_s,\alpha_{s1},\alpha_{s2})=\frac{3G_1}{2}(\varepsilon_s-\alpha_{s1})^2+\frac{3G_2}{2}(\varepsilon_s-\alpha_{s2})^2+\frac{H}{2}\alpha_{s2}^2 \tag{21.70}$$

and the dissipation function (rate dependent behaviour):

$$d(\dot{\alpha}_{s1},\dot{\alpha}_{s2})=\eta\dot{\alpha}_{s1}^2+q_y|\dot{\alpha}_{s2}|\geq 0 \tag{21.71}$$

or the flow potential:

$$z(\dot{\alpha}_{s1},\dot{\alpha}_{s2})=\frac{\eta}{2}\dot{\alpha}_{s1}^2+q_y|\dot{\alpha}_{s2}|\geq 0 \tag{21.72}$$

resulting in Ziegler's orthogonality condition (21.69) (for $\dot{\alpha}_{s2}\geq 0$):

$$3G_1(\varepsilon_s-\alpha_{s1})=\eta\dot{\alpha}_{s1} \qquad 3G_2(\varepsilon_s-\alpha_{s2})-H\alpha_{s2}=q_y \tag{21.73}$$

which together with the expression for stresses:

$$q=\frac{\partial f}{\partial \varepsilon_s}=3G_1(\varepsilon_s-\alpha_{s1})+3G_2(\varepsilon_s-\alpha_{s2}) \tag{21.74}$$

define the constitutive behaviour of the two-layer models described in the Section 21.4, thus confirming their thermomechanical consistency.

Appendices

Appendix A
Undrained Soil Behaviour

A.1 Undrained Shear Strength

The undrained shear strength of clay s_u can be derived in the laboratory from unconsolidated undrained (UU) triaxial tests, or in situ from vane tests or CPT (cone penetration test). Normally consolidated to slightly overconsolidated clays have a tendency to contract during drained loading. Since the water is incompressible, in undrained shearing this results in positive excess pore water pressures. The corresponding effective stress path to failure in $p' - q$ triaxial stress space curves to the left from vertical (Figure A.1a) and hits the failure envelope at a much lower failure deviatoric stress $q_f = 2s_u$ than in drained shearing (Figure A.1b). Therefore, this experimentally determined s_u appears to be significantly smaller than the clay strength at the same initial effective stresses in drained shearing (slow loading). An empirical rule of thumb gives the following approximate relationship for the undrained shear strength of normally consolidated clays:

$$s_u = k\sigma'_{v0} \qquad k = 0.21 - 0.25 \tag{A.1}$$

where σ'_{v0} is the in situ geostatic vertical effective stress. The lower values of $k = 0.21$ are valid for soft marine clays. Using effective stress approach, different constitutive models predict the relationship (A.1) with a different degree of accuracy.

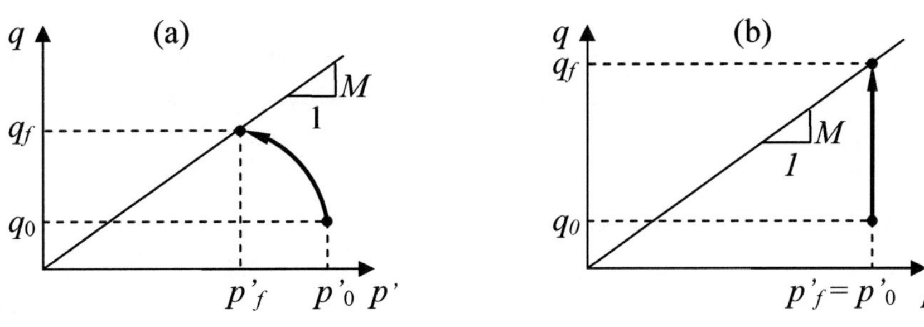

Figure A.1: Effective stress path for normally consolidated clay: (a) undrained; (b) drained.

A.2 The Mohr-Coulomb model

The Mohr-Coulomb model assumes isotropic elastic behaviour during pre-failure deformation. In undrained triaxial shear loading, when no volume change is allowed, this results in no pore water pressure being generated. The mean effective stress p' stays constant and equal to the initial mean effective stress p_0', and the corresponding effective stress path to failure in $p'-q$ triaxial stress space is vertical (Figure A.2).

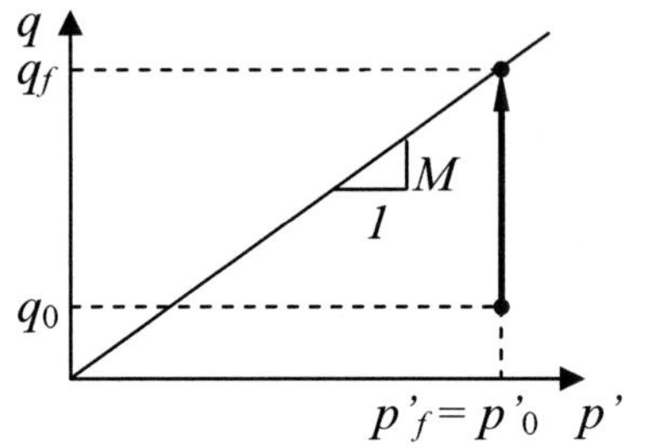

Figure A.2: Undrained effective stress path in the Mohr-Coulomb model.

The deviatoric stress at failure q_f can be then calculated as follows:

$$q_f = Mp'_f = Mp_0' \tag{A.2}$$

where

$$M = \frac{6\sin\varphi'}{3-\sin\varphi'} \tag{A.3}$$

$$p_0' = \frac{1+2K_0}{3}\sigma_{v0}' \tag{A.4}$$

where K_0 is the at rest earth pressure coefficient, which for normally consolidated clays can be estimated using the formula (Jaky, 1944):

$$K_0 = 1-\sin\varphi' \tag{A.5}$$

Substituting expressions (A.3)-(A.5) into (A.2) and noting that

$$s_u = \frac{\sigma_{1f}-\sigma_{3f}}{2} = \frac{q_f}{2} \tag{A.6}$$

we obtain the undrained shear strength estimate:

$$s_u = k\sigma_{v0}' \tag{A.7}$$

where

$$k = k_{MC} = \sin\varphi' \frac{3 - 2\sin\varphi'}{3 - \sin\varphi'} \tag{A.8}$$

For a typical range of $\varphi' = 20^\circ - 24^\circ$: $k \approx 0.30 - 0.35$, which is 40-50% higher than the value in relationship (A.1)!

A.3 The modified Cam Clay model

The modified Cam Clay model (Burland and Roscoe, 1968), in contrast, assumes plastic behaviour in pre-failure deformation using an elliptical yield surface (Figure A.3a):

$$\frac{q^2}{M^2 p'^2} + 1 = \frac{p'_c}{p'} \tag{A.9}$$

where p'_c is the pre-consolidation pressure, serving as a hardening parameter related to the specific volume via the equation of the virgin compression line (VCL, Figure A.3b):

$$V_c = N - \lambda \ln p'_c \tag{A.10}$$

The failure takes place when the stress state (p'_f, q_f) reaches the critical state line (CSL), which in the triaxial stress space (Figure A.3a) is identical to the Mohr-Coulomb failure envelope

$$q_f = Mp'_f \tag{A.11}$$

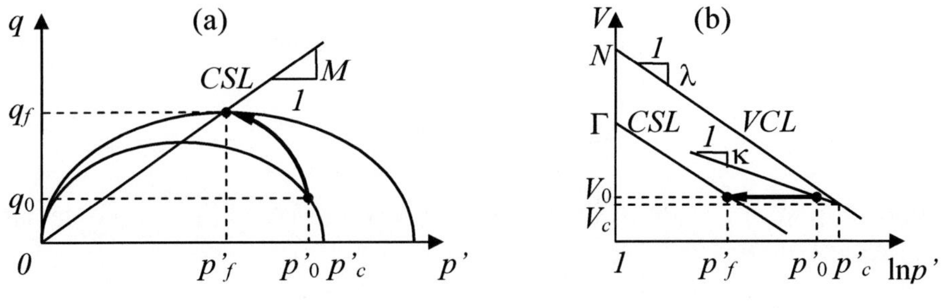

Figure A.3: Modified Cam Clay model: (a) undrained effective stress path; (b) constant volume (undrained) deformation.

In the volumetric stress-deformation space (Figure A.3b) the CSL is parallel to the VCL:

$$V = \Gamma - \lambda \ln p'_f \tag{A.12}$$

where

$$\Gamma = N - (\lambda - \kappa)\ln 2 \tag{A.13}$$

and κ defines elastic volumetric behaviour (Figure A.3b):

$$V = V_c - \kappa \ln p'/p'_c \tag{A.14}$$

The initial stress state after one-dimensional consolidation is given by

$$p'_0 = \frac{1+2K_0}{3}\sigma'_{v0} \qquad q_0 = (1-K_0)\sigma'_{v0} \tag{A.15}$$

The corresponding initial volume is found from equations (A.10) and (A.14):

$$V_0 = N - \lambda \ln p'_c - \kappa \ln p'_0/p'_c \tag{A.16}$$

where p'_c/p'_0 for the initial stress state is found from equation (A.9):

$$\frac{p'_c}{p'_0} = \frac{q_0^2}{M^2 p_0'^2} + 1 \tag{A.17}$$

During the undrained loading to failure the initial volume V_0 does not change (Figure A.3b), therefore, from equation (A.12) we obtain:

$$p'_f = e^{\frac{\Gamma - V_0}{\lambda}} \tag{A.18}$$

which after substitution of (6.13) and (6.16) turns into:

$$p'_f = p'_0\left(\frac{1}{2}\frac{p'_c}{p'_0}\right)^{1-\kappa/\lambda} \tag{A.19}$$

Substitution of equations (A.17) and (A.19) into (A.11) gives the deviatoric stress at failure:

$$q_f = Mp'_0\left(\frac{1}{2}\frac{q_0^2}{M^2 p_0'^2} + \frac{1}{2}\right)^{1-\kappa/\lambda} \tag{A.20}$$

which, using equations (A.3), (A.5), (A.9) and (A.15) can be transformed into the following relationship for the undrained shear strength:

$$s_u = k\sigma'_{v0} \tag{A.21}$$

$$k = k_{MC} \cdot \left(\frac{1}{8} \frac{\sin^2 \varphi'}{k_{MC}^2} + \frac{1}{2} \right)^{1-\kappa/\lambda} \tag{A.22}$$

where k_{MC} is calculated from equation (A.8). For a typical range of values of $\varphi' = 20^\circ - 24^\circ$ and $\kappa/\lambda = 0.2$ we obtain $k = 0.215 - 0.251$, i.e. very close to that in relationship (A.1)!

The effective undrained stress path is curved to the left (Figure A.3a) and given by:

$$\frac{q^2}{M^2 p'^2} + 1 = \left(\frac{p'_0}{p'} \right)^{\frac{1}{1-\kappa/\lambda}} \left(\frac{q_0^2}{M^2 p_0'^2} + 1 \right) \tag{A.23}$$

A.4 The original Cam Clay model

For the original Cam Clay (Schofield and Wroth, 1968) the undrained shear strength can be derived (e.g., Muir Wood, 1996) following the above procedure and simply replacing equation (A.8) of the yield surface and expression (A.13) for parameter Γ by

$$\frac{q}{Mp'} = \ln \frac{p'_c}{p'} \tag{A.24}$$

and

$$\Gamma = N - (\lambda - \kappa) \tag{A.25}$$

respectively.

The corresponding deviatoric stress at failure:

$$q_f = Mp'_0 e^{\left(\frac{q_0}{Mp'_0} - 1 \right)\left(1 - \frac{\kappa}{\lambda} \right)} \tag{A.26}$$

can be transformed using equations (A.3), (A.5), (A.6) and (A.15) into the following relationship for the undrained shear strength:

$$s_u = k\sigma'_{v0}; \qquad k = k_{MC} \cdot e^{\left(\frac{\sin \varphi'}{2k_{MC}} - 1 \right)\left(1 - \frac{\kappa}{\lambda} \right)}, \tag{A.27}$$

where k_{MC} is calculated from equation (A.8). For a typical range of values of $\varphi' = 20^\circ - 24^\circ$ and $\kappa/\lambda = 0.2$ we obtain $k = 0.212 - 0.248$, i.e. again close to that in relationship (A.1)!

The effective undrained stress path is curved to the left and given by:

$$\frac{q}{Mp'} = \frac{\ln(p'_0/p')}{1-\kappa/\lambda} + \frac{q_0}{Mp'_0} \tag{A.28}$$

A.5 Summary

For stability and bearing capacity analysis in saturated normally consolidated clays it is necessary to determine stresses in soil earth pressures under undrained loading conditions. Total stress analysis is probably the easiest way to calculate these stresses. The effective stress analysis will in principle produce the same result, provided the chosen soil model is capable of accurate prediction of the undrained shear strength. We have shown that the Mohr-Coulomb model overpredicts the undrained shear strength by as much as 50%. In contrast, both the original and modified Cam Clay models produce very close predictions, both within 1% from the experimental data, in spite of the fact that their analytical expressions look very different.

Appendix B
Finite Element Implementation

by Carlo Rabaiotti

B.1 Introduction

In the following sections we explore the main steps for implementing an elementary elastic-plastic constitutive model in a finite element code. A more detailed description of the implementation of elasto-plasticity in finite elements can be found in Owen and Hinton (1980), Potts and Zdravkovic (1999), Zienkiewicz and Taylor (2000), Bathe (2006), etc..

In Chapter 6 we introduced the method of finite elements. We have shown how all the steps of the method reduce to the solution of a system of equations given by the equilibrium condition of the nodal forces and the stress state in the body together with surface and body forces.

If we adopt a linear elastic constitutive model for describing the mechanical behaviour of the body, the system reduces to a system of linear equations, which can be solved exactly (the exact solution can be obtained for the system of equations, not for the BVP). A standard solution method for such a system of linear equations is the Gauss elimination method, with some modification, e.g., LU decomposition.

If a non-linear constitutive model (e.g., elastic-plastic) is adopted, firstly, the equilibrium has to be written in incremental form; secondly, the solution can be obtained only iteratively. This has, of course, a cost: we have to make approximations in integrating the stress-strain relationship and obtaining a solution of the system of linear equations through iterations. Last but not least, we need to constrain the plastic stress increments to the yield surface, which also comes at cost of accuracy.

On the other hand, it is an immensely powerful tool for advanced design in geotechnical engineering, which complements and but does not substitute conventional design.

B.2 Finite elements solution algorithm

The governing equation for the finite element solution a linear boundary value problem is obtained from the weak form of the equilibrium, derived from the principle of virtual work. In Chapter 6 we obtained:

$$\{\delta w_n\}^T\{q_n^e\}=\int_{V^e}\{\delta w_n\}^T\left[\left[B^e\right]^T[D]\left[B^e\right]\{w_n\}+\left[N^e\right]^T b^e(z)\right]dV+$$
$$+\int_{V^e}\{\delta w_n\}^T\left[N^e\right]^T t^e(z)dS \tag{B.1}$$

Since this equation should hold for any virtual displacements $\{\delta w_n\}$ at constant nodal displacements w_n, equation (B.1) can be rewritten as:

$$\{q_n^e\}=\int_{V^e}\left[\left[B^e\right]^T[D]\left[B^e\right]\{w_n\}+\left[N^e\right]^T b^e(z)\right]dV+\int_{S^e}\left[N^e\right]^T t^e(z)dS \tag{B.2}$$

By assembling equations (B.2) for all elements we obtain:

$$\{F_n\}=[K]\{w_n\}+\sum_e\{f_{bn}^e\}+\sum_e\{f_{sn}^e\} \tag{B.3}$$

where

$$K=\sum_e[B]^T[D][B]dV \tag{B.4}$$

When the components of the stiffness matrix $[K]$ are constants, equation (B.3) represents a system of linear equations. There are several methods to solve this system (e.g. Gaussian elimination) and generally an exact solution can be obtained. If components of $[K]$ are not constant (e.g., in an elastic-plastic stiffness matrix), the system has to be written in an incremental form and solved iteratively.

The most often used method to solve the system of non-linear equation is the Newton-Raphson algorithm. The solution scheme can be described in the following manner (NB: from now on n denotes the actual calculation step during numerical integration and i the iteration step of the general solution algorithm):

$$r(w_n)=\{F_n\}-\left([K(w_n)]\{w_n\}+\sum_e\{f_{bn}^e\}+\sum_e\{f_{sn}^e\}\right)=\{F_n\}-\{R_n\} \tag{B.5}$$

Where $r(w_n)$ is the residual error at the calculation step n, which should be close to zero.

First, we approximate $r(w_n)$ in equation (B.5) with the first two terms of the Taylor series expansion for a displacement $\overline{w}_n$ at the iteration *i*:

$$r^*\left(\overline{w}_n\right) = r\left(\overline{w}_n^i\right) + \left|\frac{\partial r(w_n)}{\partial w_n}\right|_{w_n=\overline{w}_n^i} \left(\overline{w}_n - \overline{w}_n^i\right) \tag{B.6}$$

Now we can find $\overline{w}_n^{i+1}$, where this function becomes zero:

$$r^*\left(\overline{w}_n^{i+1}\right) = r\left(\overline{w}_n^i\right) + \left|\frac{\partial r(w_n)}{\partial w_n}\right|_{w_n=\overline{w}_n^i} \left(\overline{w}_n^{i+1} - \overline{w}_n^i\right) = 0 \tag{B.7}$$

It follows then:

$$\overline{w}_n^{i+1} = \overline{w}_n^i + \frac{r\left(\overline{w}_n^i\right)}{\left|\frac{\partial r(w_n)}{\partial w_n}\right|_{w_n=\overline{w}_n^i}} \tag{B.8}$$

where

$$\frac{\partial r(w_n)}{\partial w_n} = -\left(K(w_n) + \frac{\partial K(w_n)}{\partial w_n} w_n\right) \tag{B.9}$$

and

$$T = K(w_n) + \frac{\partial K(w_n)}{\partial w_n} w \tag{B.10}$$

Equation (B.10) represents the tangent stiffness matrix. We repeat the procedure until the residual error has reached a predefined tolerance. The procedure is schematically illustrated in Figures B.1 and B.2.

Since the iterative procedure cannot last infinitely, we need to establish a convergence (stop) criterion. There are three main convergence criteria which can be adopted:

1. Displacement criterion

$$\left\|\left\{\Delta\overline{w}_n^i\right\}\right\|_2 \le e_D \left\|\left\{\Delta\overline{w}_n^i\right\} - \left\{\Delta\overline{w}_{n-1}^i\right\}\right\|_2 \tag{B.11}$$

This means the incremental displacement after certain iteration becomes so small that no more iterations are necessary. The brackets $\| \; \|_2$ mean Euclidean norm, e_D is the adopted tolerance.

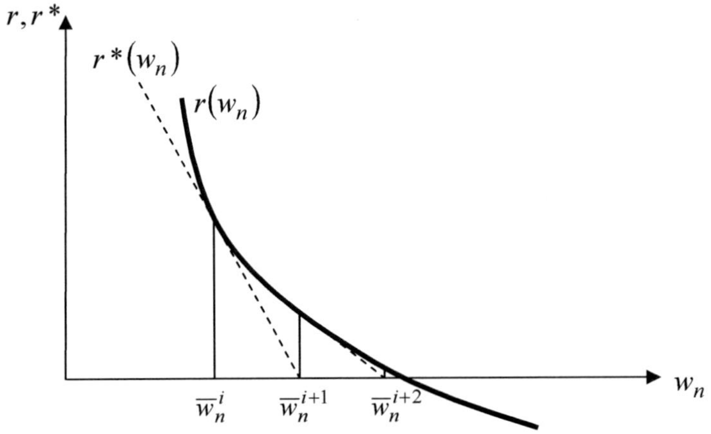

Figure B.1: Approximation of the function $r(w_n)$ by its first derivative. The second iteration point is already much closer to the zero of the error function than the previous one!

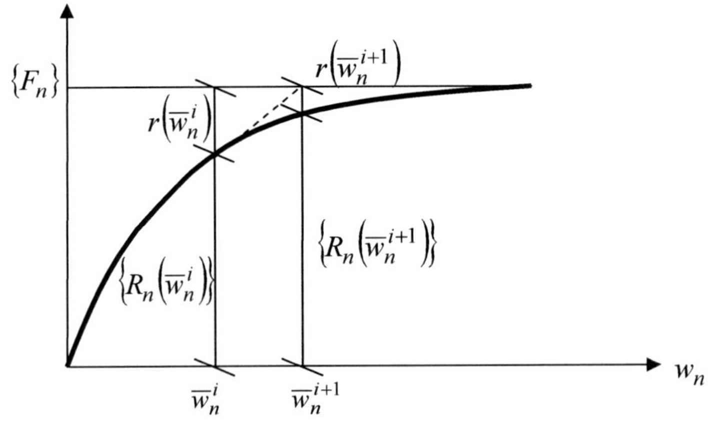

Figure B.2: Standard representation of the iteration procedure with the Newton-Raphson algorithm. The residual is represented the difference between $\{R_n(\overline{w}_n^i)\}$ (internal stress state, body and surface forces) and the nodal forces.

2. Out-of-balance force

$$\left\| \{R_n^i\} - \{F_n\} \right\|_2 \leq e_F \left\| \{R_{n-1}\} - \{F_n\} \right\|_2 \tag{B.12}$$

This means that the residual vector of the out-of-balance force $r(\overline{w}_n^i)$ is sufficiently small.

3. Internal energy criterion

$$\left\{\Delta \overline{w}_n^i\right\}\left(\left\{R_n^i\right\}-\left\{F_n\right\}\right)\leq e_E\left\{\Delta \overline{w}_n^i\right\}\left(\left\{R_{n-1}\right\}-\left\{F_n\right\}\right) \tag{B.13}$$

In this case we set a tolerance for the work done by the residual vector of the out-of-balance force $r\left(\overline{w}_n^i\right)$.

These criteria can be used individually or combined. It is obvious that the chosen tolerances are a compromise between accuracy, and computational time.

B.3 Invariants and general stress space

Before formulating the elastic-plastic constitutive model in general stress space, it is useful to recall the advantages of the formulation of constitutive models via geotechnical (triaxial) stress and strain invariants. In the following example this is done on the basis of a hyperelastic constitutive model (see Chapter 10), defined by the strain energy function. The main advantage of a constitutive model formulated with geotechnical stress and strain tensor invariants is that it can be easily generalised to a general stress-strain state. The geotechnical strain tensor invariants correspond to the following general strain formulation:

$$\varepsilon_v = \varepsilon_{ii} = \varepsilon_{11} + \varepsilon_{22} + \varepsilon_{33} \tag{B.14}$$

$$\varepsilon_s = \frac{2}{3}\sqrt{3I_{2D}} = \frac{2}{3}\sqrt{3\left(I_2 - \frac{I_1^2}{6}\right)} = \frac{2}{3}\sqrt{3\left(\frac{1}{2}\varepsilon_{ij}\varepsilon_{ij} - \frac{\varepsilon_{ii}^2}{6}\right)} \tag{B.15}$$

The general stress state can be derived from the strain energy function *f* as:

$$\sigma_{ij} = \frac{\partial f\left(\varepsilon_s,\varepsilon_v\right)}{\partial \varepsilon_{ij}} = \frac{\partial f}{\partial \varepsilon_s}\frac{\partial \varepsilon_s}{\partial \varepsilon_{ij}} + \frac{\partial f}{\partial \varepsilon_v}\frac{\partial \varepsilon_v}{\partial \varepsilon_{ij}} \tag{B.16}$$

In the following expressions (B.17)-(B.18) we obtain the partial derivatives of the triaxial strain invariants in the general strain space:

$$\frac{\partial \varepsilon_v}{\partial \varepsilon_{ij}} = \frac{\partial \varepsilon_{ii}}{\partial \varepsilon_{ij}} = \frac{\partial\left(\varepsilon_{ij}\delta_{ij}\right)}{\partial \varepsilon_{ij}} = \delta_{ij} \tag{B.17}$$

$$\frac{\partial \varepsilon_s}{\partial \varepsilon_{ij}} = \frac{2}{3\varepsilon_s}\left(\varepsilon_{ij} - \frac{\varepsilon_{mm}\delta_{ij}}{3}\right) = \frac{2\cdot e_{ij}}{3\cdot \varepsilon_s} \tag{B.18}$$

After substituting equations (B.17) and (B.18) into (B.16) we obtain the following general expression for the stress tensor:

$$\sigma_{ij} = \frac{\partial f}{\partial \varepsilon_s} \frac{2 \cdot e_{ij}}{3 \cdot \varepsilon_s} + \frac{\partial f}{\partial \varepsilon_v} \delta_{ij} \tag{B.19}$$

If we adopt a linear elastic constitutive model we can write:

$$f = \frac{3}{2} G \cdot \varepsilon_s^2 + \frac{K}{2} \cdot \varepsilon_v^2 \tag{B.20}$$

Substituting equation (B.19) into (B.20) we obtain the classical form of linear elasticity in general stress-strain space.

$$\sigma_{ij} = 2G \cdot e_{ij} + K \cdot \varepsilon_v \delta_{ij} \tag{B.21}$$

Generalisation to elastic-plastic models based on triaxial stress-strain invariants and hyperplastic formulation can be obtained similarly. This, however, requires the definition of the dissipation potential satisfying the Second Law of Thermodynamics.

B.4 General stress-strain formulation for a linear elastic-perfectly plastic model

In this section we present the derivation of a constitutive model based on linear elasticity and a Drucker-Prager failure criterion with non-associated flow in the general stress-strain space. We adopt the classical formalism of the plasticity theory. The five main components, which describe the constitutive model are:

- Decomposition of incremental strains:

$$d\varepsilon_{ij} = d\varepsilon_{ij}^e + d\varepsilon_{ij}^p \tag{B.22}$$

- Linear elastic constitutive model:

$$D_{ijkl} = \left(K - \frac{2G}{3} \right) \delta_{ij} \delta_{kl} + 2G \delta_{ik} \delta_{jl} \tag{B.23}$$

$$d\sigma_{ij} = D_{ijkl} d\varepsilon_{kl}^e \tag{B.24}$$

- Consistency condition:

$$dF = \frac{\partial F}{\partial \sigma_{ij}} d\sigma_{ij} = 0 \tag{B.25}$$

- Drucker-Prager failure surface:

$$F = \sqrt{J_{2D}} - \alpha_1 J_1 - K_1 = 0 \tag{B.26}$$

- Plastic potential:

$$P = \sqrt{J_{2D}} - \alpha_2 J_1 - K_2 = 0 \tag{B.27}$$

- Non associated flow rule:

$$d\varepsilon_{kl}^{p} = d\lambda \frac{\partial P}{\partial \sigma_{kl}} \tag{B.28}$$

The elastic-plastic constitutive relationship is derived by combining equations (B.24) with (B.22) and (B.28):

$$d\sigma_{ij} = D_{ijkl}\left(d\varepsilon_{kl} - d\lambda \frac{\partial P}{\partial \sigma_{kl}} \right) \tag{B.29}$$

Subsequently $d\lambda$ can be obtained by substituting equation (B.29) in the consistency condition (B.25):

$$d\lambda = \frac{\dfrac{\partial F}{\partial \sigma_{ij}} D_{ijkl} d\varepsilon_{kl}}{\dfrac{\partial F}{\partial \sigma_{ij}} D_{ijkl} \dfrac{\partial P}{\partial \sigma_{kl}}} \tag{B.30}$$

The partial derivatives of the failure surface and the plastic potential in equation (B.30) are given by:

$$\frac{\partial F}{\partial \sigma_{ij}} = \frac{\partial F}{\partial \sqrt{J_{2D}}} \frac{\partial \sqrt{J_{2D}}}{\partial \sigma_{ij}} + \frac{\partial F}{\partial J_1} \frac{\partial J_1}{\partial \sigma_{ij}} \tag{B.31}$$

$$\frac{\partial P}{\partial \sigma_{ij}} = \frac{\partial P}{\partial \sqrt{J_{2D}}} \frac{\partial \sqrt{J_{2D}}}{\partial \sigma_{ij}} + \frac{\partial P}{\partial J_1} \frac{\partial J_1}{\partial \sigma_{ij}} \tag{B.32}$$

Using equations (3.29) and (3.38), the partial derivatives of the stress invariants are obtained as follows:

$$\frac{\partial J_1}{\partial \sigma_{ij}} = \frac{\partial (\sigma_{ij} \delta_{ij})}{\partial \sigma_{ij}} = \delta_{ij} \tag{B.33}$$

$$\frac{\partial\sqrt{J_{2D}}}{\partial\sigma_{ij}} = \frac{1}{2\sqrt{J_{2D}}}\frac{\partial}{\partial\sigma_{ij}}\left(J_2 - \frac{1}{6}J_1^2\right) =$$
$$= \frac{1}{2\sqrt{J_{2D}}}\frac{\partial}{\partial\sigma_{ij}}\left[\frac{1}{2}\sigma_{ij}\sigma_{ij} - \frac{1}{6}\left(\sigma_{ij}\delta_{ij}\right)^2\right] = \frac{1}{2\sqrt{J_{2D}}}\left(\sigma_{ij} - \frac{1}{3}\sigma_{ij}\delta_{ij}\delta_{kl}\right) \tag{B.34}$$

so that

$$\frac{\partial\sqrt{J_{2D}}}{\partial\sigma_{ij}} = \frac{s_{ij}}{2\sqrt{J_{2D}}} \tag{B.35}$$

Substituting equations (B.33) and (B.35) into (B.31) and (B.32) we obtain:

$$\frac{\partial F}{\partial\sigma_{ij}} = \frac{\partial F}{\partial\sqrt{J_{2D}}}\frac{s_{ij}}{2\sqrt{J_{2D}}} + \frac{\partial F}{\partial J_1}\delta_{ij} \tag{B.36}$$

$$\frac{\partial P}{\partial\sigma_{ij}} = \frac{\partial P}{\partial\sqrt{J_{2D}}}\frac{s_{ij}}{2\sqrt{J_{2D}}} + \frac{\partial P}{\partial J_1}\delta_{ij} \tag{B.37}$$

In order to derive $d\lambda$, equation (B.30) can be divided into three main terms which can be calculated separately:

$$d\lambda = \frac{\overbrace{\underbrace{\frac{\partial F}{\partial\sigma_{ij}}D_{ijkl}}_{a}\,d\varepsilon_{kl}}^{b}}{\underbrace{\frac{\partial F}{\partial\sigma_{ij}}D_{ijkl}\frac{\partial P}{\partial\sigma_{kl}}}_{c}}$$

First we derive an explicit equation for (a). Combining equations (B.36) and (B.24) we obtain:

$$\frac{\partial F}{\partial\sigma_{ij}}D_{ijkl} = \left(\frac{\partial F}{\partial\sqrt{J_{2D}}}\frac{s_{ij}}{2\sqrt{J_{2D}}} + \frac{\partial F}{\partial J_1}\delta_{ij}\right)\left[\left(K - \frac{2G}{3}\right)\delta_{ij}\delta_{kl} + 2G\delta_{ik}\delta_{jl}\right] \tag{B.38}$$

By observing that:

$$s_{ij}\delta_{ij} = \sigma_{ij}\delta_{ij} - \frac{1}{3}\sigma_{ii}\delta_{ij}\delta_{ij} = \sigma_{ii} - \sigma_{ii} = 0 \tag{B.39}$$

The following equation is obtained:

$$\frac{\partial F}{\partial \sigma_{ij}} D_{ijkl} = \frac{\partial F}{\partial \sqrt{J_{2D}}} \frac{s_{ij}}{2\sqrt{J_{2D}}} 2G\delta_{ik}\delta_{jl} + \frac{\partial F}{\partial J_1} \delta_{ij}\delta_{ij}\delta_{kl}\left(K - \frac{2G}{3}\right) + \frac{\partial F}{\partial J_1} 2G\delta_{ij}\delta_{ik}\delta_{jl} \tag{B.40}$$

Equation (B.40) can be simplified as follows:

$$\frac{\partial F}{\partial \sigma_{ij}} D_{ijkl} = \frac{\partial F}{\partial \sqrt{J_{2D}}} \frac{G s_{kl}}{\sqrt{J_{2D}}} + \frac{\partial F}{\partial J_1} 3K\delta_{kl} \tag{B.41}$$

Secondly an explicit expression for (b) can be derived by using equation (B.41):

$$\begin{aligned}\frac{\partial F}{\partial \sigma_{ij}} D_{ijkl} d\varepsilon_{kl} &= \left(\frac{\partial F}{\partial \sqrt{J_{2D}}} \frac{G s_{kl}}{\sqrt{J_{2D}}} + \frac{\partial F}{\partial J_1} 3K\delta_{kl} \right) d\varepsilon_{kl} = \\ &= \frac{\partial F}{\partial \sqrt{J_{2D}}} \frac{G s_{kl} d\varepsilon_{kl}}{\sqrt{J_{2D}}} + \frac{\partial F}{\partial J_1} 3K d\varepsilon_{kk}\end{aligned} \tag{B.42}$$

Finally, an explicit expression for (c) is obtained by adopting equation (B.41) for the first two terms and equation (B.37) for the third term:

$$\begin{aligned}\frac{\partial F}{\partial \sigma_{ij}} D_{ijkl} \frac{\partial P}{\partial \sigma_{kl}} &= \left(\frac{\partial F}{\partial \sqrt{J_{2D}}} \frac{G s_{kl}}{\sqrt{J_{2D}}} + \frac{\partial F}{\partial J_1} 3K\delta_{kl} \right) \frac{\partial P}{\partial \sigma_{kl}} = \\ &= \left(\frac{\partial F}{\partial \sqrt{J_{2D}}} \frac{G s_{kl}}{\sqrt{J_{2D}}} + \frac{\partial F}{\partial J_1} 3K\delta_{kl} \right)\left(\frac{\partial P}{\partial \sqrt{J_{2D}}} \frac{s_{kl}}{2\sqrt{J_{2D}}} + \frac{\partial P}{\partial J_1} \delta_{kl} \right) = \\ &= \left(\frac{\partial F}{\partial \sqrt{J_{2D}}} \frac{\partial P}{\partial \sqrt{J_{2D}}} \frac{G s_{kl}}{\sqrt{J_{2D}}} \frac{s_{kl}}{2\sqrt{J_{2D}}} + \frac{\partial F}{\partial J_1} \frac{\partial P}{\partial J_1} 3K\delta_{kk} \right)\end{aligned} \tag{B.43}$$

Noting that: $\frac{1}{2} s_{kl} s_{kl} = J_{2D}$ and $\delta_{kk} = 3$, equation (B.43) can be simplified:

$$\frac{\partial F}{\partial \sigma_{ij}} D_{ijkl} \frac{\partial P}{\partial \sigma_{kl}} = \frac{\partial F}{\partial \sqrt{J_{2D}}} \frac{\partial P}{\partial \sqrt{J_{2D}}} G + \frac{\partial F}{\partial J_1} \frac{\partial P}{\partial J_1} 9K \tag{B.44}$$

It follows that:

$$d\lambda = \frac{\dfrac{\partial F}{\partial\sqrt{J_{2D}}}\dfrac{Gs_{kl}d\varepsilon_{kl}}{\sqrt{J_{2D}}} + \dfrac{\partial F}{\partial J_1}3Kd\varepsilon_{kk}}{\dfrac{\partial F}{\partial\sqrt{J_{2D}}}\dfrac{\partial P}{\partial\sqrt{J_{2D}}}G + \dfrac{\partial F}{\partial J_1}\dfrac{\partial P}{\partial J_1}9K} \tag{B.45}$$

For the Drucker-Prager failure criterion (B.26) and non-associated flow rule (B.28), equation (B.45) becomes:

$$d\lambda = \frac{\dfrac{Gs_{kl}d\varepsilon_{kl}}{\sqrt{J_{2D}}} + 3K\alpha_1 d\varepsilon_{kk}}{G + 9K\alpha_1\alpha_2} \tag{B.46}$$

By combining equations (B.29), (B.23) and (B.46) we obtain the stress-strain relation for the Drucker-Prager material with non-associated flow rule:

$$d\sigma_{ij} = \left[\left(K - \frac{2G}{3}\right)\delta_{ij}\delta_{kl} + 2G\delta_{ik}\delta_{jl}\right]\left[d\varepsilon_{kl} - d\lambda\left(\frac{s_{kl}}{2\sqrt{J_{2D}}} - \alpha_2\delta_{kl}\right)\right] \tag{B.47}$$

where $d\lambda$ is given by equation (B.46), or after some simplifications:

$$d\sigma_{ij} = 2Gde_{ij} + Kd\varepsilon_{kk}\delta_{ij} - d\lambda\left(\frac{G}{\sqrt{J_{2D}}}s_{ij} - 3K\alpha_2\delta_{ij}\right) \tag{B.48}$$

Substituting the index values with numbers from 1 to 3 into equation (B.48) we obtain the 9 components of the stress tensor, e.g.:

$$d\sigma_{11} = 3Kd\varepsilon_{11} + \left(\frac{3K\alpha_1 d\varepsilon_{11}}{G + 9K\alpha_1\alpha_2}\right)(9K\alpha_2) \tag{B.49}$$

For the implementation in the Newton-Raphson algorithm, the elastic-plastic stiffness matrix is needed. Combining equations (B.29) and (B.30):

$$d\sigma_{ij} = D_{ijkl}d\varepsilon_{kl} - \frac{D_{ijmn}\dfrac{\partial F}{\partial\sigma_{pr}}D_{prkl}\dfrac{\partial P}{\partial\sigma_{kl}}}{\dfrac{\partial F}{\partial\sigma_{pr}}D_{prkl}\dfrac{\partial P}{\partial\sigma_{kl}}}d\varepsilon_{kl} \tag{B.50}$$

so that

$$D_{ijkl}^{ep} = D_{ijkl} - \frac{D_{ijmn} \frac{\partial F}{\partial \sigma_{pr}} D_{prkl} \frac{\partial P}{\partial \sigma_{kl}}}{\frac{\partial F}{\partial \sigma_{pr}} D_{prkl} \frac{\partial P}{\partial \sigma_{kl}}} \tag{B.51}$$

For the Drucker-Prager model with non-associated flow rule we obtain:

$$D_{ijkl}^{ep} = \left(K - \frac{2G}{3}\right)\delta_{ij}\delta_{kl} + 2G\delta_{ik}\delta_{jl} - \frac{\left(G\frac{s_{ij}}{\sqrt{J_{2D}}} - 3K\alpha_1\delta_{ij}\right)\left(G\frac{s_{kl}}{\sqrt{J_{2D}}} - 3K\alpha_2\delta_{kl}\right)}{G + 9\alpha_1\alpha_2 K} \tag{B.52}$$

Substituting the index values with numbers from 1 to 3 we obtain the 36 components of the elastic-plastic stiffness matrix, e.g.:

$$D_{1111}^{ep} = 9K + 12G - \frac{(9K)^2 \alpha_1\alpha_2}{G + 9\alpha_1\alpha_2 K} \tag{B.53}$$

B.5 Numerical implementation

Finite element procedures calculate the stress-strain response incrementally and integrate them on the prescribed stress or strain path. Implementing a constitutive model in a FE code is not trivial, and requires a significant effort in validation within one element tests, for which a closed form solution exists. In the following paragraphs we will illustrate the main "components" of a user-defined model.

B.5.1 Incremental formulation

First of all, the incremental stress-strain relationship and the stiffness matrix have to be implemented: i.e., the expressions (B.48) and (B.52) for the stress-strain relations and the stiffness matrix, respectively.

One could wonder why to write both of them, since they are intrinsically related. The reason is that FE programs need the stress-strain relation for calculating the next increment, while the stiffness matrix is required for the equilibrium iteration according to the Newton-Raphson algorithm (equation B.10).

B.5.2 Loading state

For implementing an elastic-plastic constitutive model, it is crucial to define whether the stress state is in the elastic or in the plastic region (Figure B.3). This is done by calculating the value of the failure surface (or the yield function in case of hardening) at the end of the load step n:

a) if $F(\sigma_{ij} + d\sigma_{ij}) < 0$, the strains are elastic;

b) if $F(\sigma_{ij} + d\sigma_{ij}) > 0$, the strains are plastic and $d\sigma_{ij}$ is not correct.

Since expression b) is not admissible, we should adopt a scalar scaling factor β, such that:

$$F(\sigma_{ij} + \beta \cdot d\sigma_{ij}) = 0 \tag{B.54}$$

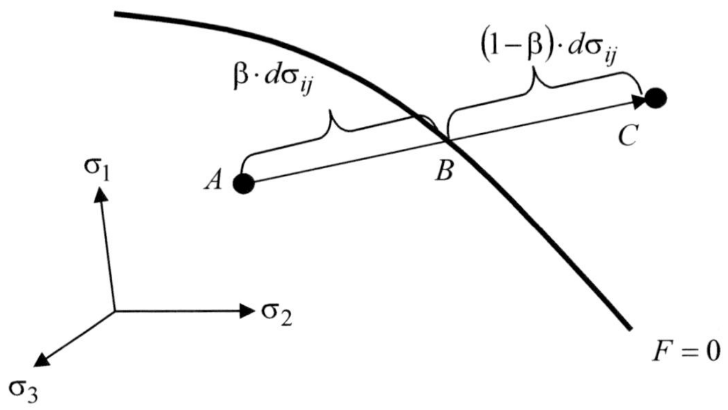

Figure B.3: Elastic and plastic stress increments. The plastic stress increment has to be redefined so that it belongs to the failure surface.

The strains can be also decomposed:

a) Elastic strains: $d\varepsilon_{ij}^{e} = \beta \cdot d\varepsilon_{ij}$

b) Plastic strains: $d\varepsilon_{ij}^{p} = (1-\beta) \cdot d\varepsilon_{ij}$

The factor β can be obtained with an iterative approach based on the Newton-Raphson method or the secant iteration scheme:

Newton-Raphson method:

$$\beta^{n+1} = \beta^{n} + \frac{F(\sigma_{ij} + \beta^{n} \cdot d\sigma_{ij})}{\left. \dfrac{\partial F(\sigma_{ij} + \beta^{n} \cdot d\sigma_{ij})}{\partial \beta} \right|_{\beta = \beta^{n}}} \tag{B.55}$$

Secant method:

$$\beta^{n+1} = \beta^n + \frac{F\left(\sigma_{ij} + \beta^n \cdot d\sigma_{ij}\right)}{F\left(\sigma_{ij} + \beta^n \cdot d\sigma_{ij}\right) - F\left(\sigma_{ij} + \beta^{n-1} \cdot d\sigma_{ij}\right)}\left(\beta^n - \beta^{n-1}\right) \tag{B.56}$$

B.5.3 Numerical integration: Modified Euler integration scheme with error control

In this section a particular integration scheme based on the explicit forward Euler method will be explored. Several more efficient methods (e.g., implicit methods, like the backward Euler method) can be found in the literature. For simplicity we only present one example of an explicit method, whose derivation is quite straightforward: the modified Euler integration scheme with error control (Sloan, 1987). The adopted formalism has been slightly modified, so that it fits into our context.

First the plastic strain increment is defined:

$$d\varepsilon_{ij}^p = (1-\beta) \cdot d\varepsilon_{ij} \tag{B.57}$$

In order to perform numerical integration of the non-linear stress path based on the strain increments, we have to divide the total strain increment into a number of substeps:

$$d\tilde{\varepsilon}_{ij}^p = d\tilde{\varepsilon}_{ij} = \Delta T \cdot d\varepsilon_{ij}^p = \Delta T \cdot (1-\beta) \cdot d\varepsilon_{ij} \tag{B.58}$$

where: $0 < \Delta T \leq 1$.

The stress increment associated to $d\tilde{\varepsilon}_{ij}$ can be obtained using equation (B.47) as follows:

$$d\tilde{\sigma}_{ij}^1 = 2G d\tilde{e}_{ij} + K d\tilde{\varepsilon}_{kk}\delta_{ij} - d\lambda\left(\frac{G}{\sqrt{J_{2D}}}s_{ij} - 3K\alpha_2\delta_{ij}\right) \tag{B.59}$$

$$d\lambda = \frac{\dfrac{G s_{kl} d\tilde{\varepsilon}_{kl}}{\sqrt{J_{2D}}} + 3K\alpha_1 d\tilde{\varepsilon}_{kk}}{G + 9K\alpha_1\alpha_2} \tag{B.60}$$

Departing from the new stress state defined by $\sigma_{ij} + \beta d\sigma_{ij} + d\tilde{\sigma}_{ij}^1$, we can now apply a second strain increment $d\tilde{\varepsilon}_{ij}$ (in case of perfect plasticity – it will be the same increment defined by equation B.58) and derive a second stress increment $d\tilde{\sigma}_{ij}^{2}$:

$$d\tilde{\sigma}_{ij}^{2} = 2Gd\tilde{e}_{ij} + Kd\tilde{\varepsilon}_{kk}\delta_{ij} - d\lambda\left(\frac{G}{\sqrt{J_{2D}}}\left(s_{ij} + \beta ds_{ij} + d\tilde{s}_{ij}^{1}\right) - 3K\alpha_{2}\delta_{ij}\right) \quad \text{(B.61)}$$

$$d\lambda = \frac{\frac{G\left(s_{kl} + \beta s_{kl} + d\tilde{s}_{kl}^{1}\right)d\tilde{\varepsilon}_{kl}}{\sqrt{J_{2D}}} + 3K\alpha_{1}d\tilde{\varepsilon}_{kk}}{G + 9K\alpha_{1}\alpha_{2}} \quad \text{(B.62)}$$

The stress increment, which will be used in the numerical integration process, is defined as:

$$d\sigma_{ij}^{ec} = \frac{1}{2}\left(d\tilde{\sigma}_{ij}^{1} + d\tilde{\sigma}_{ij}^{2}\right) \quad \text{(B.63)}$$

The error is defined as the difference between the first stress increment and the chosen incremental stress:

$$Err = d\sigma_{ij}^{ec} - d\tilde{\sigma}_{ij}^{1} = \frac{1}{2}\left(d\tilde{\sigma}_{ij}^{2} - d\tilde{\sigma}_{ij}^{1}\right) \quad \text{(B.64)}$$

The method is interpreted graphically in Figures B.4, B.5 and B.6.

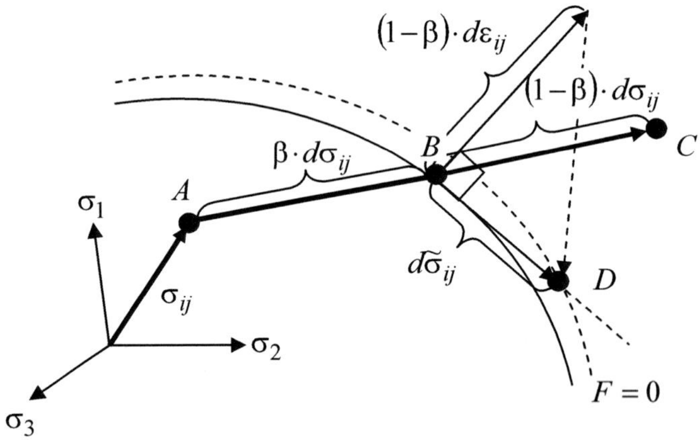

Figure B.4: Calculation of the stress increment caused by the plastic strain component (with associated flow rule) in 1 step with $\Delta T = 1.0$. The dotted curve shows the drift of the failure surface.

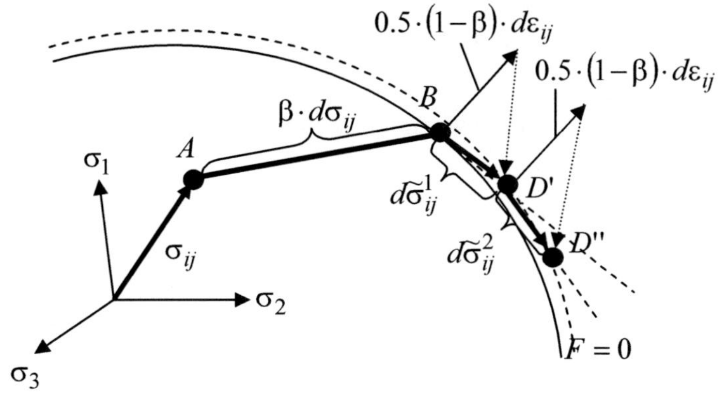

Figure B.5: Calculation of the stress increment in two steps with $\Delta T = 0.5$.

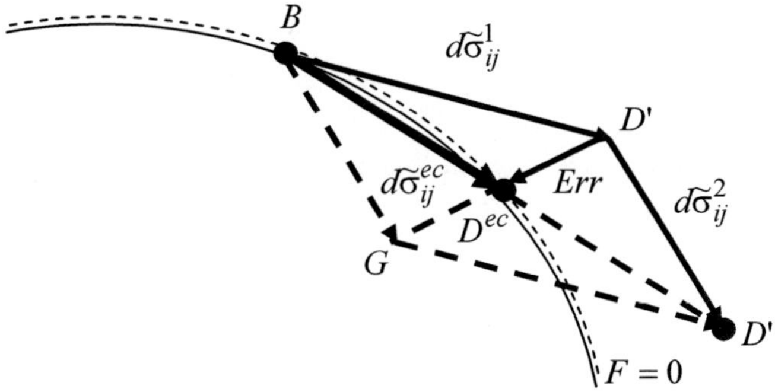

Figure B.6: Calculation of the corrected stress increment $d\tilde{\sigma}_{ij}^{ec}$ and error *Err* represented by the corresponding half-diagonals of the parallelogram $BD'D''G$.

In Figure B.4, vector AB is the stress increment caused by the elastic part of the strain increment, BC is the not admissible stress increment caused by the plastic strain increment (B.57), which has to be corrected and brought back to the failure surface (point *D*). Normally, in order to perform this correction, the plastic strain increment (B.57) has to be divided into sub-increments (B.58), so that the correct stress increment vector *BD* does not deviate too far from the failure surface. If the strains were applied in one step with $\Delta T = 1$ (see Figure B.4), this deviation could be significant and could also accumulate during further stress increments initiating from D. Figure B.5 demonstrates that if the original plastic strain increment (B.57) was applied in two steps, with $\Delta T = 0.5$, the two 2 incremental stress vectors BD' ($d\tilde{\sigma}_{ij}^1$) and $D'D''$ ($d\tilde{\sigma}_{ij}^2$), derived from equations (B.59) and (B.61), respectively, would follow the failure surface much closer.

The correction can become even more accurate and better controlled if we apply additional iteration demonstrated in Figure B.6. The vector BD^{ec} ($d\tilde{\sigma}_{ij}^{ec}$), obtained from equation (B.63) by averaging between the two incremental stress vectors $BD'(d\tilde{\sigma}_{ij}^1)$ and $D'D''$ ($d\tilde{\sigma}_{ij}^2$), brings the stress state D^{ec} significantly closer to the failure surface $F = 0$. In turn, the vector $D'D^{ec}$ represents the error (B.64) due to the linearization of the failure surface curvature. If the failure surface is linear, the error Err, represented by the vector $D'D^{ec}$ becomes equal to zero.

For a non-linear failure surface a tolerance criterion for the error is required. This can be achieved by normalizing $D'D^{ec}$ (Err) by the current stress state:

$$R = \frac{\|Err\|}{\left\|\sigma_{ij} + \beta \cdot d\sigma_{ij} + d\tilde{\sigma}_{ij}^{ec}\right\|} \leq TOL \tag{B.65}$$

If the increment ΔT is still too large, it can be reduced according to the following strategy:

$$\Delta T_{new} = \gamma \cdot \Delta T \tag{B.66}$$

$$\gamma = 0.8 \cdot \left[\frac{TOL}{R}\right]^{1/2} \tag{B.67}$$

Once the algorithm has converged, the stresses can be updated as follows:

$$\sigma_{ij} = \sigma_{ij} + \beta \cdot d\sigma_{ij} + d\tilde{\sigma}_{ij}^{ec} \tag{B.68}$$

One of the main advantages of this error control procedure is that it automatically reduces the drift of the yield surface during the stress update, as it can be observed in Figures B.4, B.5 and B.6 (dotted curve). It cannot, however, totally eliminate this drift and for highly non-linear constitutive models the consistency condition has to be checked again at the end of the procedure.

References

Bathe, K.L. (2006) *Finite element procedures in engineering analysis*, Prentice Hall, Englewood Cliffs, New Jersey.

Burland, J.B. (1989) "Ninth Laurits Bjerrum Memorial Lecture: Small is beautiful - the stiffness of soils at small strains." *Canadian Geotechnical Journal*, Vol. 26(4), 499-516.

Dafalias, Y.F. and Herrmann, L.R. (1982) "Bounding surface formulation of soil plasticity", in G. N. Pande and O. C. Zienkiewicz (eds), *Soil Mechanics: Transient and Cyclic Loads*, Wiley, New York, 253-282.

Desai, C. S. and Siriwardane, H. J. (1984) *Constitutive Laws for engineering materials: with emphasis on geologic materials*, Prentice Hall, Englewood Cliffs, New Jersey.

Chen, W.F. and Han, D. (1988) *Plasticity for Structural Engineers*, Springer Verlag, New York.

Houlsby, G.T. (1986) "A general failure criterion for frictional and cohesive materials", *Soils and Foundations*, Vol. 26(2), 97-101.

Houlsby, G.T. and Puzrin, A.M. (2006) *Principles of Hyperplasticity. An approach to plasticity based on thermodynamics*, Springer-Verlag, London.

Jaky, J. (1944) "The coefficient of earth pressure at rest." *Journal of the Society of Hungarian Engineers and Architects*, Budapest, 355–358.

Konder, R. L. (1963) "Hyperbolic stress-strain response: cohesive soils." *Journal of the Soil Mechanics and Foundations Division, ASCE*, Vol. 89, 115-143.

Masing, G. (1926) "Eigenspannungen und Verfestigung beim Messing." *Proc. of 2nd International Congress of Applied Mechanics*, Zurich, Switzerland, 332–335.

Matsuoka, H. and Nakai, T. (1974) "Stress-deformation and strength characteristics of soil under three different principal stresses." *Proc. JSCE*, Vol. 232, 59-70.

Mroz, Z., Norris, V. A., and Zienkiewicz, O. C. (1979) "Application of an anisotropic hardening model in the analysis of elasto-plastic deformation of soil." *Géotechnique*, Vol. 29(1), 1–34.

Mroz, Z. and Norris, V.A. (1982) "Elastoplastic and viscoplastic constitutive models for soils with application to cyclic loading." *Soil Mechanics -*

Transient and Cyclic Loads, ed. G.N. Pande and O.C. Zienkiewicz, Wiley, 173-218.

Potts, D.M. and Zdravkovic, L. (1999) *Finite element analysis in geotechnical engineering*, Thomas Telford, London.

Puzrin, A.M. and Burland J.B. (1996) "A logarithmic stress-strain function for rocks and soils." *Geotechnique*, Vol. 46(1), 57-164.

Puzrin, A.M. and Burland J.B. (1998) "Nonlinear model of small strain behaviour of soils." *Geotechnique*, Vol. 48(2), 217-233.

Puzrin, A.M. and Burland, J.B. (2000) "Kinematic hardening plasticity formulation of small strain behaviour of soils." *Int. J. of Num. and Anal. Methods in Geomechanics*, Vol. 24, 753-781.

Puzrin, A. M. and Shiran, A. (2000) "Effect of a constitutive relationship on seismic response of soils. Part I: Constitutive modelling of cyclic behaviour of soils." *Soil Dynamics and Earthquake Engineering*, Vol. 19, 305-318.

Puzrin, A.M., Houlsby, G.T. and Burland, J.B. (2001) "Thermomechanical formulation of small strain model for overconsolidated clays." *Proceedings of the Royal Society: Mathematical, Physical and Engineering Sciences*, Vol. 457(2006), 425-440.

Pyke, R. (1979) Nonlinear soil model for irregular cyclic loadings, *J. Geotech. Eng. Div., ASCE,* 105, 715–726.

Ramberg, W. and Osgood, W.R. (1943) "Description of stress-strain curves by three parameters." *National Advisory Committee for Aeronautics, Washington, D.C., Technical Note No. 902.*

Roscoe, K.H. and Burland, J.B. (1968) "*On the generalised behaviour of 'wet' clay." Engineering Plasticity*, ed. Heyman, J. and Leckie, F.A., Cambridge University Press, 535-610.

Schofield, A.N and Wroth, C.P. (1968) *Critical state soil mechanics*, McGraw Hill, London.

Sloan, S.W. (1987) "Substepping schemes for the numerical relations integration of elastoplastic stress-strain." *International Journal for Numerical Methods in Engineering*, Vol. 24, 893-911.

Zienkiewicz, O.C. and Taylor, R.L. (2000). *The finite element method.* Butterworth-Heinemann, Oxford.